commutative ring theory

LECTURE NOTES IN PURE AND APPLIED MATHEMATICS

1. *N. Jacobson,* Exceptional Lie Algebras
2. *L.-Å. Lindahl and F. Poulsen,* Thin Sets in Harmonic Analysis
3. *I. Satake,* Classification Theory of Semi-Simple Algebraic Groups
4. *F. Hirzebruch, W. D. Newmann, and S. S. Koh, Differentiable Manifolds and Quadratic Forms*
5. *I. Chavel,* Riemannian Symmetric Spaces of Rank One
6. *R. B. Burckel,* Characterization of C(X) Among Its Subalgebras
7. *B. R. McDonald, A. R. Magid, and K. C. Smith,* Ring Theory: Proceedings of the Oklahoma Conference
8. *Y.-T. Siu,* Techniques of Extension on Analytic Objects
9. *S. R. Caradus, W. E. Pfaffenberger, and B. Yood,* Calkin Algebras and Algebras of Operators on Banach Spaces
10. *E. O. Roxin, P.-T. Liu, and R. L. Sternberg,* Differential Games and Control Theory
11. *M. Orzech and C. Small,* The Brauer Group of Commutative Rings
12. *S. Thomier,* Topology and Its Applications
13. *J. M. Lopez and K. A. Ross,* Sidon Sets
14. *W. W. Comfort and S. Negrepontis,* Continuous Pseudometrics
15. *K. McKennon and J. M. Robertson,* Locally Convex Spaces
16. *M. Carmeli and S. Malin,* Representations of the Rotation and Lorentz Groups: An Introduction
17. *G. B. Seligman,* Rational Methods in Lie Algebras
18. *D. G. de Figueiredo,* Functional Analysis: Proceedings of the Brazilian Mathematical Society Symposium
19. *L. Cesari, R. Kannan, and J. D. Schuur,* Nonlinear Functional Analysis and Differential Equations: Proceedings of the Michigan State University Conference
20. *J. J. Schäffer,* Geometry of Spheres in Normed Spaces
21. *K. Yano and M. Kon,* Anti-Invariant Submanifolds
22. *W. V. Vasconcelos,* The Rings of Dimension Two
23. *R. E. Chandler,* Hausdorff Compactifications
24. *S. P. Franklin and B. V. S. Thomas,* Topology: Proceedings of the Memphis State University Conference
25. *S. K. Jain,* Ring Theory: Proceedings of the Ohio University Conference
26. *B. R. McDonald and R. A. Morris,* Ring Theory II: Proceedings of the Second Oklahoma Conference
27. *R. B. Mura and A. Rhemtulla,* Orderable Groups
28. *J. R. Graef,* Stability of Dynamical Systems: Theory and Applications
29. *H.-C. Wang,* Homogeneous Branch Algebras
30. *E. O. Roxin, P.-T. Liu, and R. L. Sternberg,* Differential Games and Control Theory II
31. *R. D. Porter,* Introduction to Fibre Bundles
32. *M. Altman,* Contractors and Contractor Directions Theory and Applications
33. *J. S. Golan,* Decomposition and Dimension in Module Categories
34. *G. Fairweather,* Finite Element Galerkin Methods for Differential Equations
35. *J. D. Sally,* Numbers of Generators of Ideals in Local Rings
36. *S. S. Miller,* Complex Analysis: Proceedings of the S.U.N.Y. Brockport Conference
37. *R. Gordon,* Representation Theory of Algebras: Proceedings of the Philadelphia Conference
38. *M. Goto and F. D. Grosshans,* Semisimple Lie Algebras
39. *A. I. Arruda, N. C. A. da Costa, and R. Chuaqui,* Mathematical Logic: Proceedings of the First Brazilian Conference
40. *F. Van Oystaeyen,* Ring Theory: Proceedings of the 1977 Antwerp Conference
41. *F. Van Oystaeyen and A. Verschoren,* Reflectors and Localization: Application to Sheaf Theory
42. *M. Satyanarayana,* Positively Ordered Semigroups
43. *D. L Russell,* Mathematics of Finite-Dimensional Control Systems
44. *P.-T. Liu and E. Roxin,* Differential Games and Control Theory III: Proceedings of the Third Kingston Conference, Part A
45. *A. Geramita and J. Seberry,* Orthogonal Designs: Quadratic Forms and Hadamard Matrices
46. *J. Cigler, V. Losert, and P. Michor,* Banach Modules and Functors on Categories of Banach Spaces

47. *P.-T. Liu and J. G. Sutinen,* Control Theory in Mathematical Economics: Proceedings of the Third Kingston Conference, Part B
48. *C. Byrnes,* Partial Differential Equations and Geometry
49. *G. Klambauer,* Problems and Propositions in Analysis
50. *J. Knopfmacher,* Analytic Arithmetic of Algebraic Function Fields
51. *F. Van Oystaeyen,* Ring Theory: Proceedings of the 1978 Antwerp Conference
52. *B. Kadem,* Binary Time Series
53. *J. Barros-Neto and R. A. Artino,* Hypoelliptic Boundary-Value Problems
54. *R. L. Sternberg, A. J. Kalinowski, and J. S. Papadakis,* Nonlinear Partial Differential Equations in Engineering and Applied Science
55. *B. R. McDonald,* Ring Theory and Algebra III: Proceedngs of the Third Oklahoma Conference
56. *J. S. Golan,* Structure Sheaves Over a Noncommutative Ring
57. *T. V. Narayana, J. G. Williams, and R. M. Mathsen,* Combinatorics, Representation Theory and Statistical Methods in Groups: YOUNG DAY Proceedings
58. *T. A. Burton,* Modeling and Differential Equations in Biology
59. *K. H. Kim and F. W. Roush,* Introduction to Mathematical Consensus Theory
60. *J. Banas and K. Goebel,* Measures of Noncompactness in Banach Spaces
61. *O. A. Nielson,* Direct Integral Theory
62. *J. E. Smith, G. O. Kenny, and R. N. Ball,* Ordered Groups: Proceedings of the Boise State Conference
63. *J. Cronin,* Mathematics of Cell Electrophysiology
64. *J. W. Brewer,* Power Series Over Commutative Rings
65. *P. K. Kamthan and M. Gupta,* Sequence Spaces and Series
66. *T. G. McLaughlin,* Regressive Sets and the Theory of Isols
67. *T. L. Herdman, S. M. Rankin III, and H. W. Stech,* Integral and Functional Differential Equations
68. *R. Draper,* Commutative Algebra: Analytic Methods
69. *W. G. McKay and J. Patera,* Tables of Dimensions, Indices, and Branching Rules for Representations of Simple Lie Algebras
70. *R. L. Devaney and Z. H. Nitecki,* Classical Mechanics and Dynamical Systems
71. *J. Van Geel,* Places and Valuations in Noncommutative Ring Theory
72. *C. Faith,* Injective Modules and Injective Quotient Rings
73. *A. Fiacco,* Mathematical Programming with Data Perturbations I
74. *P. Schultz, C. Praeger, and R. Sullivan,* Algebraic Structures and Applications: Proceedings of the First Western Australian Conference on Algebra
75. *L Bican, T. Kepka, and P. Nemec,* Rings, Modules, and Preradicals
76. *D. C. Kay and M. Breen,* Convexity and Related Combinatorial Geometry: Proceedings of the Second University of Oklahoma Conference
77. *P. Fletcher and W. F. Lindgren,* Quasi-Uniform Spaces
78. *C.-C. Yang,* Factorization Theory of Meromorphic Functions
79. *O. Taussky,* Ternary Quadratic Forms and Norms
80. *S. P. Singh and J. H. Burry,* Nonlinear Analysis and Applications
81. *K. B. Hannsgen, T. L. Herdman, H. W. Stech, and R. L. Wheeler,* Volterra and Functional Differential Equations
82. *N. L. Johnson, M. J. Kallaher, and C. T. Long,* Finite Geometries: Proceedings of a Conference in Honor of T. G. Ostrom
83. *G. I. Zapata,* Functional Analysis, Holomorphy, and Approximation Theory
84. *S. Greco and G. Valla,* Commutative Algebra: Proceedings of the Trento Conference
85. *A. V. Fiacco,* Mathematical Programming with Data Perturbations II
86. *J.-B. Hiriart-Urruty, W. Oettli, and J. Stoer,* Optimization: Theory and Algorithms
87. *A. Figa Talamanca and M. A. Picardello,* Harmonic Analysis on Free Groups
88. *M. Harada,* Factor Categories with Applications to Direct Decomposition of Modules
89. *V. I. Istrăţescu,* Strict Convexity and Complex Strict Convexity
90. *V. Lakshmikantham,* Trends in Theory and Practice of Nonlinear Differential Equations
91. *H. L. Manocha and J. B. Srivastava,* Algebra and Its Applications
92. *D. V. Chudnovsky and G. V. Chudnovsky,* Classical and Quantum Models and Arithmetic Problems
93. *J. W. Longley,* Least Squares Computations Using Orthogonalization Methods
94. *L. P. de Alcantara,* Mathematical Logic and Formal Systems
95. *C. E. Aull,* Rings of Continuous Functions
96. *R. Chuaqui,* Analysis, Geometry, and Probability
97. *L. Fuchs and L. Salce,* Modules Over Valuation Domains

98. *P. Fischer and W. R. Smith,* Chaos, Fractals, and Dynamics
99. *W. B. Powell and C. Tsinakis,* Ordered Algebraic Structures
100. *G. M. Rassias and T. M. Rassias,* Differential Geometry, Calculus of Variations, and Their Applications
101. *R.-E. Hoffmann and K. H. Hofmann,* Continuous Lattices and Their Applications
102. *J. H. Lightbourne III and S. M. Rankin III,* Physical Mathematics and Nonlinear Partial Differential Equations
103. *C. A. Baker and L, M. Batten,* Finite Geometrics
104. *J. W. Brewer, J. W. Bunce, and F. S. Van Vleck,* Linear Systems Over Commutative Rings
105. *C. McCrory and T. Shifrin,* Geometry and Topology: Manifolds, Varieties, and Knots
106. *D. W. Kueker, E. G. K. Lopez-Escobar, and C. H. Smith,* Mathematical Logic and Theoretical Computer Science
107. *B.-L. Lin and S. Simons,* Nonlinear and Convex Analysis: Proceedings in Honor of Ky Fan
108. *S. J. Lee,* Operator Methods for Optimal Control Problems
109. *V. Lakshmikantham,* Nonlinear Analysis and Applications
110. *S. F. McCormick,* Multigrid Methods: Theory, Applications, and Supercomputing
111. *M. C. Tangora,* Computers in Algebra
112. *D. V. Chudnovsky and G. V. Chudnovsky,* Search Theory: Some Recent Developments
113. *D. V. Chudnovsky and R. D. Jenks,* Computer Algebra
114. *M. C. Tangora,* Computers in Geometry and Topology
115. *P. Nelson, V. Faber, T. A. Manteuffel, D. L. Seth, and A. B. White, Jr.,* Transport Theory, Invariant Imbedding, and Integral Equations: Proceedings in Honor of G. M. Wing's 65th Birthday
116. *P. Clément, S. Invernizzi, E. Mitidieri, and I. I. Vrabie,* Semigroup Theory and Applications
117. *J. Vinuesa,* Orthogonal Polynomials and Their Applications: Proceedings of the International Congress
118. *C. M. Dafermos, G. Ladas, and G. Papanicolaou,* Differential Equations: Proceedings of the EQUADIFF Conference
119. *E. O. Roxin,* Modern Optimal Control: A Conference in Honor of Solomon Lefschetz and Joseph P. Lasalle
120. *J. C. Díaz,* Mathematics for Large Scale Computing
121. *P. S. Milojević,* Nonlinear Functional Analysis
122. *C. Sadosky,* Analysis and Partial Differential Equations: A Collection of Papers Dedicated to Mischa Cotlar
123. *R. M. Shortt,* General Topology and Applications: Proceedings of the 1988 Northeast Conference
124. *R. Wong,* Asymptotic and Computational Analysis: Conference in Honor of Frank W. J. Olver's 65th Birthday
125. *D. V. Chudnovsky and R. D. Jenks,* Computers in Mathematics
126. *W. D. Wallis, H. Shen, W. Wei, and L. Zhu,* Combinatorial Designs and Applications
127. *S. Elaydi,* Differential Equations: Stability and Control
128. *G. Chen, E. B. Lee, W. Littman, and L. Markus,* Distributed Parameter Control Systems: New Trends and Applications
129. *W. N. Everitt,* Inequalities: Fifty Years On from Hardy, Littlewood and Pólya
130. *H. G. Kaper and M. Garbey,* Asymptotic Analysis and the Numerical Solution of Partial Differential Equations
131. *O. Arino, D. E. Axelrod, and M. Kimmel,* Mathematical Population Dynamics: Proceedings of the Second International Conference
132. *S. Coen,* Geometry and Complex Variables
133. *J. A. Goldstein, F. Kappel, and W. Schappacher,* Differential Equations with Applications in Biology, Physics, and Engineering
134. *S. J. Andima, R. Kopperman, P. R. Misra, J. Z. Reichman, and A. R. Todd,* General Topology and Applications
135. *P Clément, E. Mitidieri, B. de Pagter,* Semigroup Theory and Evolution Equations: The Second International Conference
136. *K. Jarosz,* Function Spaces
137. *J. M. Bayod, N. De Grande-De Kimpe, and J. Martínez-Maurica,* p-adic Functional Analysis
138. *G. A. Anastassiou,* Approximation Theory: Proceedings of the Sixth Southeastern Approximation Theorists Annual Conference
139. *R. S. Rees,* Graphs, Matrices, and Designs: Festschrift in Honor of Norman J. Pullman
140. *G. Abrams, J. Haefner, and K. M. Rangaswamy,* Methods in Module Theory

141. *G. L. Mullen and P. J.-S. Shiue*, Finite Fields, Coding Theory, and Advances in Communications and Computing
142. *M. C. Joshi and A. V. Balakrishnan*, Mathematical Theory of Control: Proceedings of the International Conference
143. *G. Komatsu and Y. Sakane*, Complex Geometry: Proceedings of the Osaka International Conference
144. *I. J. Bakelman*, Geometric Analysis and Nonlinear Partial Differential Equations
145. *T. Mabuchi and S. Mukai*, Einstein Metrics and Yang–Mills Connections: Proceedings of the 27th Taniguchi International Symposium
146. *L. Fuchs and R. Göbel*, Abelian Groups: Proceedings of the 1991 Curaçao Conference
147. *A. D. Pollington and W. Moran*, Number Theory with an Emphasis on the Markoff Spectrum
148. *G. Dore, A. Favini, E. Obrecht, and A. Venni*, Differential Equations in Banach Spaces
149. *T. West*, Continuum Theory and Dynamical Systems
150. *K. D. Bierstedt, A. Pietsch, W. Ruess, and D. Vogt*, Functional Analysis
151. *K. G. Fischer, P. Loustaunau, J. Shapiro, E. L. Green, and D. Farkas*, Computational Algebra
152. *K. D. Elworthy, W. N. Everitt, and E. B. Lee*, Differential Equations, Dynamical Systems, and Control Science
153. *P.-J. Cahen, D. L. Costa, M. Fontana, and S.-E. Kabbaj*, Commutative Ring Theory
154. *S. C. Cooper and W. J. Thron*, Continued Fractions and Orthogonal Functions: Theory and Applications
155. *P. Clément and G. Lumer*, Evolution Equations, Control Theory, and Biomathematics

Additional Volumes in Preparation

commutative ring theory

proceedings of the Fès international conference

edited by

Paul-Jean Cahen
Université d'Aix-Marseille III
Marseille, France

Douglas L. Costa
University of Virginia
Charlottesville, Virginia

Marco Fontana
Terza Università degli Studi di Roma
Rome, Italy

Salah-Eddine Kabbaj
Université S. M. Ben Abdellah
Fès, Morocco

Marcel Dekker, Inc. **New York • Basel • Hong Kong**

Library of Congress Cataloging-in-Publication Data

Commutative ring theory : proceedings of the Fès international conference / edited by Paul-Jean Cahen ... [et al.].
p. cm. -- (Lecture notes in pure and applied mathematics ; v. 153)
Proceedings of the International Conference on Commutative Ring Theory, held at Fès, Morocco, Apr. 20-24, 1992.
Includes bibliographical references.
ISBN 0-8247-9170-3
1. Commutative rings--Congresses. I. Cahen, Paul-Jean II. International Conference on Commutative Ring Theory (1992 : Fès, Morocco) III. Series.
QA251.3.C66 1993
512'.4--dc20 93-35648
CIP

The publisher offers discounts on this book when ordered in bulk quantities. For more information, write to Special Sales/Professional Marketing at the address below.

This book is printed on acid-free paper.

MARCEL DEKKER, INC.
270 Madison Avenue, New York, New York 10016

Current printing (last digit):
10 9 8 7 6 5 4 3 2 1

PRINTED IN THE UNITED STATES OF AMERICA

Preface

The present volume contains the proceedings of the International Conference on Commutative Ring Theory held at Fès, Morocco, April 20-24, 1992.

This book consists of a collection of articles of original research, and in some cases survey articles, selected from those submitted by the conference participants. Certain of the papers not actually delivered at the conference are nonetheless based on results presented there. The principal subjects treated are: spectra of rings, chain conditions and dimension theory, fiber products, group rings and semigroup rings, class groups, graded rings and integer-valued polynomials. In addition, connections with and applications to topological algebra and algebraic geometry are also represented.

The goal of this work is to set forth some recent developments in Commutative Algebra with particular interest given to the theory of rings and ideals. It is aimed primarily at researchers in algebra. Instructors and students at the graduate level should also find these articles to be of interest.

This Conference was the first international encounter in Commutative Ring Theory held in Morocco. It was organized by the Faculty of Sciences of Fès with the collaboration of the University of Aix-Marseille III, the University of Lyon I, the University of Rome "La Sapienza" and the University of Virginia at Charlottesville. The Conference was made possible by the financial support of the Moroccan Mathematical Society, the Faculty of Sciences of Fès and also the University of S.M. Ben Abdellah, the Faculty of Sciences of Tétouan, the Ecole Supérieure of Technology of Fès, the Municipal Council of the City of Fès and the Fès-Saïs Association.

The editors wish to express their thanks to professors M. Saghi and A. Atmani, respectevely Dean and Vice-Dean of the Faculty of Sciences of Fès, M. Guennoun, chairman of the Mathematics Department, and N. Elyacoubi, member of the Moroccan Mathematical Society council for their never-failing encouragement and warm support in helping the local committee with the preparation of the Conference.

Paul-Jean Cahen
Douglas L. Costa
Marco Fontana
Salah-Eddine Kabbaj

Contents

Contributors

V. BARUCCI, Dipartimento di Matematica, Università di Roma "La Sapienza", Roma, Italy

M. BOULAGOUAZ, Département de Mathématiques et d'Informatique, Université de Fès, Fès, Morocco

P.-J. CAHEN, Laboratoire de Mathématiques Fond. et Appl., Université d'Aix-Marseille III, Marseille, France

J.-L. CHABERT, Institut Supérieur des Sciences et Techniques,Université de Picardie, Saint Quentin, France

R. CHIBLOUN, Département de Mathématiques, Faculté des Sciences , Meknès, Morocco

D. COSTA, Department of Mathematics, University of Virginia, Charlottesville, U.S.A.

H. DICHI, Département de Mathématiques, Université d'Abidjan, Abidjan, Côte d'Ivoire

D. DOBBS, Department of Mathematics, University of Tennessee, Knoxville, U.S.A.

O. ECHI, Département de Mathématiques, Faculté des Sciences de Sfax, Sfax, Tunisia

A. FACCHINI, Dipartimento di Matematica, Università di Udine, Udine, Italy

M. FONTANA, Dipartimento di Matematica, Terza Università degli Studi di Roma, Roma, Italy

S. GABELLI, Dipartimento di Matematica, Università di Roma "La Sapienza", Roma, Italy

R. GILMER, Department of Mathematics, Florida State University, Tallahassee, U.S.A.

F. GIROLAMI, Dipartimento di Matematica, Università di Roma "La Sapienza", Roma, Italy

E. HOUSTON, Department of Mathematics, University of North Carolina, Charlotte, U.S.A.

A. IDELHADJ, Département de Mathématiques, Faculté des Sciences de Tétouan, Tétouan, Morocco

L. IZELGUE, Département de Mathématiques et Informatique, Université de Fès, Fès, Morocco

S. KABBAJ, Département de Mathématiques et Informatique, Université de Fès, Fès, Morocco

A. KAIDI, Département de Mathématiques, Faculté des Sciences de Rabat, Rabat, Morocco

G. KELLER, Department of Mathematics, University of Virginia, Charlottesville, U.S.A.

T. LUCAS, Department of Mathematics, University of North Carolina, Charlotte, U.S.A.

H. MATSUMURA, Department of Mathematics, Nagoya University, Nagoya, Japan

A. MICALI, Département de Mathématiques, Université de Montpellier II, Montpellier, France

J. MOTT, Department of Mathematics, Florida State University, Tallahassee, U.S.A.

J.-I. NISHIMURA, Department of Mathematics, Kyoto University, Kyoto, Japan

G. PICAVET, Département de Mathématiques, Université de Clermont II, Aubière, France

M. PICAVET-L'HERMITTE, Département de Mathématiques, Université de Clermont II, Aubière, France

D. SANGARE, Département de Mathématiques, Université d'Abidjan, Abidjan, Côte d'Ivoire

M. ZAFRULLAH, Department of Mathematics, Winthrop College, Rock Hill, S. Carolina, U.S.A.

S. ZARZUELA, Departament d'Àlgebra i Geometria, Universitat de Barcelona, Barcelona, Spain

1
Seminormal Mori Domains

VALENTINA BARUCCI Dipartimento di Matematica, Università di Roma 'La Sapienza', P.le A. Moro, 2 - 00185 Roma, Italy

INTRODUCTION

A Mori domain is a domain such that the ascending chain condition holds for integral divisorial ideals. Thus the first examples of Mori domains are Noetherian domains. It is well known that a completely integrally closed Mori domain is a Krull domain . However, as M.Roitman has shown, the complete integral closure A* of a Mori domain A is not necessarily a Krull domain. Indeed A* need be neither completely integrally closed nor Mori (cf.[R]).The main result of this paper is the following: if A is a seminormal Mori domain , then A* is a Krull domain.

Recall that a domain A is seminormal if, for any element x in the quotient field of A, whenever $x^2, x^3 \in A$, then $x \in A$ (cf.[S]).

In Section 1 we define, for any Mori domain A, a sequence $\{A_i; i \geq 0\}$ of overrings of A. This algorithmic construction is similar to that given in [B2], but it is more general (in fact we do not exclude the case $A:A^* = (0)$). The first part of the paper extends to this 'new' sequence many of the properties of the sequence defined in [B2]. This sequence of overrings is analyzed in Section 2 in the seminormal case, in order to show that (in this case) $\cup\{A_i; i \geq 0\}$ is the complete integral closure of A and it is a Krull domain (cf. Theorem 2.9)

Throughout the paper the strongly divisorial ideals play an important role. Recall that a non-zero ideal I of a domain A is divisorial if $I = A:(A:I) = I_v$, is strong if $A:I = I:I$ and is strongly divisorial if it is strong and divisorial. It turns out that the number of strongly divisorial prime ideals of a Mori domain A is a measure of how far is A from being completely integrally closed. More precisely, in the seminormal case, in passing in our sequence from A_i to A_{i+1}, the number of strongly divisorial prime ideals decreases and becomes zero in $\cup\{A_i; i \geq 0\}$, which in fact turns out to be completely integrally closed.

If A is a seminormal Mori domain and $\{A_i; i \geq 0\}$ is the associated sequence of overrings, we show in Section 3 that, descending in the sequence, A_i is obtained from A_{i+1} by glueing over a suitable set of (strongly divisorial) prime ideals of A.Thus finally we get, in the integral case, a generalization of a result of C.Traverso [T] about seminormal Noetherian rings (cf.Corollary 3.2).

Throughout the paper, with ideal we mean an integral ideal.

We will use often the following well known facts in a Mori domain A. If I and J are non-zero fractional ideals of A and if S is a multiplicative closed subset of A, then $S^{-1}(A:I) = S^{-1}A : S^{-1}I$. Moreover, if $P \in \mathrm{Spec}(A)$, $P \neq (0)$, then P is strong if and only if A_P is not a DVR (cf.[G, Proposition 1.2]). Finally, if $(0) \neq P \in \mathrm{Spec}(A)$ and if $P \cap S = \emptyset$ (where S is a multiplicatite closed subset of A), then P is divisorial if and only if $S^{-1}P$ is divisorial and P is strong if and only if $S^{-1}P$ is strong.

1. THE ASSOCIATED SEQUENCE OF OVERRINGS

If A is a Mori domain we denote, as in [BG1], by $D_m(A)$ the set of maximal divisorial ideals of A. The elements of $D_m(A)$ are prime ideals and, if $P \in D_m(A)$, either A_P is a DVR or P is strongly divisorial (cf. [BG1, Proposition (2.1) and Theorem (2.5)]). Let us denote by $\mathcal{I}(A)$ the set $\{P \in D_m(A)$ such that A_P is a DVR$\}$ and by $\mathcal{S}(A)$ the set $\{P \in D_m(A)$ such that P is strongly divisorial$\}$.

Given a Mori domain A, we can construct a sequence of Mori overrings of A

$$A = A_0 \subset A_1 \subset \ldots \subset A_i \subset \ldots \qquad (*)$$

setting, for each $i \geq 0$,

$$A_{i+1} = \cap\{(A_i)_P;\ P \in \mathcal{I}(A_i)\} \cap \{(A_i)_P : P(A_i)_P;\ P \in \mathcal{S}(A_i)\}.$$

Notice that, for each $P \in \mathcal{S}(A_i)$, $P(A_i)_P$ is a strongly divisorial ideal of the Mori domain $(A_i)_P$. Thus $(A_i)_P : P(A_i)_P = P(A_i)_P : P(A_i)_P$ is still a Mori domain (cf.[B1, Corollary 11]). Therefore, for each $i \geq 0$, A_i is a Mori domain because it is an intersection with finite character of Mori domains (cf.[RD, I, Theorem 2]).

REMARK 1.1. a) If $A:A^* \neq (0)$, this construction coincides with that given in [B2, Section 1]. In fact in this case $\mathcal{S}(A_i)$ is empty or finite by [B2, Proposition 1.5]. If $\mathcal{S}(A_i)$ is empty, $A_{i+1} = \cap\{(A_i)_P;\ P \in \mathcal{I}(A_i) = D_m(A_i)\} = A_i$ as in the definition in [B2]. On the other hand, if

$\mathcal{S}(A_i) = \{Q_1, \ldots Q_n\}$ and if $R_i = Q_1 \cap \ldots \cap Q_n$, then , for each $P \in \mathcal{I}(A_i)$, $(A_i)_P = R_i(A_i)_P$ and for each j, $j = 1,\ldots,n$, $Q_j(A_i)_{Q_j} = R_i(A_i)_{Q_j}$. Hence in this case

$$A_{i+1} = \cap\{(A_i)_P:R_i(A_i)_P \,;\, P \in \mathcal{I}(A_i)\} \cap \{(A_i)_{Q_j}:R_i(A_i)_{Q_j} \,;\, j=1,\ldots,n\} =$$

$$\cap\{(A_i)_P:R_i(A_i)_P \,;\, P \in D_m(A_i)\} = \cap\{(A_i)_P \,;\, P \in D_m(A_i)\}:R_i = A_i:R_i.$$

For examples of this type see [B2, Examples 1.9].

b) If $A:A^* = (0)$, then, for some i, $A_i : A_{i+1} = (0)$. In this case A_i has an infinite number of strong maximal divisorial ideals, otherwise, as in a), $A_i : A_{i+1} = R_i. \neq (0)$, where R_i is the intersection of the strong maximal divisorial ideals of A_i. For such a Noetherian domain see for example [N, Appendix, Example 8]. A class of Mori domains of this type is constructed in [BGR2].

LEMMA 1.2. *Let A be a Mori domain and let* $B \subset C$ *be two consecutive domains of the associated sequence (*) constructed above. If* $Q \in \mathcal{S}(B)$ *and* $S = B\setminus Q$, *then* $S^{-1}C = B_Q:QB_Q$

Proof. Notice first that, for each $P \in D_m(B)$, $P \neq Q$, we have $B:Q \subset B_P$. In fact if $x \in B:Q$ and $y \in Q\setminus P$, then $xy \in B$ and so $x \in B_P$. Thus, for each $P \in D_m(B)$, $P \neq Q$, we have $B_Q:QB_Q = S^{-1}(B:Q) \subset S^{-1}B_P \subset S^{-1}(B_P: PB_P)$.

Now $S^{-1}C = \cap\{S^{-1}B_P \,;\, P \in \mathcal{I}(B)\} \cap \{S^{-1}(B_P:PB_P)\,,\ P \in \mathcal{S}(B)\}$. For each $P \in \mathcal{I}(B)$, B_P is a DVR and $S^{-1}B_P$ is the quotient field of B. On the other hand $S^{-1}(B_Q:QB_Q) = B_Q:QB_Q$ and , if $P \in \mathcal{S}(B)$, $P \neq Q$, as we just proved, $B_Q:QB_Q \subset S^{-1}(B_P:PB_P)$. Thus $S^{-1}C = B_Q:QB_Q$.

PROPOSITION 1.3. *Let A be a Mori domain and let* $B \subset C$ *be two consecutive domains of the associated sequence (*). If* $J \subset C$ *is a fractional ideal of* B *then* $B:(B:J) \subset C:(C:J)$.

Proof. First we show that $J_v = B:(B:J) \subset C$. Since B is a Mori domain, by [BG1, Proposition 2.2,c)], we have that $J_v = \cap\,\{(JB_P)_v \,;\, P\in D_m(B)\}$. If $P \in \mathcal{I}(B)$, then $(JB_P)_v \subset B_P$, because $J \subset C \subset B_P$, for each $P \in \mathcal{I}(B)$; if $P\in \mathcal{S}(B)$, then, setting $S = B\setminus P$, we have $JB_P = S^{-1}J \subset S^{-1}C = PB_P:PB_P$ (cf. Lemma 1.2 for the last equality). Hence, since $PB_P:PB_P$ is a fractional divisorial ideal of B_P, $(JB_P)_v \subset PB_P:PB_P$. Thus $J_v \subset \cap\{B_P;\ P\in \mathcal{I}(B)\} \cap \{PB_P:PB_P \,;\, P \in \mathcal{S}(B)\} = C$.

Suppose now $z \in C:J$. Then $zJ \subset C$. Hence $z(B:(B:J)) = B:(B:zJ) \subset C$, by the previous part of the proof applied to the ideal $J' = zJ \subset C$, which is the desired result.

COROLLARY 1.4. *Let A be a Mori domain and let* $B \subset C$ *be two consecutive domains of the associated sequence (*). If* H *is a divisorial ideal of* C, *then* $H \cap B$ *is a divisorial ideal of* B. *Moreover if* H *is a strongly divisorial prime ideal of* C, *then* $H \cap B$ *is a strongly divisorial (prime) ideal of* B.

Proof. Since C is an overring of B we can assume $h \neq (0)$. By Proposition 1.3, $B:(B:h) \subset C:(C:h) \subset C:(C:H) = H$. On the other hand $B:(B:h) \subset B$ and so $B:(B:h) \subset H \cap B = h$ and h is divisorial.

Now suppose that H is a strongly divisorial prime ideal of C. If $h = H \cap B$ is not strong, then B_h is a DVR and, since $B_h \subset C_H$, also C_H is a DVR, a contradiction because HC_H is a strongly divisorial prime ideal of C. Thus h is a strongly divisorial prime ideal of B.

For the convenience of the reader, we state the following result which is partially known:

LEMMA 1.5. *Let* H *be a strongly divisorial ideal of a domain* D *and let* $E = (D:H)$. *If* p *is a prime ideal of* D *and if* $p \not\supset H$, *then there exists exactly one prime ideal* P *of* E *lying over* p *(i.e. such that* $P \cap D = p$*) and for this* P *we have* $D_p = E_P$. *Moreover if* I *is an ideal of* D, *if* $I \subset p$, *and if* J *is an ideal of* E *lying over* I, *then* $J \subset P$.

Proof. It is well known that for any $p \in \text{Spec}(D)$, $p \not\supset H$, there exists exactly one prime ideal P of E lying over p, $P = (p:H)$ and $D_p = E_P$ (cf. [B2, p.104-105]). Let us show now that $J \subset P$. If $y \in J \setminus P$, then $yh \notin p$, for some $h \in H$. Thus $yh \notin I$. On the other hand H is an ideal of E, so $yh \in H \subset D$. Therefore $yh \notin J$, which is a contradiction because J is an ideal of E.

PROPOSITION 1.6. *Let A be a Mori domain and let* $B \subset C$ *be two consecutive domains of the associated sequence (*). If* p *is a divisorial prime ideal of* B *and* $p \notin \mathcal{S}(B)$, *then there exists exactly one prime ideal* P *of* C *lying over* p *and*

1) $B_p = C_P$;

2) if p *is strongly divisorial, then* P *is strongly divisorial;*

3) if $p \in \mathcal{I}(B)$, *then* $P \in \mathcal{I}(C)$.

Moreover if I *is a divisorial ideal of* B, *if* $I \subset p$ *and if* J *is a divisorial ideal of* C *lying over* I, *then* $J \subset P$ *and* $JC_P = IB_p$.

Proof. Since $p \notin \mathcal{S}(B)$, we have two possibilities:

a) $p \subsetneq m$, for some $m \in \mathcal{S}(B)$;

b) $p \in \mathcal{I}(B)$.

In case a), setting $S = B \setminus m$, we have $S^{-1}B \subset S^{-1}C = S^{-1}B:S^{-1}m$ (cf. Lemma 1.2), and $S^{-1}p \subsetneq S^{-1}m$ are divisorial prime ideals of $S^{-1}B$. Appying Lemma 1.5 to $S^{-1}B \subset S^{-1}C$, we get that there is exactly one prime ideal of $S^{-1}C$ lying over $S^{-1}p$; so there is exactly one prime ideal P of C lying over p and $B_p = S^{-1}B_{S^{-1}p} = S^{-1}C_{S^{-1}P} = C_P$. Moreover, if p is strongly divisorial, then, by [B2, Proposition 1.10] $S^{-1}P$ is strongly divisorial, hence P is strongly divisorial too.

In case b), set $P = p B_p \cap C$. P is a prime ideal of C lying over p. Since $B_p \subset C_P$ and B_p is a DVR, we have that $C_P = B_p$ is a DVR. Moreover if P' is another prime ideal of C lying over p, then, as before, $C_{P'} = B_p$ and $P' = P'C_{P'} \cap C = p B_p \cap C = P$. Since $p \in I(B)$, P is also maximal divisorial, because P is the only prime ideal of C lying over p, p is maximal divisorial in B and the contraction of a divisorial ideal of C is a divisorial ideal of B (cf. Corollary 1.4). Thus $P \in I(C)$.

Let I be a divisorial ideal of B, $I \subset p$, and let J be a divisorial ideal of C lying over I.

In case a), if $S = B \setminus m$, applying Lemma 1.5, we get $S^{-1}J \subset S^{-1}P$ and so $J \subset P$. In case b) suppose $J \not\subset P$. Let $P_1,\dots,P_n$ be the maximal divisorial ideals of C containing J. Setting for any i, i = 1,...,n, $p_i = P_i \cap B$, we have that $p \not\subset p_i$. We want to show that for any i, i = 1,...,n, $IB_{p_i} \cap B \not\subset p$. Let $x \in \mathrm{rad}(IB_{p_i}) \cap B$. Then $x^n \in IB_{p_i} \cap B$, for some $n \in \mathbb{N}$. If $IB_{p_i} \cap B \subset p$, then $x \in p$ and so $\mathrm{rad}(IB_{p_i}) \cap B \subset p$, a contradiction, because $\mathrm{rad}(IB_{p_i})$ is a finite intersection of primes, $p \not\subset p_i$ and $\mathrm{ht}(p) = 1$. Thus $IB_{p_1} \cap \dots \cap IB_{p_n} \cap B \not\subset p$ and so $IB_{p_1} \cap \dots \cap IB_{p_n} \cap B \not\subset IB_p \cap B$.

Therefore we have $I \subset IB_p \cap IB_{p_1} \cap \dots \cap IB_{p_n} \cap B \subsetneq IB_{p_1} \cap \dots \cap IB_{p_n} \cap B$. On the other hand $I = J \cap B = JC_{P_1} \cap \dots \cap JC_{P_n} \cap B \supset IB_{p_1} \cap \dots \cap IB_{p_n} \cap B$ and so we get a contradiction.

Finally we have to prove that (in both cases a) and b)) $JC_P = IB_p$. Setting $T = B \setminus p$, we have $JC_P = J B_p = T^{-1}J = T^{-1}P \cap T^{-1}J = T^{-1}p \cap T^{-1}J = T^{-1}(p \cap J) = T^{-1}I = IB_p$.

COROLLARY 1.7. *Let* A *be a Mori domain and let* $B \subset C$ *be two consecutive domains of the associated sequence (*). Suppose that* I *is a divisorial ideal of* B *and that, for any maximal divisorial ideal* p *of* B *containing* I, $p \in I(B)$. *Then there exists exactly one divisorial ideal* J *of* C *lying over* I *and, for any maximal divisorial ideal* P *of* C *containing* J, $P \in I(C)$.

Proof. Let $p_1,\dots,p_n$ be the maximal divisorial ideals of B containing I. Since $p_1,\dots,p_n \in I(B)$, by Proposition 1.6, there exist $P_1,\dots,P_n \in I(C)$, such that for any i, i = 1,...,n, P_i lies over p_i. Notice that the maximal divisorial ideals of C containing any divisorial ideal J of C lying over I are precisely $P_1,\dots,P_n$. As a matter of fact, by Proposition 1.6, $J \subset P_i$ and if $J \subset P \in D_m(C)$, then $I \subset P \cap B = p$ is a divisorial ideal of B (cf. Corollary 1.4) and whenever $p \neq p_i$, i.e. $p \notin I(B)$, then $I \subset p \subset m \in S(B)$. This is a contradiction.

Set $J = IB_{p_1} \cap \dots \cap IB_{p_n} \cap C$. Then J is a divisorial ideal of C lying over I. On the other hand, any divisorial ideal J' of C lying over I is such that for any i, i = 1,...,n, $J'C_{P_i} = IB_{p_i}$ and so $J' = J'C_{P_1} \cap \dots \cap J'C_{P_n} \cap C = IB_{p_1} \cap \dots \cap IB_{p_n} \cap C = J$.

2. THE MAIN RESULT

We consider now the seminormal case.

LEMMA 2.1. *Let* H *be a strongly divisorial ideal of a domain* D *and let* E = (D:H). *If* D *is seminormal, if* H *is a radical ideal of* E *and if* E *is a Mori domain, then* E *is seminormal.*

Proof. We know that $E = \cap\{ E_P ; P \in D_m(E)\}$ (cf.[BG1, Proposition 2.2, b)]). Thus it is enough to show that, for each $P \in D_m(E)$, E_P is seminormal. Let $P \in D_m(E)$; if $P \not\supset H$ and $p = P \cap D$, then $E_P = D_p$ is seminormal (cf. Lemma 1.5). On the other hand, if $P \supset H$, then, by [B2, Lemma 3.1], P is not a strongly divisorial ideal of E. Thus $P \in \mathcal{I}(E)$ and E_P is a DVR, which is trivially seminormal.

PROPOSITION 2.2. *Let* A *be a seminormal Mori domain and let (*) be the associated sequence. Then, for each* $i \geq 0$, A_i *is a seminormal (Mori) domain and, for any* $Q \in \mathcal{S}(A_i)$, *the ideal* $Q(A_i)_Q$ *is a radical ideal of* $(A_i)_Q:Q(A_i)_Q$.

Proof. Since A is seminormal, any localization A_P is also seminormal. Consider $Q \in \mathcal{S}(A)$. Since A_Q is seminormal, QA_Q is a radical ideal of the ring $A_Q:QA_Q$ by [BGR, Lemma 3.9]. So, by Lemma 2.1, $A_Q:QA_Q$ is seminormal and $A_1 = \cap\{A_P; P \in \mathcal{I}(A)\} \cap \{A_Q:QA_Q; Q \in \mathcal{S}(A)\}$ is also seminormal. The same argument applied to $A_i \subset A_{i+1}$ proves (by induction) the Proposition.

LEMMA 2.3. *Let* A *be a seminormal Mori domain and let* $B \subset C$ *be two consecutive domains of the associated sequence (*). Then:*

a) If Q *is a strongly divisorial prime ideal of* C, *then* $q = Q \cap B$ *is not a maximal divisorial ideal of* B.

b) the map $Q \rightarrow Q \cap B$ *gives a one to one inclusion preserving correspondence between the set of all the strongly divisorial prime ideals of* C *and the set of strongly divisorial prime ideals of* B *that are not maximal divisorial.*

Proof. a): If q is a maximal divisorial ideal of B, then, by Corollary 1.4, $q \in \mathcal{S}(B)$. In this case, setting $S = B \setminus q$, by Lemma 1.2, we have that $S^{-1}C = S^{-1}B : S^{-1}q$, where $S^{-1}q$ is a strongly divisorial ideal of $S^{-1}B$. Since Q is a strongly divisorial ideal of C, $S^{-1}Q$ is a strongly divisorial ideal $S^{-1}C$. Moreover, by Proposition 2.2, $S^{-1}q$ is a radical ideal of $S^{-1}C$ and this contradicts [B2, Lemma 3.1].

b): by a) and Proposition 1.6, because the contraction of a strongly divisorial ideal of C is a strongly divisorial ideal of B (cf. Corollary 1.4).

PROPOSITION 2.4. *Let* A *be a seminormal Mori domain and let* $B \subset C$ *be two consecutive domains of the associated sequence (*). Then for any divisorial ideal* I *of* B, *there exists an integer* m(I) *such that the number of divisorial ideals of* C *lying over* I *is at most* m(I).

Proof. Consider $(IC)_v = C:(C:I)$. If $(IC)_v \cap B \supsetneq I$, then there are no divisorial ideals of C lying over I and the Proposition trivially holds. Suppose that $(IC)_v \cap B = I$. Let $P_1,...,P_r$ be the maximal divisorial ideals of C containing $(IC)_v$ and such that $p_i = P_i \cap B \notin \mathcal{S}(B)$. Let $Q_1,...,Q_s$ be the maximal divisorial ideals of C containing $(IC)_v$ and such that $Q_j \cap B \in \mathcal{S}(B)$. By Lemma 2.3 a), for any j, j = 1,...,s, $Q_j \in \mathcal{I}(C)$ and so C_{Q_j} is a DVR.

Let J be a divisorial ideal of C lying over I. By Proposition 1.6, $J \subset P_i$, for any i, i = 1,...,r, and $JC_{P_i} = IB_{p_i}$. On the other hand, if $J \subset P \in D_m(C)$, then $P = P_i$, for some i, i = 1,...,r, or $P = Q_j$, for some j, j = 1,...,s. By [BG1,Proposition 2.2(c)], we have $J = JC_{P_1} \cap ... \cap JC_{P_r} \cap JC_{Q_1} \cap ... \cap JC_{Q_s} \cap C$. Since, for any j, j = 1,...,s, $C_{Q_j} = (V_j,M_j)$ is a DVR and since $JC_{Q_j} = JV_j \supset IV_j = M_j^{k_j}$, we have $JC_{Q_j} = M_j^{h_j}$, for some h_j, $0 \le h_j \le k_j$. Thus, for the generic divisorial ideal J of C lying over I, we have at most (k_j+1) choices for the components JC_{Q_j}. On the other hand, as we have seen before, the components $JC_{P_i} = IB_{p_i}$ are uniquely determined by I. Therefore there are at most $(k_1+1)...(k_s+1)$ divisorial ideals J of C lying over I and the Proposition is proved.

EXAMPLE 2.5. Consider the seminormal Mori (in fact Noetherian) domain $A = k + (X^2-1)k[X]$, where k is a field. In this case $\mathcal{S}(A) = \{q = (X^2-1)k[X]\}$ and the associated sequence (*) is simply $A_0 = A \subset A_1 = k[X]$. Consider the divisorial ideal $I = X^2k[X] \cap q = X^2(X^2-1)k[X]$ of A. The maximal divisorial ideals of k[X] containing $(Ik[X])_v = I$ are $P_1 = Xk[X]$, $Q_1 = (X+1)k[X]$ and $Q_2 = (X-1)k[X]$. Notice that $P_1 \cap A \notin \mathcal{S}(A)$ and that, for j = 1,2, $Q_j \cap A = q \in \mathcal{S}(A)$. Setting $k[X]_{(X+1)} = (V_1,M_1)$ and $k[X]_{(X-1)} = (V_2,M_2)$, we have $J = JV_1 \cap JV_2 \cap Jk[X]_{(X)} \cap k[X]$. Since, for j = 1,2, $IV_j = M_j \subset JV_j$, we have two choices for each component JV_j, while the component $Jk[X]_{(X)} = Ik[X]_{(X)} = X^2k[X]_{(X)}$ is uniquely determined. Thus we have four divisorial ideals of k[X] lying over I, namely $J_1 = X^2(X^2-1)k[X]$, $J_2 = X^2(X+1)k[X]$, $J_3 = X^2(X-1)k[X]$, $J_4 = X^2k[X]$.

LEMMA 2.6. *Let* A *be a seminormal Mori domain and let (*) be the associated sequence. Let* I *be a divisorial ideal of* A_i *and let* k *be the maximum of the lengths of chains of strongly divisorial prime ideals of* A_i *containing* I. *If* J *is a divisorial ideal of* A_{i+k+1} *lying over* I, *then any maximal divisorial ideal of* A_{i+k+1} *containing* J *is in* $\mathcal{I}(A_{i+k+1})$.

Proof. Suppose that $J \subset P \in \mathcal{S}(A_{i+k+1})$. Applying Lemma 2.3 b) (k+1) times, we get that there is in A_i a chain of length (k+1) of strongly divisorial prime ideals containing $I = J \cap A_i$, a contradiction.

PROPOSITION 2.7. *Let* A *be a seminormal Mori domain and let* A_i *be any domain of the associated sequence (*). Then, for any divisorial ideal I of* A_i *there exists an integer* m'(I) *such that any strictly ascending chain of divisorial ideals of* A_{i+h} *(for each integer* h, $h \geq 0$*) lying over* I *contains at most* m'(I) *terms.*

Proof. Let k be the maximum of the lengths of chains of strongly divisorial prime ideals of A_i containing I. Applying Proposition 2.4 (k+1) times, we get that there exists an integer m'(I) such that the number of divisorial ideals of A_{i+k+1} lying over I is at most m'(I). In particular any strictly ascending chain of divisorial ideals of A_{i+k+1} lying over I contains at most m'(I) terms. So, if $h \leq k+1$, the Proposition is easily proved. On the other hand, by Lemma 2.6, if J is a divisorial ideal of A_{i+k+1} lying over I, then any maximal divisorial ideal of A_{i+k+1} containing J is in $\mathcal{I}(A_{i+k+1})$. Thus, by Corollary 1.7, for each $h > k+1$, in A_{i+h} there is exactly one divisorial ideal lying over J.

In conclusion, for each integer h, $h \geq 0$, any strictly ascending chain of divisorial ideals of A_{i+h} lying over I contains at most m'(I) terms.

COROLLARY 2.8. *Let* A *be a seminormal Mori domain and let (*) be the associated sequence. Then* $\cup\{A_i\,;\, i \geq 0\}$ *is a Mori domain.*

Proof. By Proposition 2.7 , we can apply Theorem 2.4 of [BGR2]. Notice that, by Corollary 1.4 and Proposition 1.1 of [BGR2], condition a) of [BGR2, Theorem 2.4] is satisfied in our case.

THEOREM 2.9. *Let* A *be a seminormal Mori domain and let (*) be the associated sequence. Then* $\cup\{A_i\,;\, i \geq 0\} = A^*$ *is a Krull domain.*

Proof. Since, for any i, $A_i \subset A^*$, we have that $\cup\{A_i\,;\, i \geq 0\} \subset A^*$. To prove the opposite inclusion, it is enough to show that $D = \cup\{A_i\,;\, i \geq 0\}$ is completely integrally closed. Since we know that D is a Mori domain (cf. Corollary 2.8), this is equivalent to show that D does not have strongly divisorial prime ideals (cf.[B1, Corollary 14]). Suppose that P is a strongly divisorial prime ideal of D. Then, by Proposition 1.1 and Lemma 2.1 of [BGR2], for any i, $p_i = P \cap A_i$ is a divisorial prime ideal of A_i. Moreover p_i is strongly divisorial (otherwise $(A_i)_{p_i}$ is a DVR and also D_P is a DVR, a contradiction). Fix an index i and apply Lemma 2.6 to the divisorial ideal p_i of A_i. For some k, any maximal divisorial ideal of A_{i+k} containing $P \cap A_{i+k}$, that lies over p_i, is in $\mathcal{I}(A_{i+k})$. On the other hand, as we observed before, $P \cap A_{i+k}$ is a strongly divisorial prime ideal of A_{i+k}. So we get a contradiction. Therefore D is a completely integrally closed Mori domain, i.e. a Krull domain.

As in [B2], we call the supremum of the lengths of all chains of strongly divisorial prime ideals of an integral domain A the *strong dimension* of A.

COROLLARY 2.10. *Let* A *be a seminormal Mori domain and suppose that the strong dimension* s *of* A *is finite. Then the sequence (*) is stationary at* $A_{s+1} = A^*$ *and* A^* *is a Krull domain.*

Proof. By Theorem 2.9, it is enough to show that the sequence (*) is stationary at A_{s+1}, i.e. that $\mathcal{S}(A_{s+1}) = \emptyset$. If $P \in \mathcal{S}(A_{s+1})$, applying Lemma 2.3 b) (s+1) times, we get that there is in A a chain of strongly divisorial prime ideals of lenght (s+1), a contradiction.

3. GLUEINGS

Recall the following definitions. If $A \subset B$ are two rings, if $p \in \operatorname{Spec} A$ and $I = \operatorname{rad}(pB)$ lies over p, then the *ring obtained from* B *by glueing over* p is the pullback of the following diagram

$$\begin{array}{ccc} & & k(p) \\ & & \updownarrow \\ B & \xrightarrow{\ f\ } & S^{-1}B/S^{-1}I \end{array}$$

where $S = A \setminus p$, f is the composition of the canonical maps $B \longrightarrow S^{-1}B \longrightarrow S^{-1}B/S^{-1}I$, $k(p) = A_p/pA_p$ and $k(p) \longrightarrow S^{-1}B/S^{-1}I$ is the canonical immersion.

If $\{p_i;\ i \in \Omega\}$ is a set of prime ideals of A and for each $i \in \Omega$, $\operatorname{rad}(p_iB)$ lies over p_i, then the *ring obtained from* B *by glueing over* $\{p_i;\ i \in \Omega\}$ is the ring $\cap\{B_i;\ i \in \Omega\}$, where, for each $i \in \Omega$, B_i is the ring obtained from B by glueing over p_i.

Before stating our next result, we need the following notation. If A is a seminormal Mori domain and if (*) is the associated sequence of overrings, then, for any i, denote by $\mathcal{P}_i(A) = \mathcal{P}_i$ the set of $q \in \operatorname{Spec} A$ such that $q = Q \cap A$, for some $Q \in \mathcal{S}(A_i)$. By Corollary 1.4, any $\mathcal{P}_i$ is a set of strongly divisorial prime ideals of A. Conversely, if p is a strongly divisorial prime ideal of A and $p \subset p_1 \subset \ldots \subset p_i$ is a chain of strongly divisorial prime ideals of A of maximum length, then, by Lemma 2.3 b), $p = P \cap A$ for some $P \in \mathcal{S}(A_i)$. Thus $\mathcal{P} = \cup \{\mathcal{P}_i ; i \geq 0\}$ is the set of strongly divisorial prime ideals of A .

THEOREM 3.1. *Let* A *be a seminormal Mori domain and let* (*) *be the associated sequence. Then* A_i *is obtained from* A_{i+1} *by glueing over the set* $\mathcal{P}_i$.

Proof. By Proposition 2.2, for each i, and for each $Q \in \mathcal{S}(A_i)$, if $S = A_i \setminus Q$, we have that $S^{-1}Q$ is a radical ideal of $(A_i)_Q$: $Q(A_i)_Q = S^{-1}A_{i+1}$ (cf. Lemma 1.2). Thus $(A_i)_Q \cap A_{i+1}$ is obtained from A_{i+1} by glueing over Q and, since $A_i = \cap\{(A_i)_Q \cap A_{i+1} ; Q \in \mathcal{S}(A_i)\}$, A_i is obtained from A_{i+1} by glueing over $\mathcal{S}(A_i)$.

Let $Q \in \mathcal{S}(A_i)$ and $q = Q \cap A_{i-1}$. By Corollary 1.4, q is a strongly divisorial prime ideal of A_{i-1} , but it is not maximal divisorial because A is seminormal (cf. Lemma 2.3 a)). Let $q \subsetneq m \in \mathcal{S}(A_{i-1})$. Set $\Sigma = A_{i-1} \setminus m$. By Lemma 1.2, we have that $\Sigma^{-1}A_i = \Sigma^{-1}A_{i-1} : \Sigma^{-1}m$, where $\Sigma^{-1}m$ is a strongly divisorial ideal of $\Sigma^{-1}A_{i-1}$ and $\Sigma^{-1}q \supsetneq \Sigma^{-1}m$. Applying [B2, Lemma 3.5], we get that the ring obtained from $\Sigma^{-1}A_{i+1}$ by glueing over $\Sigma^{-1}Q$ coincides with the ring obtained from $\Sigma^{-1}A_{i+1}$ by glueing over $\Sigma^{-1}q$. Therefore the ring obtained from A_{i+1} by glueing over Q coincides with the ring obtained from A_{i+1} by glueing over $q \in \mathrm{Spec}(A_{i-1})$. Iterating this argument, we get that A_i is obtained from A_{i+1} by glueing over the prime ideals q of Spec(A) such that $q = Q \cap A$, for some $Q \in \mathcal{S}(A_i)$, that is by glueing over $\mathcal{P}_i$.

In [T] Traverso has proved that if A is a seminormal Noetherian ring, such that the integral closure A' is a finitely generated A-module, then there exists a chain of overrings of A, $A = A_0 \subset A_1 \subset \dots \subset A_n = A'$, where A_i is obtained from A_{i+1} by glueing over a prime ideal of A. Moreover it is well known that, if A is an integral domain, then A' is a Krull domain (Mori-Nagata Theorem).

We can state a similar result for seminormal Mori domains of finite (strong) dimension. For such a domain A, the complete integral closure A*, which coincides in the Noetherian case with the integral closure A', is also a Krull domain as we have just proved.

COROLLARY 3.2. *Let* A *be a seminormal Mori domain of finite strong dimension* s. *Then there exists a sequence* $A = A_0 \subset A_1 \subset \dots \subset A_{s+1} = A^*$ *of overrings of* A *such that* A_i *is obtained from* A_{i+1} *by glueing over the set* $\mathcal{P}_i$.

Proof. Apply Corollary 2.10 and Theorem 3.1.

EXAMPLES 3.3. Examples of seminormal Mori domains obtained with a finite number of glueings from a Krull domain of the type $k[X_1,...,X_n]$ (where k is a field) may be found in [B2, Examples 3.12].

Notice that Corollary 3.2 holds in particular for any seminormal Noetherian domain A of finite dimension, without the hypothesis that the integral closure A' = A* of A is a finitely generated A-module.

Since in a Noetherian domain, any prime ideal has a finite height, we can give a 'local version' of Corollary 3.2 for any Noetherian domain:

COROLLARY 3.4. *Let* A *be a seminormal Noetherian domain. Then, for any* $P \in D_m(A)$, *there exists a sequence*

$$A_P = A_0(P) \subset A_1(P) \subset \dots \subset A_{s(P)+1}(P) = (A_P)'$$

of overrings of A_P *such that* $A_i(P)$ *is obtained from* $A_{i+1}(P)$ *by glueing over the set* $\mathcal{P}_i(A_P)$.

Proof. For any $P \in D_m(A)$, A_P is a seminormal Mori (in fact Noetherian) domain of finite strong dimension s(P). So we can apply Corollary 3.2. Since A_P is Noetherian, the complete integral closure $(A_P)^*$ coincides with the integral closure $(A_P)'$.

Notice that Corollary 3.4 can also be formulated for any maximal ideal of A.

REFERENCES

[B1] Barucci, V.(1986). Strongly divisorial ideals and complete integral closure of an integral domain, J. Algebra, 99: 132-142.

[B2] Barucci, V.(1989). A Lipman's type construction, glueings and complete integral closure, Nagoya Math. J., 113: 99-119.

[BG1] Barucci,V., Gabelli, S. (1987). How far is a Mori domain from being a Krull domain?, J. Pure Appl. Alg., 45: 101-112.

[BGR] Barucci,V., Gabelli, S., Roitman, M. (1992) On Semi-Krull domains, J. Algebra 145: 306-328.

[BGR2] Barucci, V., Gabelli, S., Roitman, M.(1991).On the class group of a strongly Mori domain, Comm.Alg. (to appear).

[Fn] Fontana, M.,(1980). Topologically defined classes of commutative rings, Annali Mat. Pura Appl., 123: 331-355.

[F] Fossum, R. (1973). The divisor class group of a Krull domain, Springer-Verlag.

[G] Gabelli, S. (1987). On divisorial ideals in polynomial rings over Mori domains, Comm. Alg., 15: 2349-2370.

[HOP] Heinzer, W., Ohm, J., Pendleton, R.L. (1970). On integral domains of the form $\cap D_P$, P minimal, J. Reine Angew. Math., 241: 147-159.

[N] Nagata, M. (1962). Local rings, Interscience, New York.

[RD] Raillard (Dessagnes), N. (1975). Sur les anneaux de Mori, C.R. Acad. Sc. Paris., 280: 1571-1573.

[R] Roitman, M. (1990). On the complete integral closure of a Mori domain, J. Pure Appl. Alg., 66: 55-79.

[S] Swan, R. (1980). On seminormality, J. Algebra, 67: 210-229.

[T] Traverso, C. (1970). Seminormality and Picard group, Ann Sc. Norm. Sup. Pisa, 24: 585-595.

2
Maximality Properties in Numerical Semigroups, with Applications to One-Dimensional Analytically Irreducible Local Domains

Valentina Barucci[*,†]
Dipartimento di Matematica
Università di Roma "La Sapienza"
00185 Roma
ITALY

David E. Dobbs[*,**]
Department of Mathematics
University of Tennessee
Knoxville, TN 37996-1300
U.S.A.

Marco Fontana[*,†]
Facoltà di Scienze
Terza Università degli studi di Roma
00146 Roma
ITALY

1. Introduction

Suppose that A is a Noetherian local (commutative integral) domain which is analytically irreducible and of Krull dimension at most 1; as we noted in [BDF2, Remark 5(b)], this is equivalent to supposing that A is a Noetherian conducive domain, in the sense of [BDF1]. Thus (apart from the trivial case of a field A), we are supposing that the integral closure of A is a DVR, say $(V, \mathbf{m})$, which is finite over A. Suppose also that A contains the residue field, k, of V. As noted by Kunz [K], the conductor $(A : V) = \mathbf{m}^{g+1}$, where g is the Frobenius number of the numerical value semigroup S of A. Building on

[*]Supported in part by NATO Collaborative Research Grant CRG 900113
[**]Supported in part by Dipartimento di Matematica, Università di Roma "La Sapienza"
[†]Supported in part by Ministero dell' Università e delle Ricerca Scientifica e Tecnologica (60% Fund)

work of Fröberg-Gottlieb-Häggkvist [FGH] for the case $V = k[[X]]$, we showed in [BDF2, Theorem 4] that if g is odd, then A is maximal inside V with respect to the property $(A : V) = \mathbf{m}^{g+1}$ if and only if A is a Gorenstein domain. It is natural to ask which rings A (and associated semigroups S) are characterized by the corresponding maximality condition in case g is even. We answer this motivating question, by introducing a class of domains of Cohen-Macaulay type 2 (dubbed "Kunz domains" in honor of [K]) and a corresponding class of semigroups (dubbed "pseudo-symmetric," since Gorenstein A corresponds to symmetric S). This announcement deals with these new concepts and their relations to various classes of rings (Gorenstein; Arf, in the sense of [L]) and various related classes of semigroups.

Let A be as above. The condition that A be maximal inside V with respect to $(A : V) = \mathbf{m}^{g+1}$ (regardless of whether g is odd or even) is characterized in Corollary 19 in terms of the so-called type sequence of A. Moreover, the "conducive" property allows us to view each overring of A, apart from the quotient field, as a fractional ideal of A. Theorem 26 asserts that (even if k is not inside A) if A is a Gorenstein domain and $g \geq 2$, then $B \mapsto I = (A : B)$ gives an order-reversing bijection between the integral overrings of A and the strongly divisorial ideals of A; in this bijection, B is Gorenstein if and only if I is stable.

The previous result suggests a connection with Arf domains (that is, domains in which each integrally closed ideal is stable). We obtain several characterizations of the Arf Gorenstein domains A of the above type. In particular, Theorem 23 asserts that if A is as above, then A is an Arf Gorenstein domain $\Leftrightarrow$ each integral overring of A is a Gorenstein domain $\Leftrightarrow$ A is Gorenstein and of maximal embedding dimension. A key tool is Theorem 22, in which the "Arf" property for rings A of the above type is characterized by the coincidence of certain canonical sequences of integral overrings of A. (The rings in one of these sequence are defined by blowing-up, as in [L]; the rings in the other sequences arise as duals of relevant ideals.)

Insofar as possible, we first develop matters semigroup-theoretically, introducing numerical semigroup analogues of the above (and other) ring-theoretic concepts. Despite the symmetric $\leftrightarrow$ Gorenstein and pseudo-symmetric $\leftrightarrow$ Kunz parallels, there are some surprises. For instance, although "Arf" and "maximal embedding dimension" semigroups are of interest, they do not *by themselves* serve to characterize the correspondingly named ring-theoretic notions. The "missing link" involves comparing the semigroup–and ring-theoretic notions of "type" and "embedding dimension": see Proposition 21 and Theorem 22.

Let A, V, S, g be as above. Theorem 24 states that $(A, \mathbf{m}_A)$ is Kunz of maximal embedding dimension $\Leftrightarrow$ A is of Cohen-Macaulay type 2 and $\mathbf{m}_A^{-1}$ is a Gorenstein domain. In conjunction with earlier semigroup-theoretic material, this leads to a characterization of the Kunz (resp., Gorenstein) domains A of maximal embedding dimension: these correspond to the semigroups $S = \langle 3, \frac{g}{2}+3, g+3 \rangle$ for even $g \equiv 1, 2 \pmod 3$ (resp., $S = \langle 2, g+2 \rangle$ for odd g). In this way, we believe that the "Kunz" and "pseudo-symmetric" concepts, in addition to answering the question left open in [FGH] and [BDF2], establish a theory within what Brown-Curtis [BC,p. 234] term the "sporadic" type 2 numerical semigroups which are not of almost maximal length.

Full proofs, related examples, and additional results will appear elsewhere.

2. Results

By a *numerical semigroup*, or simply a *semigroup*, S, we mean an additive submonoid of the monoid $\mathbb{N}$ of all natural numbers (including 0). It is well known that each semigroup S is finitely generated; that is, there exist $a_1, a_1, \ldots, a_\nu \in S$ such that $S = \langle a_1, a_2, \ldots, a_\nu \rangle := \{\sum_{i=1}^{\nu} a_i n_i : n_i \in \mathbb{N}\}$. When $\nu \geq 2$, we always suppose that $a_i < a_{i+1}$ for $1 \leq i \leq \nu - 1$, that $GCD(a_1, a_2, \ldots, a_\nu) = 1$ [or, equivalently, that $\mathbb{N} \setminus S$ is finite] and that $a_i \notin \langle a_1, \ldots, a_{i-1}, \widehat{a_i}, a_{i+1}, \ldots, a_\nu \rangle$. Under these conditions, it is easy to see that $\nu \leq a_1$. The least positive integer belonging to S is called the *multiplicity of S* and is denoted by $\mu = \mu(S)$. The cardinality of an irredundant set of generators of S is called the *embedding dimension of S* and is denoted by $e = e(S)$. With the above assumptions, $\mu(S) = a_1$ and $e(S) = \nu$. Therefore, for each semigroup S that we consider,

$$e(S) \leq \mu(S).$$

Also, we let $g = g(S)$ denote the *Frobenius number of S*, in the sense of [Br]; that is, the greatest integer which does not belong to S. When $S \neq \mathbb{N}$, it is often useful to consider the role of the integer $n = n(S) := \text{Card}\,(S \cap \{0, 1, \ldots, g(S)\})$; for convenience, we set $n(\mathbb{N}) = 0$.

A *relative ideal* of a semigroup S is a nonempty subset H of $\mathbb{Z}$ such that $H + S \subseteq H$ and $H + d \subseteq S$ for some $d \in S$. A relative ideal of S which is contained in S is simply called an *ideal* of S. It is straightforward to show that if H, K and L are relative ideals of S, then $H + K$, $kH(= H + H + \cdots + H, k$ summands for $k \geq 1; = \{0\}$ for $k = 0)$ and $(H - K) := \{z \in \mathbb{Z} : z + K \subseteq H\}$ are also relative ideals of S. If H is a relative ideal of S, then the relative ideal $(S - H)$ is called *the dual of H with respect to S* and is denoted by H^*. A relative ideal H of S such that $H^{**} = H$ is called a *bi-dual ideal of S*.

If $x_1, \ldots, x_r \in \mathbb{Z}$, then the relative ideal of S generated by these elements will be denoted by $(x_1, x_2, \ldots, x_r)$. If $r = 1$ and $x = x_1$, then $(x) = x + S$ is called the *relative principal ideal (of S) generated by x*. It is easy to see that every (relative) principal ideal is bidual and that $H^{**} = \cap \{x + S : x + S \supseteq H\}$.

For every relative ideal I of S, $(I - I)$ is a numerical semigroup, and it is the largest semigroup having I as a (relative) ideal.

Let S be a semigroup. The ideal $M = M(S) := \{x \in S : x \neq 0\}$ is called the *maximal ideal of S*. It is easy to see that $(M - M) = (S - M)$. Let $T(S)$ denote the (finite) set $M^* \setminus S$, and let $t = t(S)$, called the *type of S*, denote the cardinality of $T(S)$. If I is an ideal of S and $\{i_k : 1 \leq k \leq \sigma\}$ is an irredundant set of generators of I, then we notice that $hI = \{\sum_{k=1}^{\sigma} \alpha_k i_k : \alpha_k \in \mathbb{N}, \sum_{k=1}^{\sigma} \alpha_k \geq h\}$, and so

$$hI - (h+1)I = \left\{ \sum_{k=1}^{\sigma} \alpha_k i_k : \alpha_k \in \mathbb{N}, \sum_{k=1}^{\sigma} \alpha_k = h \right\}.$$

In particular, $M \setminus 2M = \{a_1, \ldots, a_\nu\}$, whence $\text{Card}(M \setminus 2M) = e(S)$.

Suppose that $g \in \mathbb{Z}$ satisfies $g \geq -1$ and $g \neq 0$. Then we can consider the set

$$\mathcal{S}_g := \{S : S \text{ is a numerical semigroup, } g(S) = g\},$$

which is partially ordered under set-theoretic inclusion. It is easy to see, by Zorn's Lemma, that $\mathcal{S}_g$ has at least one maximal element. Also, it is clear that $\mathcal{S}_g = \{\mathbb{N}\}$ if and only if $g = -1$.

The following two results are well known, and admit straightforward proofs.

LEMMA 1. *Let $g \in \mathbb{N}$ be odd. Then, for any semigroup $S \in \mathcal{S}_g$, the following are equivalent:*

(i) *S is maximal in $\mathcal{S}_g$;*
(ii) *The map $S \cap \{0, 1, \ldots, g\} \to (\mathbb{N} \setminus S) \cap \{0, 1, \ldots, g\}$, $s \mapsto g - s$, is a bijection;*
(iii) *$n(S) = (g+1)/2$;*
(iv) *$T(S) = \{g\}$ (or, equivalently, $M^* = \langle S, g \rangle$);*
(v) *$t(S) = 1$.* □

LEMMA 2. *Let $g \in \mathbb{N}$ be even. Then, for any semigroup $S \in \mathcal{S}_g$, the following are equivalent:*

(i) *S is maximal in $\mathcal{S}_g$;*
(ii) *the map $S \cap \left\{0, 1, \ldots, \widehat{g/2}, \ldots, g\right\} \to (\mathbb{N} \setminus S) \cap \left\{0, 1, \ldots, \widehat{g/2}, \ldots, g\right\}$, $s \mapsto g - s$, is a bijection;*
(iii) *$n(S) = g/2$;*
(iv) *$T(S) = \{g/2, g\}$ (or, equivalently, $M^* = \langle S, g/2 \rangle$).* □

A numerical semigroup satisfying the equivalent conditions of Lemma 1 (respectively, Lemma 2) is called a *symmetric* (respectively, a *pseudo-symmetric*) *semigroup*. The relevance of symmetric semigroups to the motivating question mentioned in the Introduction is clear; the relevance of pseudo-symmetric semigroups will appear in Proposition 17(a) and Theorem 18. It is important that, although symmetric subgroups are characterized by having type 1, the pseudo-symmetric semigroups are only one particular kind of semigroup having type 2. It is classically known that any semigroup S minimally generated by two relatively prime elements, $S = \langle a, b \rangle$, is symmetric, with $g(S) = ab - a - b$.

Our next goal is to sharpen the notion of "type," in order to characterize, in Proposition 3, the maximal elements of the set $\mathcal{S}_g$.

Let $S = \{0 = s_0, s_1, \ldots, s_{n-1}, s_n = g+1, \to\}$ be a given semigroup (where the symbol "$\to$" means that all subsequent natural numbers belong to S) with $s_i < s_{i+1}$ for $0 \le i \le n = n(S)$ and $g = g(S)$. For each $i \ge 0$, we can consider the ideal $S_i := \{s \in S : s \ge s_i\}$ and, hence, the relative ideal $S(i) := S_i^* = (S - S_i)$. For $0 \le i \le n$, it is easy to see that $S(i) = (S_i - S_i)$ and, thus, $S(i)$ is a semigroup. Moreover, $S(0) = S$, $S(1) = (M(S) - M(S))$, and $S(n) = \mathbb{N}$.

For $1 \le i \le n$, let $T_i(S)$ denote the (finite) set $S(i) \setminus S(i-1)$, and let $t_i = t_i(S)$ denote the number of elements of $T_i(S)$. In particular, $T_1(S) = T(S)$ and $t_1(S) = t(S)$. We call $(t_i(S) : 1 \le i \le n)$ the *type sequence of* S. It is obvious that

$$g(S) + 1 - n(S) = \sum_{i=1}^{n(S)} t_i(S)$$

is an integer reflecting the number of "gaps" of S inside $\mathbb{N}$; this integer is also called the *degree of singularity of* S and is denoted by $\delta = \delta(S)$.

If $1 \leq i \leq n(S)$, it is easy to verify the following:

$$g(S(i)) = g(S) - s_i;$$
$$1 \leq t_i(S) \leq t(S);$$
$$2n(S) \leq g(S) + 1 \leq n(S)(t(S) + 1); \quad \text{and}$$
$$t(S) \leq g(S) + 2 - 2n(S).$$

Using the above information, we can prove the following result. It will lead, in Corollary 19, to one answer for the motivating question that was left open in [FGH] and [BDF2].

PROPOSITION 3. *Let $1 \leq g \in \mathbb{N}$ and let $S \in \mathcal{S}_g$. Then S is maximal in $\mathcal{S}_g$ if and only if its type sequence is $(t(S), 1, \ldots, 1)$ and $t(S) \leq 2$.* □

We next use the notion of bi-dual ideal to characterize the symmetric and the pseudo-symmetric semigroups.

PROPOSITION 4. *Let S be a numerical semigroup, with maximal ideal M and Frobenius number g. Then:*

(a) *S is symmetric if and only if $I^{**} = I$ for every nonzero relative ideal I of S.*
(b) *S is pseudo-symmetric if and only if g is even, $g/2 \in (M - M)$, and $I^{**} = I$ for every relative ideal I of S such that $g/2 \in (I - I)$.* □

For each ideal I of a numerical semigroup S and for each integer $h \geq 0$, we can consider the semigroup $(hI - hI)$. Since $S \subseteq (hI - hI) \subseteq ((h+1)I - (h+1)I) \subseteq \mathbb{N}$, we may introduce the numerical semigroup $L(S, I) := \cup\{(hI - hI) : h \geq 0\}$, called (in honor of [L]) the *Lipman semigroup of S with respect to I*, or the *semigroup obtained from S by blowing-up I*.

If i_1 is the least positive integer in I, then one may show the following:

$$L(S, I) = (hI - hI), \text{ for some } h \geq 1;$$
$$(h+1)I = hI + i_1, \text{ for some } h \geq 1;$$
$$L(S, I) = \{z - ki_1 : z \in kI, k \geq 1\};$$
$$I + L(S, I) = i_1 + L(S, I); \text{ and}$$

if T is a semigroup such that $S \subseteq T \subseteq \mathbb{N}$ and $I + T = i + T$ for some $i \in I$, then $L(S, I) \subseteq T$.

From the previous considerations, one can show

PROPOSITION 5. *If $h \geq 1$ and i_1 is the least positive integer in the ideal I, then the following are equivalent:*

(i) *$I + (hI - hI) = i_1 + (hI - hI)$;*
(ii) *$Card(hI \setminus (h+1)I) = i_1$;*
(iii) *$(h+1)I = hI + i_1$;*
(iv) *$(hI - hI) = ((h+1)I - (h+1)I)$.* □

By analogy with [L], we say that an ideal I of a semigroup S is *stable* if I is a principal ideal of $(I - I)$. Stable ideals are characterized as follows:

PROPOSITION 6. *Let S be a numerical semigroup, and let I be an ideal with least positive element i_1. Then the following are equivalent:*

(i) *I is stable;*
(ii) *$Card(I \setminus 2I) = i_1$;*
(iii) *$2I = I + i_1$;*
(iv) *$(I - I) = L(S, I)$.* □

If $M = M(S)$ is the maximal ideal of a semigroup S, then the semigroup $L(S) := L(S, M)$ obtained from S by blowing-up M is called the *Lipman semigroup of S*. Since $S = S(1)$ implies that $(hM - hM) = S(1)$ for every $h \geq 1$, then

$$S = L(S) \Leftrightarrow S = S(1) \Leftrightarrow S = \mathbb{N} \Leftrightarrow g(S) = -1.$$

Since $S(1) \subseteq L(S)$ by definition, our next goal is to (use the concept of stability to) characterize the semigroups such that $S(1) = L(S)$.

PROPOSITION 7. *Let $S \neq \mathbb{N}$ be a numerical semigroup, with maximal ideal $M = M(S)$. Then the following are equivalent:*

(i) *$S(1) = L(S)$;*
(ii) *$e(S) = \mu(S)$;*
(iii) *$2M = M + s_1$ (where $s_1 = \mu(S)$);*
(iv) *$S(1) = S_1 - s_1$;*
(v) *$n(S(1)) = n(S) - 1$;*
(vi) *$S(i) = S(1)(i-1)$, for $1 \leq i \leq n(S)$;*
(vii) *if $n(S) \geq 2$, then $t_i(S) = t_{i-1}(S(1))$ for $2 \leq i \leq n(S)$;*
(viii) *M is stable.* □

As we remarked earlier, $e(S) \leq \mu(S)$ in general. For this reason, a numerical semigroup satisfying the equivalent conditions of Proposition 7 is called a *semigroup of maximal embedding dimension.*

Besides the ascending chain of numerical semigroups

$$(\mathbf{S}(\cdot)) \qquad S = S(0) \subset S(1) \subset \cdots \subset S(n) = \mathbb{N}$$

already considered, we can study two other interesting chains of semigroups obtained by duals and blowing-ups, respectively:

$$(\mathbf{B(S)}) \qquad S =: B_0 \subseteq B(B_0) =: B_1 \subseteq \cdots \subseteq B(B_h) =: B_{h+1} \subseteq \cdots$$

$$(\mathbf{L(S)}) \qquad S =: L_0 \subseteq L(L_0) =: L_1 \subseteq \cdots \subseteq L(L_h) =: L_{h+1} \subseteq \cdots$$

where, for a numerical semigroup T with maximal ideal N, we denote the numerical semigroup $(N - N) = (T - N) = T(1)$ by $B(T)$ and the Lipman semigroup of T by $L(T)$. Since $B_1 = B(S) = S(1)$, we see that

$$S = B(S) \Leftrightarrow S = L(S)$$

Moreover, since $\mathbb{N} \setminus S$ is finite (or empty), the sequences $(\mathbf{B(S)})$ and $(\mathbf{L(S)})$ are stationary.

In effect, Proposition 7 characterized the semigroups S such that $S(1) = B_1(S) = L_1(S)$ (that is, the semigroups of maximal embedding dimension). Next, we characterize the particular semigroups of maximal embedding dimension for which the chains $(\mathbf{S}(\cdot))$, $(\mathbf{B(S)})$, and $(\mathbf{L(S)})$ coincide.

THEOREM 8. *If S is a numerical semigroup and $n = n(S)$, then the following conditions are equivalent:*

(i) *The sequences* **(B(S))** *and* **(L(S))** *coincide;*
(ii) *The sequences* **(S(·))** *and* **(B(S))** *coincide;*
(iii) *The sequences* **(S(·))**, **(B(S))** *and* **(L(S))** *coincide;*
(iv) $e(S(i)) = \mu(S(i))$, *for* $0 \leq i \leq n(S)$*;*
(v) $n(S(i)) = n(S) - i$, *for* $0 \leq i \leq n(S)$*;*
(vi) $S(i)(j) = S(i + j)$, *for* $0 \leq i + j \leq n(S)$*;*
(vii) $t_j(S(i)) = t_{i+j}(S)$, *for* $1 \leq i + j \leq n(S)$*;*
(viii) $S = \{s_0, s_1 = \mu(S), \ldots, s_h = \sum_{i=1}^{h} \mu(S(i-1)), \ldots, s_n = \sum_{i=1}^{n} \mu(S(i-1)), \rightarrow\}$*;*
(ix) S_i *is a stable ideal, for every* $0 \leq i \leq n(S)$*.* □

A semigroup satisfying the equivalent conditions of Theorem 8 is called an *Arf semigroup.* Of course, each Arf semigroup is of maximal embedding dimension. In Theorems 9 and 11, we characterize the Arf semigroups which are either symmetric or pseudo-symmetric. Ring-theoretic applications of Theorems 8–12 appear in Theorems 22–26.

A semigroup which is either symmetric or pseudo-symmetric need not be of maximal embedding dimension and, hence, need not be an Arf semigroup. Similarly, it is easy to give explicit examples showing that an Arf semigroup (and, hence, a semigroup of maximal embedding dimension) need not be symmetric or pseudo-symmetric.

The next three theorems characterize the symmetric (respectively, pseudo-symmetric) semigroups that are Arf semigroups or semigroups of maximal embedding dimension.

THEOREM 9. *For a numerical semigroup S, the following are equivalent:*

(i) S *is symmetric and Arf;*
(ii) S *is symmetric and has maximal embedding dimension;*
(iii) $S = \langle 2, g + 2 \rangle$, *with* $g = g(S)$ *an odd integer,* $g \geq -1$*;*
(iv) $2 \in S$ *(or, equivalently,* $S = \langle 2, m \rangle$, *with* $m \geq 1$ *and* m *odd);*
(v) *Each semigroup* T, $S \subseteq T \subseteq \mathbb{N}$, *is symmetric and Arf;*
(vi) *Each semigroup* T, $S \subseteq T \subseteq \mathbb{N}$, *is symmetric.* □

THEOREM 10. *For a numerical semigroup S with $n(S) \geq 1$, the following are equivalent:*

(i) S *is pseudo-symmetric and has maximal embedding dimension;*
(ii) $S = \langle 3, \frac{g}{2} + 3, g + 3 \rangle$, *with* $g = g(S)$ *even and* $g \equiv 1, 2$ *(mod 3);*
(iii) $t(S) = 2$ *and* $S(1)$ *is a symmetric semigroup.* □

THEOREM 11. *If S is a numerical semigroup, then the following are equivalent:*

(i) S *is pseudo-symmetric and Arf;*
(ii) S *is either* $\langle 3, 4, 5 \rangle$ *or* $\langle 3, 5, 7 \rangle$*;*
(iii) $t(S) = 2$, $S(1)$ *is symmetric, and* $n(S) \leq 2$*.* □

We saw in Theorem 9 that a semigroup S is symmetric and Arf if and only if each semigroup T, $S \subseteq T \subseteq \mathbb{N}$, is symmetric. The purpose of the following results is to characterize, using the ideals of an arbitrary semigroup$S$, all symmetric semigroups T such that $S \subseteq T \subseteq \mathbb{N}$.

A nonzero ideal I of a semigroup S is called a *strong ideal of* S if $I = I + I^*$, or equivalently, if $(I - I) = (S - I)$; and I is called a *strongly bidual ideal of* S if $I^{**} = I$ and I is strong.

THEOREM 12. *Let S be a symmetric semigroup. View*

$$\mathcal{O}(S) := \{T : T \text{ is a semigroup, } S \subseteq T \subseteq \mathbb{N}\} \text{ and}$$

$\mathcal{D}(S) := \{I : I$ *is a strongly bi-dual ideal of* $S\}$ *as posets under set-theoretic inclusion. Define the function*

$$\varphi : \mathcal{O}(S) \to \mathcal{D}(S) \text{ by } T \mapsto (S - T).$$

Then:

(a) *φ is an order-reversing bijection, with inverse map given by $I \mapsto (I - I)$.*
(b) *If $\Sigma(S) := \{T : T \in \mathcal{O}(S),\ T$ is symmetric$\}$ and if $\mathcal{S}(S) := \{I : I \in \mathcal{D}(S)$ and I is stable$\}$, then the map $\varphi\,|_{\Sigma(S)}: \Sigma(S) \to \mathcal{S}(S)$ is an order-reversing bijection.* □

The bijection φ defined in Theorem 12 must not be confused with a bijection σ between $\mathcal{O}(S) := \{T : T \text{ is a semigroup, } S \subseteq T \subseteq \mathbb{N}\}$ and $\mathcal{S}t(S) := \{I : I \text{ is a stable ideal of } S\}$ which holds for *every* numerical semigroup S. To prepare for a description of σ in Proposition 13(b), note that if $T \in \mathcal{O}(S)$, we can consider $y := \inf\{x \in S : x \text{ nonzero, } x + T \subset S\}$ and the ideal $J_T := y + T$, which is an ideal of both S and T. It is easy to see that J_T is a stable ideal of S.

PROPOSITION 13. *Let S be a numerical semigroup. Then:*

(a) *$\mathcal{O}(S) = \{L(S, I) : I$ is a stable ideal of $S\} = \{(I - I) : I$ is a stable ideal of $S\}$.*
(b) *The map $\sigma : \mathcal{O}(S) \to \mathcal{S}t(S)$, $T \mapsto J_T$, is a bijection, with inverse map $\mathcal{S}t(S) \to \mathcal{O}(S)$ given by $J \mapsto (J - J)$.* □

Let I be a nonzero ideal of a semigroup S, and let i_1 be the least positive integer in I. Then we say that I is a *complete ideal of* S if $I = (i_1 + \mathbb{N}) \cap S$. Thus, I is a complete ideal of S if and only if $I = S_k$ for some k, $0 \leq k$.

The concepts of complete ideal and stable ideal lead, in Theorems 14 and 15, to new characterizations of the semigroups studied in Theorems 8 and 9.

THEOREM 14. *A numerical semigroup S is Arf if and only if each complete ideal of S is stable.* □

THEOREM 15. *A numerical semigroup S is Arf and symmetric if and only if each ideal of S is stable.* □

The remainder of this announcement deals with analogues and applications of the preceding material to certain local domains. Let $(V, \mathbf{m} = \pi V, k)$ be a rank 1 discrete valuation domain with maximal ideal $\mathbf{m}$ and residue field k. Let K denote the quotient field of V, let v be the (discrete) valuation on K^* associated to V and, for each subset B of K, let $v(B)$ denote the image under v of the set of nonzero elements of B. Let A be a subring of V such that the integral closure A' (of A in K) coincides with V. We always assume that the conductor $\mathbf{f} = (A : V)$ is nonzero and we let $g = g(A)$ denote the integer such that $\mathbf{f} = \mathbf{m}^{g+1} = \pi^{g+1}V$. The above conditions on A are equivalent to requiring that

A is a Noetherian conducive domain, or that (apart from the trivial case of a field) A is a one-dimensional Noetherian local analytically irreducible domain (cf. [BDF1], [BDF2]). Given A, let $\mathbf{m}_A = \mathbf{m} \cap A$ be the maximal ideal of A. We denote by $\mathcal{V} = \mathcal{V}(V)$ the class of all Noetherian conducive domains A such that $A' = V$ and $A/\mathbf{m}_A \simeq V/\mathbf{m} = k$. We call $v(A)$ the *value semigroup associated to* A (i.e., the numerical semigroup obtained as the image under v of the nonzero elements of A).

Let $A \in \mathcal{V}$ and let $S = v(A)$ be the value semigroup of A. It is not difficult to prove that $g(A) = g(S)$. If $n = n(S)$ and $S = \{0 = s_0, s_1, \ldots, s_n = g+1, \rightarrow\}$, we can consider the following ideals of A:

$$\mathbf{a}_k = \mathbf{a}_k(A) := \mathbf{m}^{s_k} \cap A, \quad 0 \leq k \leq n.$$

(Clearly, $\mathbf{a}_0 = A$, $\mathbf{a}_1 = \mathbf{m}_A$, and $\mathbf{a}_n = \mathbf{f} = (A : V)$.) Since one sees easily that

$$\mathbf{a}_k^{-1} := (A : \mathbf{a}_k) = (\mathbf{a}_k : \mathbf{a}_k) \quad \text{for} \quad 0 \leq k \leq n,$$

we may also consider the following fundamental chain of overrings of A:

$$\textbf{(A}(\cdot)\textbf{)} \qquad A =: A(0) \subset A(1) := \mathbf{a}_1^{-1} \subset A(2) := \mathbf{a}_2^{-1} \subset \cdots \subset A(n) := \mathbf{a}_n^{-1} = V$$

Let ℓ_A (or, simply, ℓ) denote the length of an A-module. It is well known that:

$$n(A) := \ell(A/\mathbf{f}) = n(S); \quad \text{and}$$
$$g(A) + 1 = \ell(V/\mathbf{f}) = g(S) + 1.$$

We call the sequence of natural numbers $(t_i(A) := \ell(A(i)/A(i-1)) : 1 \leq i \leq n)$ the *type sequence of* A, and we call the natural number $t(A) := t_1(A)$ the *type of* A. Since A is a one-dimensional Noetherian domain, A is Cohen-Macaulay and $t(A)$ coincides with the CM-type of A.

PROPOSITION 16. *Let $A \in \mathcal{V}$. Then $t(A) \leq t(v(A))$; and $t(A) = t(v(A))$ if and only if $v((A : \mathbf{m}_A)) = (v(A) - v(\mathbf{m}_A))$.* □

It is obvious that:

$$\ell(V/A) = \sum_{i=1}^{n} t_i(A)$$

where the integer $\ell(V/A)$ is called *the degree of singularity of* A and is denoted by $\delta(A)$. It is clear that:

$$\delta(A) = \ell(V/\mathbf{f}) - \ell(A/\mathbf{f}) = g(S) + 1 - n(S) = \delta(S).$$

It is well known [Ms, Propositions 2 and 3] that

$$1 \leq t_i(A) \leq t(A), \quad \text{for} \quad 1 \leq i \leq n,$$

and so we immediately obtain

(1) $\ell(A/\mathbf{f}) \leq \delta(A) \leq \ell(A/\mathbf{f})t(A)$;
(2) $t(A) \leq \ell(V/\mathbf{f}) - 2\ell(A/\mathbf{f}) + 1$.

In our setting, *a ring A of $\mathcal{V}$ is a Gorenstein domain if and only if* $\ell(V/\mathbf{f}) = 2\ell(A/\mathbf{f})$ (or, equivalently, $\ell(V/A) = \ell(A/\mathbf{f})$). Therefore, for a Gorenstein domain $A \in \mathcal{V}$, the first inequality of (1) is an equality; and this happens if and only if $t(A) = 1$. We say that a ring A of $\mathcal{V}$ is a *Kunz domain* if $\ell(V/\mathbf{f}) = 2\ell(A/\mathbf{f}) + 1$. Evidently, by (2), a Kunz domain A is a particular kind of domain having Cohen-Macaulay type $t(A) = 2$.

Kunz [K] proved that $A \in \mathcal{V}$ is Gorenstein if and only if its value semigroup $v(A)$ is symmetric. We next state the analogue when $v(A)$ is pseudo-symmetric.

PROPOSITION 17. *Let $A \in \mathcal{V}$. Then*

(a) *A is a Kunz domain if and only if its value semigroup $v(A)$ is pseudo-symmetric.*
(b) *If A is either Gorenstein or Kunz, then $t(A) = t(v(A))$.* □

Consider the class of domains $\mathcal{A} := \{A \in \mathcal{V}(V) : k \subset A\} = \cup\{\mathcal{A}_g : g \geq -1\}$, where $\mathcal{A}_g := \{A \text{ domain} : k \subset A \subset V,\ A' = V,\ (A : V) = \pi^{g+1}V\}$. Generalizing a result by Fröberg, Gottlieb and Häggkvist [FGH], we proved in [BDF2] that: *$A \in \mathcal{A}$ is a Gorenstein domain if and only if A is maximal inside $\mathcal{A}_g$ for some odd integer g.* In a similar way, we can now prove that:

THEOREM 18. *Let $A \in \mathcal{A}$. Then A is a Kunz domain if and only if A is maximal inside $\mathcal{A}_g$ for some even integer g.* □

COROLLARY 19. *Let $g \geq -1$ and $A \in \mathcal{A}$. Then A is maximal in $\mathcal{A}_g$ if and only if its type sequence is $(t(A), 1, \ldots, 1)$ with $t(A) \leq 2$.* □

Proposition 4 can be translated ring theoretically. It is well known that a 1-dimensional local domain A is Gorenstein if and only if each nonzero (fractional) ideal of A is *divisorial* (cf. [Bas, Theorem 6.3]). We can infer that $A \in \mathcal{V}$ is Gorenstein if and only if each overring of A is divisorial (as a fractional ideal of A). On the other hand, an analogue for Kunz domains is stated next.

PROPOSITION 20. *If $A \in \mathcal{V}$ and $g = g(A)$, then the following are equivalent:*

(i) *A is Kunz;*
(ii) *g is even, $(\mathbf{m}_A : \mathbf{m}_A)$ has an element of valuation $\frac{g}{2}$, and each fractional ideal $\mathbf{a}$ of A, such that $(\mathbf{a} : \mathbf{a})$ has an element of valuation $\frac{g}{2}$, is divisorial.* □

Let $A \in \mathcal{V}$. We denote by $e(A)$ the *embedding dimension of A*; that is, $e(A) = \ell(\mathbf{m}_A/\mathbf{m}_A^2)$. Notice that if $v(A)$ is the value semigroup of A, then

$$e(A) \leq e(v(A)),$$

and it can happen that this inequality is strict.

Next, we introduce the ring $L(A) := \cup\{(\mathbf{m}_A^k : \mathbf{m}_A^k) : k \geq 0\}$, obtained from A *by blowing up its maximal ideal*; $L(A)$ is called the Lipman ring of A (cf. [L, p.651]).

Since $L(A) \subseteq V$, it is easy to see that $L(A)$ is a finitely generated A-module and that $L(A) = (\mathbf{m}_A^h : \mathbf{m}_A^h)$ for all $h >> 0$. Moreover, it is well known that $\mathbf{m}_A L(A)$ is principal; an element $x \in \mathbf{m}_A$ such that $xL(A) = \mathbf{m}_A L(A)$ is called a *transversal generator* of $\mathbf{m}_A$. Notice that if x is a transversal generator of $\mathbf{m}_A$ then, for $h >> 0$, $x\mathbf{m}_A^h = \mathbf{m}_A^{h+1}$ [L, Proposition 1.1 and Lemma 1.8]. If $\mathbf{a}$ is an ideal of A, we say that $\mathbf{a}$ is *stable* when $\mathbf{a}$ is principal in $(\mathbf{a} : \mathbf{a})$ [L, Definition 1.3 and Lemma 1.11].

Let $\mu(A)$ denote the multiplicity of the ring A (cf. [Mt, §14]). In the above situation, it is known (cf. [L, Corollary 1.6, Lemma 1.8, Theorem 1.9]) that

$$\begin{aligned}\mu(A) &= \ell(L(A)/\mathbf{m}_A L(A)) = \ell(\mathbf{m}_A^h/\mathbf{m}_A^{h+1}) \quad \text{for } h >> 0 \\ &= \ell(A/xA) \quad \text{for } x \text{ a transversal generator of } \mathbf{m}_A.\end{aligned}$$

Moreover,

$$e(A) \le e(v(A)) \le \mu(v(A)) = \mu(A).$$

Since, for every $R \in \mathcal{V}$, $n(R) = n(v(R))$ coincides with the length of the chain $(\mathbf{R}(\cdot))$, it is clear that:

$$n(A(1)) \le n(A) - 1.$$

We also have the inequality $g(A(1)) \ge g(A) - \mu(A)$. The next result, an analogue of Proposition 7, addresses when these inequalities are equalities.

PROPOSITION 21. *If $A \in \mathcal{V}$, then the following are equivalent:*

(i) $A(1) = L(A)$;
(ii) $A = A + \mathbf{m}_A L(A)$ *(i.e., A is obtained from $L(A)$ by gluing over $\mathbf{m}_A$ the primary ideals belonging to $\mathbf{m}_A L(A)$, in the sense of [T]);*
(iii) $e(A) = \mu(A)$;
(iv) $\mathbf{m}_A^2 = x\mathbf{m}_A$ *for some nonzero (transversal) element* $x \in \mathbf{m}_A$;
(v) $\mathbf{m}_A = xA(1)$ *for some nonzero element $x \in \mathbf{m}_A$ (i.e., $\mathbf{m}_A$ is stable);*
(vi) *if* $n(A) \ge 1$, *then* $n(A(1)) = n(A) - 1$ *and* $g(A(1)) = g(A) - \mu(A)$;
(vii) *the value semigroup $v(A)$ has maximal embedding dimension and* $t(A) = t(v(A))$;
(viii) *the value semigroup $v(A)$ has maximal embedding dimension and* $e(A) = e(v(A))$. □

When the equivalent conditions of Proposition 21 hold, we say that $A \in \mathcal{V}$ is a *ring of maximal embedding dimension*. An important class of such A will be characterized in Theorem 22.

Besides the ascending chain of overrings:

$$(\mathbf{A}(\cdot)) \qquad A = A(0) \subset A(1) \subset \cdots \subset A(n) = V,$$

where $n = n(A) = \ell(A/\mathbf{f})$, we can study two other interesting ascending chains of overrings:

$$(\mathbf{B(A)}) \qquad A =: B_0 \subseteq B_1 := B(B_0) \subseteq B_2 := B(B_1) \subseteq \cdots \subseteq B_h := B(B_{h-1}) \subseteq \cdots \subseteq V$$

$$(\mathbf{L(A)}) \qquad A =: L_0 \subseteq L_1 := L(L_0) \subseteq L_2 := L(L_1) \subseteq \cdots \subseteq L_h := L(L_{h-1}) \subseteq \cdots \subseteq V$$

where for a (local) overring $(R, \mathbf{m}_R)$ of A with $A \subseteq R \subseteq V$, we have denoted the overring $(\mathbf{m}_R : \mathbf{m}_R) = R(1)$ by $B(R)$ and the Lipman overring of R by $L(R)$.

By analogy with Theorem 8, our next goal is to characterize the rings $A \in \mathcal{V}(V)$ for which the chains $(\mathbf{A}(\cdot))$, $(\mathbf{B(A)})$ and $(\mathbf{L(A)})$ coincide. These rings A are exactly the *Arf rings* (cf. [L, Theorem 2.2], [A]).

THEOREM 22. *For $A \in \mathcal{V}(V)$, the following are equivalent:*

(i) *the sequences* $(\mathbf{B(A)})$ *and* $(\mathbf{L(A)})$ *coincide;*
(ii) *the sequences* $(\mathbf{A(\cdot)})$ *and* $(\mathbf{B(A)})$ *coincide;*
(iii) *the sequences* $(\mathbf{A(\cdot)})$, $(\mathbf{B(A)})$ *and* $(\mathbf{L(A)})$ *coincide;*
(iv) $e(A(i)) = \mu(A(i))$, *for* $0 \leq i \leq n(A)$;
(v) $n(A(i)) = n(A) - i$ *and* $g(A(i)) = g(A) - \sum_{k=0}^{i-1} \mu(A(k))$, *for* $0 \leq i \leq n(A)$;
(vi) $A(i)(j) = A(i+j)$ *for* $0 \leq i+j \leq n(A)$, *and* $v(A(i)) = v(A)(i)$ *for* $0 \leq i \leq n(A)$;
(vii) *if* $n(A) \geq 1$, *then* $t_j(A(i)) = t_{i+j}(A)$ *for* $1 \leq i+j \leq n(A)$, *and* $v(A(i)) = v(A)(i)$ *for* $0 \leq i \leq n(A)$;
(viii) $v(A)$ *is an Arf semigroup and* $t(A(i)) = t(v(A)(i))$, *for* $0 \leq i \leq n(A)$;
(ix) $v(A)$ *is an Arf semigroup and* $e(A(i)) = e(v(A)(i))$, *for* $0 \leq i \leq n(A)$. □

Examples show that the semigroup properties of $v(A)$ alone cannot give satisfactory variants of conditions (vii) and (viii) in Proposition 21 or conditions (viii) and (ix) in Theorem 22.

From the previous considerations, we know that a ring A of $\mathcal{V}$ can be Gorenstein or Kunz without having maximal embedding dimension (*a fortiori*, without being Arf); and, conversely, a ring A of $\mathcal{V}$ can be Arf without being Gorenstein or Kunz.

The next four results are ring-theoretic analogues of Theorems 9–12.

THEOREM 23. *If $A \in \mathcal{V}$, then the following conditions are equivalent:*

(i) *A is Gorenstein and Arf;*
(ii) *A is Gorenstein and of maximal embedding dimension;*
(iii) *every integral overring R of A (that is, $A \subseteq R \subseteq V$) is Gorenstein and Arf;*
(iv) *every integral overring R of A (that is, $A \subseteq R \subseteq V$) is Gorenstein;*
(v) *the value semigroup $v(A)$ is symmetric and Arf.* □

THEOREM 24. *Let $A \in \mathcal{V}$, with $n(A) \geq 1$. Then the following are equivalent:*

(i) *A is Kunz and of maximal embedding dimension;*
(ii) *$t(A) = 2$ and $A(1)$ is Gorenstein;*
(iii) *the value semigroup $v(A)$ is pseudo-symmetric and of maximal embedding dimension.* □

THEOREM 25. *Let $A \in \mathcal{V}$, with $n(A) \geq 1$. Then the following are equivalent:*

(i) *A is Kunz and Arf;*
(ii) *$t(A) = 2$, $A(1)$ is Gorenstein, and $n(A) \leq 2$;*
(iii) *the value semigroup $v(A)$ is pseudo-symmetric and Arf.* □

Recall from [Bar] that a nonzero ideal $\mathbf{a}$ of an integral domain R is called a *strong ideal of R* if $\mathbf{a} = \mathbf{a}\mathbf{a}^{-1}$ (or, equivalently, if $(\mathbf{a} : \mathbf{a}) = (R : \mathbf{a})$). A *strongly divisorial ideal* $\mathbf{a}$ of an integral domain R is a nonzero ideal of R which is both strong and divisorial.

THEOREM 26. *Let $A \in \mathcal{V}$ and $g(A) \geq 2$. Suppose that A is Gorenstein. View $\mathcal{O}(A) := \{B : B$ is an overring of A, $A \subseteq B \subseteq V\}$, $\mathcal{O}_{Gor}(A) := \{B \in \mathcal{O}(A) : B$ is Gorenstein$\}$, $\mathcal{D}(A) := \{\mathbf{a} : \mathbf{a}$ is a strongly divisorial ideal of $A\}$ and $\mathcal{S}(A) := \{\mathbf{a} \in \mathcal{D}(A) : \mathbf{a}$ is stable$\}$*

as posets under inclusion. Then the map $\varphi : \mathcal{O}(A) \to \mathcal{D}(A)$, $B \mapsto (A : B)$, *is an order-reversing bijection, with inverse map* $\mathcal{D}(A) \to \mathcal{O}(A)$ *given by* $\mathbf{a} \mapsto (\mathbf{a} : \mathbf{a})$. *Moreover,* $\varphi|_{\mathcal{O}_{Gor(A)}}: \mathcal{O}_{Gor}(A) \to \mathcal{S}(A)$ *is also an order-reversing bijection.* □

We close by using the earlier material to reobtain, for the rings A of $\mathcal{V}$, some well known results.

COROLLARY 27. *Let* $A \in \mathcal{V}$. *Then* A *is Arf if and only if each integrally closed ideal of* A *is stable.* □

PROPOSITION 28. *Let* $A \in \mathcal{V}$. *Then* A *is Arf and Gorenstein if and only if each ideal of* A *is stable.* □

References

[A] C. Arf, *Une interprétation algébrique de la suite des ordres de multiplicité d'une branche algébrique*, Proc. London Math. Soc. **50** (1949), 256–287.

[Bar] V. Barucci, *Strongly divisorial ideals and complete integral closure of an integral domain*, J. Algebra **99** (1986), 132–142.

[Bas] H. Bass, *On the ubiquity of Gorenstein rings*, Math. Zeit. **82** (1963), 8–28.

[BDF1] V. Barucci, D.E. Dobbs and M. Fontana, *Conducive integral domains as pullbacks*, Manuscripta Math. **54** (1986), 261–277.

[BDF2] V. Barucci, D.E. Dobbs and M. Fontana, *Gorenstein conducive domains*, Comm. Algebra **18** (1990), 3889–3903.

[Br] A. Brauer, *On a problem of partitions*, Amer. J. Math. **64** (1942), 299–312.

[BC] W.C. Brown and F. Curtis, *Numerical semigroups of maximal and almost maximal length*, Semigroup Forum **62** (1991), 219–235.

[FGH] R. Fröberg, C. Gottlieb and R. Häggkvist, *Gorenstein rings as maximal subrings of* $k[[x]]$ *with fixed conductor*, Comm. Algebra **16** (1988), 1621–1625.

[K] E. Kunz, *The value-semigroup of a one-dimensional Gorenstein ring*, Proc. Amer. Math. Soc. **25** (1970), 748–751.

[L] J. Lipman, *Stable ideals and Arf rings*, Amer. J. Math. **93** (1971), 649–685.

[Mt] H. Matsumura, *Commutative ring theory*, Cambridge Univ. Press, 1986.

[Ms] T. Matsuoka, *On the degree of singularity of one-dimensional analytically irreducible Noetherian local rings*, J. Math. Kyoto Univ. **11** (1971), 485–494.

[T] G. Tamone, *Sugli incollamenti di ideali primari e la genesi di certe singolarità*, Boll. Un. Mat. Ital., Suppl. Algebra e Geometria **2** (1980), 243–258.

3
The Graded and Tame Extensions

M. BOULAGOUAZ Département de Mathématiques et d'Informatique, Faculté des Sciences de Fès, Université de Fès, B.P. 1796 Atlas Fès, Morocco

1 INTRODUCTION

Every valuation on a field k defines a filtration of this field. The graded ring $gr(k)$ associated to this filtration has the property that all homogeneous non-zero elements are invertible; such a ring is called graded field. The idea of using this kind of construction to study valuated fields is due to M.Krasner [8] who prefers to study the set of homogeneous elements of the graded ring, rather than the graded ring itself.

The main point of this article is to show how certain properties of some extension of valuated fields can be translated in terms of the corresponding graded fields and their fields of fractions.

We characterize defectless and tame extension of finite degree i.e extension K/k of valuated field for which: first, the residual extension $\bar{K}/\bar{k}$ is separable; second, the ramification index $(\Gamma_K : \Gamma_k)$ is not a multiple of the residual characteristic and finally,

$$[K : k] = (\Gamma_K : \Gamma_k).[\bar{K} : \bar{k}].$$

The following theorems appear respectively in section 6 and 7.

THEOREM 1 *The following statements are equivalent.*
1- K/k is tame and defectless of degree n.
2- The extension of fraction fields $frac(gr(K))/frac(gr(k))$ is separable of degree n.

THEOREM 2 *The Galois group $G_{K/k}$ of a defectless and tame extension K/k is the extension group of $Hom(\Gamma_K/\Gamma_k, \bar{K}^*)$ by the Galois group of $\bar{K}/\bar{k}$.*

Finally, we give new proofs of some results in [2].

2 GRADED FIELDS

DEFINITION 1 *Let $(\Delta, +)$ be an abelian group. A graded ring A of type Δ,*

$$A = \bigoplus_{\delta \in \Delta} A_\delta,$$

is called graded field of type Δ if all non-zero homogeneous elements are invertible.

Put

$$\Gamma_A = \{\delta \in \Delta \mid A_\delta \neq \{0\}\}.$$

Γ_A is a subgroup of Δ, called the support of the graded field A. Also, $H = \cup(A_\delta - \{0\})$ is the set of homogeneous and non-zero elements of A. Finally, for $a \in H$, we define the degree of a, denoted by $deg(a)$ as the element $\delta \in \Delta$ such that $a \in A_\delta$.
The following proposition is straightforward.

PROPOSITION 1
1) For every graded field A, the homogeneous component of degree zero, denoted by A_0, is a field.
2) For each $\delta \in \Gamma_A$ the homogeneous component A_δ is an A_0-vector space of dimension 1.

Next, when Δ is a torsion-free group it is easy to show that all graded fields of type Δ have no 0-divisors (see [[4], ch.II, P173, Prop.8]) . Hence graded commutative fields of type Δ have a field of fractions. Note at this stage that there are graded fields for which the type group is not torsion-free.
Before we continue, we introduce some notation. If $(\Delta, +)$ is an abelian group acting on a set Θ by $\delta(\theta) = \delta + \theta$, for each $(\delta, \theta) \in \Delta \times \Theta$. We say that Δ acts freely on Θ if $\delta(\theta) = \delta'(\theta)$ implies $\delta = \delta'$.
Finally, if A is a graded field of type Δ and M is a left A-module then we say that M is a graded A-module of type Θ if and only if

1) $M = \bigoplus_{\theta \in \Theta} M_\theta$.
2) for each $\delta \in \Delta$ and $\theta \in \Theta$, we have $A_\delta . M_\theta \subset M_{\delta+\theta}$.
Analogously, the support Γ_A of A acts on

$$\Gamma_M = \{\theta \in \Theta \mid M_\theta \neq \{0\}\} = support(M).$$

LEMMA 1 *Let M be a graded left A-module where A is a given graded field and N is a graded A-submodule of M. For each homogeneous $m \in M$, either $m \in N$ or $N \cap Am = \{0\}$.*

Proof: Fix m a homogeneous element of M. Set

$$I_m = \{a \in A : am \neq 0\}.$$

It's clear that I_m is an homogeneous ideal of A. Hence either $I_m = \{0\}$ or $I_m = A$. That first equality implies $Am \cap N = \{0\}$ and the second one implies $m \in N$. ■

PROPOSITION 2 *Every graded submodule N of a graded left A-module where A is a graded field, has a supplementary graded submodule in M.*

Proof: Put $F = \{P \mid P$ is a graded submodule of$N$ so that $N \cap P = \{0\}\}$. By Zorn's lemma pick a maximal element in F, say L. It is clear that $L \cap N = \{0\}$. Next, suppose $L \bigoplus N \neq M$. Pick a homogeneous element m in $M - (L \bigoplus N)$. By lemma 1, $(L \bigoplus N) \cap Am = \{0\}$, but this a contradiction with the maximality of L. So the proof is finished. ■

COROLLARY 1 *Let A be a graded field. then every left A-module is a free module.*

Proof: By Zorn lemma, there exist a maximal free system S of homogeneous elements. put N the graded submodule generated by S. Use lemma1 and Proposition2 to prove that $N = M$. ■

THEOREM 3 *Every graded module M on a graded field is a free module. Hence a basis of M can be chosen using homogeneous elements only.*

Proof: Pick $(\theta_i)_{i \in I}$ a set of representatives of different orbits of Γ_M under the action of the group Γ_A. Now, if for each $i \in I$ $(e_{ij})_{j \in J_i}$ is a basis for the A_0-vector space M_{Θ_i} then $E = (e_{ij})_{i \in I, j \in J}$ is a basis for the A-module M.
Next, put $N_i = \Sigma_{j \in J_i} A e_{ij}$. Note that $N_i \subset M_{\theta_i + \theta}$.
Now, pick W a maximal element L among subsets of J_i such that $\Sigma_{j \in L} A e_{ij}$ is a direct sum. By applying Lemma1 to $N = \Sigma A e_{ij}$, it follows that $\Sigma A e_{ij}$ is a direct sum and so is $\Sigma N_i = P$.

To show that $P = M$, pick a homogeneous element in M of degree θ, and θ_i a representative of Θ. It follows that $a \in A_{\theta_i - \theta}$, and the degree of am is θ_i. Hence,

$$am = \Sigma_{j \in J_i} a_{ij} e_{ij}, \qquad \text{i.e} \qquad m = \Sigma_{j \in J_i} a^{-1} a_{ij} e_{ij},$$

which finishes up the proof.

COROLLARY 2 *Under the same assumption of the previous theorem we have:*

$$rank\ M = \sum_{i \in I} dim_{A_0} M_{\theta_i}\ .$$

PROPOSITION 3 *If A is a graded field such that A_0 is algebraically closed and Γ a finite group, then A is isomorphic to the group ring $A_0[\Gamma]$.*

Proof:
Since Γ is a finite abelian group, Γ is a direct sum of r cyclic subgroups each generated by δ_i, of order n_i respectively. Now, since $\delta \in \Delta$ implies $\delta = \Sigma m_i \delta_i$, with $m_i \leq n_i$ it follows that

$$\Sigma m_i \delta_i + \Sigma r_i \delta_i = \Sigma \lambda_i \delta_i,$$

where $m_i + r_i = q_i n_i + \lambda_i$, where λ_i is the the rest of the euclidian division of $m_i + r_i$ by n_i.
Next fix $0 \neq e_{\delta_i} \in A_{\delta_i}$, then $e_{\delta_i}^{n_i} \in A_{n_i \delta_i} (= A_0)$. Now since A_0 is algebraically closed, pick $b_i \in A_0$ such that $b_i^{n_i} = e_{\delta_i}^{n_i}$. Hence we may assume, for all fixed e_{δ_i}, that $e_{\delta_i}^{n_i} = 1$. Thus we suppose

$$e_{\Sigma m_i \delta_i} = e_1^{m_1} e_2^{m_2} ... e_r^{m_r}\ .$$

Therefore, the resulting family e_δ is a basis for the A_0-vector space A and has the same multiplication table as the associated elements of Γ in the group algebra of Γ over A_0. ■

COROLLARY 3 *Assume Γ is finite and A is a graded field of support Γ. The following statements are equivalent.*
1- the A_0-algebra A is étale.
2- The order of Γ is not a multiple of the characteristic p of A_0.

Proof:
1) implies 2)
Assume A is étale, and denote by Ω the algebraic closure of A_0. Then the Ω-algebra $\Omega \otimes_{A_0} A$ does not have nilpotent elements [[5],chV §6 n7 th.4]. Now by proposition 3 $\Omega \otimes_{A_0} A$ can be identified whith the algebra $\Omega[\Gamma]$. If the characteristic p of A_0 divides the order of Γ, then if δ in Γ of order p it will follow that:
$1 - e_\delta \neq 0$ and $(1 - e_\delta)^p = 1 - e_{p\delta} = 0$, which is a contradiction.

2) implies 1)
suppose that the characteristic p of A_0 does not divide the order of δ. By Maschke's theorem, the algebra $\Omega[\Gamma]$ is a semi-simple algebra and hence étale since it is a commutative algebra. Hence we apply proposition 2 and [[5],chV §6, n5 corollary2 a)] to conclude.

3 THE RING OF POLYNOMIALS OVER A-GRADED FIELD

In this section, A will denote a graded field with support an abelian group Γ and

$$\Delta = \Gamma \bigotimes_{\mathbb{Z}} \mathbb{Q}.$$

For each $\theta \in \Delta$, we define a gradation of $A[X]$ of type, the subgroup generated by θ and Γ, by:

$$A[X]_\alpha = \{\Sigma a_i X^i \mid a_i \text{ homogeneous and } deg(a_i) + i\theta = \alpha\}.$$

this gradation is called a θ-gradation of $A[X]$ and the graded ring $A[X]$ is denoted by $A[X]^{(\theta)}$.

DEFINITION 2 *$P(X) \in A[X]$ is homogeneisable if there is $\theta \in \Delta$ such that $P(X)$ is homogeneous in $A[X]^{(\theta)}$.*

PROPOSITION 4 *The following statements are equivalent.*
1- $P(X) = \Sigma_{i=0}^{n} a_i X^i$ is homogeneisable.
2- for each $i = 0, ..., n$ a_i is homogneous and

$$\frac{deg(a_i) - deg(a_j)}{i - j},$$

calculated in Δ, does not depend on (i, j) such that $i \neq j$ and a_i , $a_j \neq 0$.

Proof:
1) implies 2)
This part follows by the definition of $A[X]^{(\theta)}$.
2) implies 1)
If we put

$$\theta = \frac{deg(a_i) - deg(a_j)}{i - j},$$

then $P(X)$ is clearly homogeneous in $A[X]^{(\theta)}$. ∎

PROPOSITION 5 *Let $P(X)$ be a homogeneous polynomial for a g-graduation of $A[X]$ and assume that g is of order d in Δ/Γ. If k is the greatest integer such that $kd \leq n$, a is a non zero element of A of degree dg and b is a non zero element of degree $deg(a_n) + k.d.deg(X)$, then:*

$$H(X) = \Sigma_{j=0}^{k} b^{-1} a_{n+(j-k)d} a^j X^j \in A_0[X] \text{ and } P(X) = bX^{n-kd} H(a^{-1}X^d).$$

Proof:
Since for $a_i, a_j \neq 0$, we have $(j-i)g = \deg(a_i) - \deg(a_j) \in \Gamma$, so $(i-j)$ should be a multiple of d. But since $a_n \neq 0$, all $a_i \neq 0$ are under the form a_{n-jd} and

$$Q(X) = \Sigma_{j=0}^{k} a_{n-jd} X^{n-jd} = \Sigma_{j=0}^{k} a_{n-jd} X^{n-kd} X^{d(k-j)}$$
$$= X^{n-kd} \Sigma_{j=0}^{k} a_{n-(k-j)d} X^{jd} = bX^{n-kd} \Sigma_{j=0}^{k} b^{-1} a_{n-(k-j)d} a^j (a^{-1}X^d)^j$$
$$= bX^{n-kd} H(a^{-1}X^d).$$

To finish up the proof note that $\deg(b^{-1}a_{n-(k-j)d}a^j) = 0$. ■

Next assume that A is a graded field of support Γ which is totally ordered and define v on A by

$$v(\Sigma a_{\delta_i}) = inf\{\delta_i \mid a_{\delta_i} \neq 0\}.$$

Note that v can be extented to F (the fraction field of A) by

$$v(\frac{a}{b}) = v(a) - v(b).$$

v is called *the canonical valuation* of F.

PROPOSITION 6 *Let E be an extension of F, valuated by a valuation w extending the canonical valuation v of F. If $x \in E$ is such that $P(x) = 0$ for a homogeneisable polynomial $P(X)$ in $A[X]$, then $P(X)$ is homogeneous in $A[X]^{(w(x))}$.*

Proof:
Let $P(X) = \Sigma_{i=0}^{n} a_i X^i$ be a homogeneisable polynomial in $A[X]$ such that $P(x) = 0$. By ([[5], chVI,§3, n.1 cor. of the prop.1]), there are $i \neq j$ such that

$$w(a_j) + jw(x) = w(a_i) + iw(x).$$

Now since w extends v, it follows that: $v(a_i) - v(a_j) = (j-i)w(x)$. To end the proof note that $P(X)$ is homogeneous in $A[X]^{(\theta)}$, for some θ in Δ and use the proposition 4. ■

PROPOSITION 7 *Let B a graded extension of A and assume E, the field of fractions of B, is of finite degree on F (the fraction field of A). Then the minimal polynomial of each non zero element of B_δ on F is a homogeneous element of $A[X]^\delta$.*

Proof:
Let $b \in B_\delta$ and denote $P(X) = \Sigma_{i=0}^n a_i X^i$ the minimal polynomial of b over F. Pick $a \in A$ such that $aP(X) \in A[X]$. Let $aP(X) = \Sigma P_{\delta_i}(X)$ be the decomposition of $aP(X)$ into homogeneous elements in $A[X]^{(\delta)}$. Hence, if $P_{\delta_i}(X) = \Sigma_{j=0}^n a_{ij} X^j$, we really have $a_{ij} b^j \in A[b]_{\delta_i}$. Hence $P_{\delta_i}(b)$ is in $A[b]_{\delta_i}$ and $P(b) = 0$ implies $P_{\delta_i}(b) = 0$. Now the degree of $P_{\delta_i}(X)$ is less than the degree of $P(X)$. Thus $aP(X) = cP_{\delta_i}(X)$. Since $P_{\delta_i}(X)$ has homogeneous coefficients and can be chosen an unitary polynomial, it follows then that $P(X) = P_{\delta_i}(X)$.

4 CRITERION OF SEPARABILITY OF GRADED FIELDS EXTENSION

In this section we denote by A a graded field with totally ordered support Γ_A and by F its fraction field . Also, B will denote any graded extension of A with support a subgroup of $\Delta = \Gamma_A \otimes_{\mathbb{Z}} \mathbb{Q}$. Furthermore we assume that the fraction field E is of finite degree over F, and v, the canonical valuation on E that extends the canonical valuation on F.

THEOREM 4 *The following statements are equivalent.*

- *E/F is separable.*
- *B_0/A_0 is separable and the characteristic p of A_0 does not divide the order of the group Γ_B/Γ_A.*

Assume E/F is separable. Let $M = B_0.F$, it's easy to see that
$M \simeq B_0 \otimes_{A_0} F$ as an F-algebra. Hence M is the F-scalar extension algebra of the A_0-algebra B_0. Thus M/F is separable if and only if B_0/A_0 is separable (see [[5],ch.V §15 n2 pro.3-d]).
Assume p divides $(\Gamma_B : \Gamma_A)$. Pick $\delta \in \Gamma_B - \Gamma_A$ such that $p\delta \in \Gamma_A$. The minimal polynomial $P(X)$ of b with respect to F has its coefficients in A and is homogeneous in $A[X]^{(v(b))}$ (see the proposition4, 6).
Now proposition 5 implies $P(X) = cX^i H(a^{-1}X^d)$, where $d = (\Gamma_B : \Gamma_A)$. since $P(X)$ is irreducible we have $i = 0$ and therefore $P(X)$ is a polynomial in X^p which is a contradiction.
Put $M = B_0.F$. We have E/F is separable if and only if E/M is separable and M/F is separable. Hence, B_0/A_0 is separable implies M/F is separable.
Now, on one hand B_0/A_0 is separable and so is M/F. On the other hand the M-algebra $M.B$ is a graded extension with support Γ_B/Γ_A. So the homogeneous component of degree ν is defined by $(M.B)_\nu = M.(\bigoplus_{\alpha \in \nu} B_\alpha)$. Note that the homogeneous component of degree zero is $(MB)_0 \simeq M$. So the corollary 3 implies MB/M is etale and [[4] proposition 4 §6] implies MB/M is separable. Thus E/M is separable by ([[3], prop.4 §15 n2]).

5 CASE WHERE $frac(B)/frac(A)$ IS NORMAL

Let B/A be an extension of graded fields of rank n and of supports respectively Γ_B and Γ_A. Put $E = frac(B)$, $F = frac(A)$. Thus $n = [E : F]$. We assume, in this section, that E/F is a normal extension with Galois group T. Put

$$H = \{\sigma \in T, \text{ where } B_0 \text{ is fixed by } \sigma\}.$$

H is a normal subgroup in T (since it's the kernel of the homomorphism ϕ from T into $\mathrm{Aut}(B_0)$ defined by $\phi(\sigma) = \sigma_{|B_0}$). For each $\delta \in \Gamma_B$, fix a non-zero element $e_\delta \in B_\delta$, with $e_\delta \in A$ if $\delta \in \Gamma_A$. Now we have,

$$B = \bigoplus_{\Gamma_B} B_0 e_\delta \qquad \text{and} \qquad A = \bigoplus_{\Gamma_A} A_0 e_\delta.$$

Put $S = A \otimes_{A_0} B_0$. Hence $S = \bigoplus_{\Gamma_A} B_0 e_\delta$. So if $M = frac(S)$, then it is not hard to see that the Galois group $G_{E/M}$ of E/M is H. Let $\sigma \in H$ be fixed. Then $\sigma(e_\delta) \in B_\delta$. So $\sigma(e_\delta) = b_\delta(\sigma)e_\delta$, where $b_\delta(\sigma) \in B_0$. Hence $b_\delta(\sigma)$ does not depend on the choice of e_δ (since $\sigma \in H$). Put $e_0 = 1$ and define the mapping $b(\sigma)$ from Γ_E/Γ_F into B_0^* such that

$$b(\sigma)(\delta + \Gamma_F) = b_\delta(\sigma).$$

So $b(\sigma)$ is well defined and thus $b(\sigma) \in Hom(\Gamma_E/\Gamma_F, B_0^*)$.

PROPOSITION 8 *The mapping Φ from H into $Hom(\Gamma_E/\Gamma_F, B_0^*)$ defined by $\Phi(\sigma) = b(\sigma)$ is a group isomorphism.*

Proof: Let σ and τ in H. Then

$$\sigma\tau(e_\delta) = \sigma(b_\delta(\tau)e_\delta)$$

$$= b_\delta(\tau)\sigma(e_\delta) = b_\delta(\tau)b_\delta(\sigma),$$

which shows that Φ is a homomorphism.
To show that Φ is one-one, first note that if $b(\sigma) = 1$, then $\sigma(e_\delta) = e_\delta$, for $\delta \in \Gamma_B$. But $\sigma \in H$ implies that σ fixes B_0 and the e_δ, hence σ fixes B and therefore σ fixes E.
Now if $h \in Hom(\Gamma_E/\Gamma_F, B_0^*)$, h can be viewed as element of $Hom(\Gamma_E, B_0^*)$ such that $h(\Gamma_F) = \{1\}$. Now, let σ be defined by $\sigma(e_\delta) = h(\delta)e_\delta$. It's not hard to see that $\sigma \in H$ and $\Phi(\sigma) = h \in H$. COMMENT:
The construction of Φ is well known (Bourbaki Alg.Comm. ChVI §8 exercise.11) But it is interesting to notice that, in this case, it leads to an isomorphism which would allow in section 7 to determine the Galois group of K_V/K_T of some valuated extension K/k, where K_V is the largest tame extension of k in K and K_T is the largest inertial extension of k in K.

6 GRADED FIELD ASSOCIATED WITH A VALUATED FIELD

Let k be a field equipped with a valuation v on Γ_k. $\mathcal{O}_k$, and $\mathcal{M}_k$ respectively the ring and ideal associated with the valuation v. Also, set
$\bar{k} = \mathcal{O}_k/\mathcal{M}_k$ and $\Delta = \Gamma_k \otimes_{\mathbf{Z}} \mathbf{Q}$.
For each $\delta \in \Delta$ we put

$$k_\delta = \{x \in k \mid v(x) \geq \delta\}$$

and

$$k_{\delta^+} = \{x \in k \mid v(x) > \delta\} = \mathcal{M}_k.k_\delta.$$

In particular $k_0 = \mathcal{O}_k$ et $k_{0^+} = \mathcal{M}_k$.
The sets k_{δ^+} and k_δ are clearly $\mathcal{O}_k$-modules which define a filtration on k. Next, we denote the graded ring associated by

$$gr(k) = \oplus_{\delta\in\Delta} gr(k)_\delta .$$

where

$$gr(k)_\delta = k_\delta/k_{\delta^+};$$

recall that the multiplication on $gr(k)$ is defined as follows: if $x \in k_\delta$ and y in k_ϵ, then

$$(x + k_{\delta^+})(y + k_{\epsilon^+}) = xy + k_{(\delta+\epsilon)^+} .$$

Hence, every homogeneous element of $gr(k)$ is invertible, indeed if $x \in k_\delta/k_{\delta^+}$ then $v(x) = \delta$, then $x^{-1} \in k_{-\delta}/k_{(-\delta)^+}$ and

$$(x + k_{\delta^+})(x^{-1} + k_{(-\delta)^+}) = 1 + k_{0^+} .$$

therefore the ring $gr(k)$ is a graded field of type Δ.

DEFINITION 3 $gr(k)$ *is called* **the graded field associated with the valuated field** k . *Its homogeneous component of degree 0 is the residual field i.e*

$$gr(k)_0 = \bar{k}$$

and its support is the value group of k *i.e*

$$\Gamma_{gr(k)} = \Gamma_k .$$

Now, define

$$\tilde{v} : gr(k) \to \Gamma_k \cup \{\infty\} \subset \Delta \cup \{\infty\}$$

by: if

$$a = \sum_{\delta \in \Delta} a_\delta \in gr(k),$$

then

$$\tilde{v}(a) = \inf\{\delta \mid a_\delta \neq 0\}.$$

Notice that if $gr(k)$ is a commutative ring, then it has a fraction field (since it does not have 0-divisors). Put $\tilde{k}$ the field of fractions of $gr(k)$ whenever $gr(k)$ is a commutative ring.
For $x \in k_\delta$ set

$$\tilde{x} = x + k_{v(x)^+} \in gr(k)_{v(x)};$$

so $\tilde{v}(\tilde{x}) = v(x)$.
From now, on valuated fields would be assumed to be commutative fields, hence their associated graded fields are commutative. Also, it's easy to see that $\tilde{v}$ can be extended to a valuation on $\tilde{k}$ by

$$\tilde{v}(\frac{a}{b}) = \tilde{v}(a) - \tilde{v}(b).$$

EXAMPLE:
Let k be a field equipped with a discrete valuation. Denote by π an uniformizer of k. Hence each $x \in k$ can be writen as $a\pi^{v(x)}$ for some $a \in \mathcal{O}_k$ (valuation ring of k). Put $T = \tilde{\pi}$, then each element of $gr(k)$ can be written as

$$\Sigma_{i=-m}^{n} a_i T^i, \text{ for some m and n in } \mathbb{N}.$$

therefore $gr(k) = \bar{k}[T, T^{-1}]$, $\tilde{k} = \bar{k}(T)$ and $\tilde{v}$ is the natural valuation defined on $\bar{k}(T)$, its residual field is $\bar{k}$.

PROPOSITION 9 *The residual field of $\tilde{k}$ with the valuation $\tilde{v}$ is $\bar{k}$.*

Proof: It's clear that the residual field of $\tilde{k}$ contains $gr(k)_0 = \bar{k}$. Conversely, the residue of $\frac{a}{b} \in \tilde{k}$, where $a, b \in gr(k)$ and $\tilde{v}(a) \geq \tilde{v}(b) \neq \infty$, is the same as the residue of $\frac{a_{\tilde{v}(a)}}{b_{\tilde{v}(b)}}$. This residue is zero if $\tilde{v}(a) > \tilde{v}(b)$ and is in the image of $gr(k)_0$ if $\tilde{v}(a) = \tilde{v}(b)$. ■

Let's assume now that k is a commutative field, and K is some finite extension of k of degree n. We assume that the valuation v on k can be extended to a valuation w on K. Note that $\Gamma_k \subset \Gamma_K$ and $(\Gamma_K : \Gamma_k)$ is finite. We may also look at Γ_K as a subgroup of Δ. Moreover, we identify $gr(k)$ with a subring of $gr(K)$ and $\tilde{k}$ with a subfield of $\tilde{K}$. Hence, $\tilde{w}$ is then an extension of $\tilde{v}$.

PROPOSITION 10 *Let e and f be respectively the ramification index and the residual degree of w/v. The graded ring $gr(K)$ is a graded module over $gr(k)$ of rank ef. Moreover, e and f are respectively the ramification index and residual degree of $\tilde{w}/\tilde{v}$ and*

$$[\tilde{K} : \tilde{k}] = ef.$$

Proof:
It's clear that $gr(K)$ is a graded module on $gr(k)$, and its rank may be be computed by corollary 2. Indeed, each homogeneous component $gr(K)_\delta$ which is not $\{0\}$ is a $gr(K)_0$-vector space of dimension 1, with $gr(K)_0 = \bar{K}$. Hence,

$$dim_{gr(k)_0}\, gr(K)_\delta = [\bar{K} : \bar{k}] = f.$$

On the other hand, the number of orbits of Γ_K under the action of Γ_k is $(\Gamma_K : \Gamma_k) = e$. This shows that the rank of $gr(K)$ over $gr(k)$ is ef.
Now, by definition of $\tilde{v}$, $\tilde{w}$ it's clear that the values groups of $\tilde{v}$ and $\tilde{w}$ are respectively Γ_k and Γ_K. Hence the ramification index of $\tilde{w}/\tilde{v}$ is e. Moreover, by proposition 9, the residual field of $\tilde{k}$ and $\tilde{K}$ are respectively $\bar{k}$ and $\bar{K}$. Hence, the residual degree of $\tilde{w}/\tilde{v}$ is f. Finally, since $gr(K)$ has a finite rank, the ring $gr(K) \otimes_{gr(k)} \tilde{k}$ is of finite dimension over $\tilde{k}$. Now, since $gr(K) \otimes_{gr(k)} \tilde{k}$ is a ring without 0-divisors and hence it's a field. so,

$$gr(K) \otimes_{gr(k)} \tilde{k} = \tilde{K}.$$

Thus

$$[\tilde{K} : \tilde{k}] = ef.$$

∎

COROLLARY 4 *The following statements are equivalent*

- *v is a defectless valuation and can be extented in a unique way from k to K.*
- *$[K : k] = [gr(K) : gr(k)] = [\tilde{K} : \tilde{k}]$.*

We finish this section by giving a characterization of a defectless and tame extension of finite degree i.e valuated fields extension K/k so that the residual extension $\bar{K}/\bar{k}$ is separable, with the ramification index $(\Gamma_K : \Gamma_k)$ is not a multiple of the residual characteristic and

$$[K : k] = (\Gamma_K : \Gamma_k).[\bar{K} : \bar{k}].$$

This characterization is given in terms of the extension of fraction fields and their associate graded fields.

THEOREM 5 *For any valuated extension K/k of finite degree, the following statements are equivalent,*

- *$\tilde{K}$ is a separable extension of $\tilde{k}$ of degree n.*
- *K/k is a defectless and tame extension of degree n.*

Proof:
Apply theorem 4 , with $E = \tilde{K}$, $F = \tilde{k}$ and proposition 9, 10.

7 VALUATED GALOIS EXTENSION

Let K/k be a Galois extension of commutative fields, with Galois group G of order n.

We assume that there is a valuation v on k that extends uniquely to a valuation on K. Also we suppose that K/k is defectless i.e

$$n = ef,$$

where $e = (\Gamma_K : \Gamma_k)$ is the ramification index of K/k and $f = [\bar{K} : \bar{k}]$ is the residual degree of K/k.

We define the inertial group G_T and the ramification group G_V of G by:

$$G_T = \{\sigma \in G \mid w(\sigma(x) - x) > 0\,; \forall x \in O_K\}$$

et

$$G_V = \{\sigma \in G \mid w(\sigma(x) - x) > w(x); \forall x \in K^*\}.$$

We also define the inertial field K_T of K/k and the ramification field of K/k as the invariant subfields of K/k under G_T and G_V respectively (see [7] or [6]).

It's clear that $G_V \subset G_T$ and thus $K_T \subset K_V$. Moreover the extension K_V/k is a tame extension [7]. Now, remark that for $\sigma \in G$ we define an automorphism $\tilde{\sigma}$ of $\tilde{K}$ by :

$$\tilde{\sigma}(\tilde{x}) = \widetilde{\sigma(x)}, \text{ for } \tilde{x} \in gr(K)_{v(x)},$$

$\tilde{\sigma}$ can extended by linearity to $gr(K)$, and afterwards to $\tilde{K}$ by

$$\tilde{\sigma}(\frac{a}{b}) = \frac{\tilde{\sigma}(a)}{\tilde{\sigma}(b)}.$$

Next, set $\tilde{G} = \{\tilde{\sigma} \mid \sigma \in G\}$. It's clear that $\tilde{G}$ is a group. Moreover $\widetilde{\sigma \circ \tau} = \tilde{\sigma} \circ \tilde{\tau}$ for $\sigma, \tau \in G$. The mapping $\phi : G \to \tilde{G}$ defined by

$$\phi(\sigma) = \tilde{\sigma}$$

is a group homomorphism .

Now, for $\sigma \in G_V$, we have $w(\sigma(x) - x) > \omega(x)$. So $\sigma(x) - x \in K_{\omega(x)}$ and

$$\widetilde{\sigma(x)} = \tilde{x}.$$

So, $G_V = Ker(\phi)$. Hence ϕ induces an isomorphism from G/G_V onto $\tilde{G}$.

COROLLARY 5 *K/k is a defectless and tame extension if and only if $G \simeq \tilde{G}$.*

Next, let K/k be a defectless valuated normal extension of degree n. It is well known that K_T is the largest inertial extension of k which is contained in K i.e the largest defectless extension among E/k such that $\bar{E}/\bar{k}$ is separable, $\Gamma_E = \Gamma_k$

and $E \subset K$.
Also, K_V is the largest defectless and tame extension of k contained in K i.e the largest extension E/k which is defectless such that $\bar{E}/\bar{k}$ is separable, p the characteristic of $\bar{k}$ does not divise $(\Gamma_E : \Gamma_k)$ and $E \subset K$ (see [7]).
Finally $\bar{K_V} = \bar{K_T}$ is the separable closure of $\bar{k}$ in $\bar{K}$, which we shall denote by $\bar{K}^{(sep)}$. Also

$$\Gamma_{K_V} = \{x \in \Gamma_K \text{ such that the order of } x \text{ over } \Gamma_k \text{ is relatively prime to } p\}.$$

Remark that Γ_{K_V} is a subgroup of Γ_K, and we put $\Gamma_K^{(sep)} = \Gamma_{K_V}$.

8 GALOIS GROUP OF K_V/K_T

PROPOSITION 11 *If K_V/K_T is a Galois extension of finite degree then,*

$$G_{K_V/K_T} \simeq Hom(\Gamma_K^{(sep.)}/\Gamma_k, \bar{K}^{(sep.)*}).$$

Proof:
Apply proposition 8 to

$$A = gr(K_T) = \bar{K}^{(sep)} \bigotimes_{\bar{k}} gr(k) \text{ and } B = gr(K_V),$$

so we have

$$Hom(\Gamma_{K_V}/\Gamma_k, \bar{K}^{(sep)*}) \simeq H.$$

but $B_0 = A_0 = \bar{K}^{(sep)}$ hence $H = T$. On the other hand K_V/K_T is a tame and defectless extension and by corollary 5 $T \simeq G_{K_V/K_T}$. ∎

COROLLARY 6 *If K/k is a defectless and tame extension of finite degree then its Galois group is a group extension of the group $Hom(\Gamma_K/\Gamma_k, \bar{K}^*)$ by the Galois group of $\bar{K}/\bar{k}$.*

Proof: K/k is normal and $\bar{K}/\bar{k}$ is also normal and so is K_T/k. Also, it's well know that Galois group of $\bar{K}/\bar{k}$ and of K_T/k are isomorphic whenever K/k is a tame extension. Finally, Galois theory shows that the Galois group $G_{K/k}$ of K/k is a group extension of $Hom(\Gamma_K/\Gamma_k, \bar{K}^*)$ by the Galois group of $\bar{K}/\bar{k}$. ∎

References

[1] M.Boulagouaz ; Anneaux de Krasner et extensions modérément ramifiées; thèse de troisième cycle, Juin 1983 à Paris 6

[2] M.Boulagouaz ; Corps gradué d'un corps valué; Rapport de séminaire N.194, institut de mathématique pure et appliquée de l'U.C.L, Belgique, Mai 1991

[3] N.Bourbaki; Algèbre commutative (chapitre VI)

[4] N.Bourbaki ; Algèbre(Chapitre 1 à 3), Masson , Paris (1964)

[5] N.Bourbaki ; Algèbre(Chapitre IV à V), Masson , Paris (1964)

[6] E.Artin ; Algebraic Numbers and Algebraic Functions; Gordon and Breach, 1967

[7] O. Endler;Valuation theory; Springer-Verlag, Berlin, 1972

[8] M. Krasner; Une généralisation de la notion de corps: le corpoïde. Un corpoïde remarquable de la théorie des corps valués; Comptes Rendus t.219 (1944) , pp 345–347

[9] M.Krasner;Théorie de Galois des corpoïdes commutatifs sans torsion et ses applications à la théorie de la ramification des extensions algébriques des corps valués , Université Pierre et Marie Curie, Paris (1976)

[10] O.F.G Schilling; The Theory of Valuations; Math. surveys 4, Amer. Math. Soc., Providence, R.I, 1950

[11] Van Geel et F.Van Oystayen; About graded fields, Indag. Math. 43(1981), 273-286

4
Ascending Chain Conditions and Associated Primes

PAUL-JEAN CAHEN Laboratoire de Mathématiques Fondamentales et Appliquées, Case 322, Faculté des Sciences de Saint-Jérôme, Université d'Aix Marseille III, 13 397 Marseille Cedex 13, France

ABSTRACT: It is known that associated and weakly associated primes are the same for a noetherian domain; here it is shown that a.c.c. on divisorial ideals (Mori domains) implies equality of associated and weakly associated primes for A/I, where I is divisorial. Converse implications do not hold and a.c.c. on principal ideals does not imply nor is implied by equality of associated and weakly associated primes for A/I, where I is principal.

INTRODUCTION

For classical notions and definitions on rings, we refer to [4], [10] and [13]. In this paper, we denote by A a commutative domain with unit and K its quotient field. A domain is called a *Mori domain* if the ascending chain condition holds on integral divisorial ideals (see [7], [8], [12]); Krull domains are the completely integrally closed Mori domains [12]. If M is an A-module, recall that the set Ass(M) (resp. $\mathrm{Ass}_f(M)$) of *associated primes* of M (resp. of *weakly associated primes* of M) is the set of primes that annihilate an element of M (resp. are minimal primes of the annihilator of an element of M).

Recall that for any module M of a noetherian ring, i.e. a ring with ascending chain condition (a.c.c.) on *every* ideal, $\mathrm{Ass}(M) = \mathrm{Ass}_f(M)$, therefore $\mathrm{Ass}(A/I) = \mathrm{Ass}_f(A/I)$ for *every* ideal I. In a first section we show that for every *divisorial* ideal I of a Mori domain, i.e. a domain with a.c.c. on *divisorial* ideals, then $\mathrm{Ass}(A/I) = \mathrm{Ass}_f(A/I)$.

Next sections are mostly devoted to examples. First a Krull domain (hence a Mori domain) with a (non divisorial) ideal such that $Ass_f(A/I) \neq Ass(A/I)$, then a domain such that $Ass(A/I) = Ass_f(A/I)$, for every ideal I, but which is not Mori (hence not noetherian). Lastly we provide examples showing that ascending chain condition on principal ideals is not implied nor implies $Ass(A/I) = Ass_f(A/I)$, for every principal ideal I.

1 CHAIN CONDITIONS AND ASSOCIATED PRIMES

If I and J are two fractional ideals of A, we denote by $(I : J)$ the conductor

$$(I : J) = \{ \alpha \in K \mid \alpha J \subset I \}.$$

If $x \in K$, we denote by $(I :_A x)$ the annihilator of the class of x in the A-module K/I, hence

$$(I :_A x) = A \cap (I : Ax) = \{ \alpha \in A \mid \alpha x \in I \}.$$

The following lemma is immediate (see [4] or [11, proposition 1.1]) but useful:

LEMMA 1.1: *Let A be a domain of field of fractions K, I a divisorial ideal and $x \in K$, then $(I :_A x)$ is a divisorial ideal.*

If A is a noetherian domain, then $Ass(A/I) = Ass_f(A/I)$, for any ideal I of A. The same holds for any divisorial ideal of a Mori domain:

THEOREM 1.2: *Let I be a divisorial ideal of a Mori domain A, then $Ass_f(A/I) = Ass(A/I)$.*

Proof: If $\mathbf{p} \in Ass_f(A/I)$, there is $x \in A$ such that $\mathbf{p}$ is a minimal prime of $(I :_A x)$, which is a divisorial ideal [lemma 1.1]; since A is a Mori domain, we may choose x such that $(I :_A x)$ is maximal among such annihilators contained in $\mathbf{p}$; we then prove that $(I :_A x)$ is prime, thus $(I :_A x) = \mathbf{p}$, and $\mathbf{p} \in Ass(A/I)$:
let $a,b \in A$ such that $ab \in (I :_A x)$, then $(I :_A x) \subset (I :_A bx)$,
- either $(I :_A bx) \subset \mathbf{p}$, but then $(I :_A x) = (I :_A bx)$ hence $a \in (I :_A x)$,
- or there is $c \notin \mathbf{p}$, such that $cbx \in A$, but then $(I :_A x) \subset (I :_A cx) \subset \mathbf{p}$ (since $c \notin \mathbf{p}$ and $\mathbf{p}$ is prime), but then $(I :_A x) = (I :_A cx)$ hence $b \in (I :_A x)$ ■

Principal ideals are divisorial, thus in particular $Ass_f(A/I) = Ass(A/I)$, for every principal ideal of a Mori domain; we then record the following (the proof of which is inspired by [11, corollary 2.5]).

PROPOSITION 1.3: *Let A be a domain; the following assertions are equivalent.*
i) For any principal ideal I of A, $Ass_f(A/I) = Ass(A/I)$.
ii) $Ass_f(K/A) = Ass(K/A)$.

Proof: If $x = \frac{a}{d}$ in K, then $(A :_A x) = (dA :_A a)$.
Suppose (i) and let $\mathbf{p} \in Ass_f(K/A)$; $\mathbf{p}$ is a minimal prime of $(A :_A x) = (dA :_A a)$, hence $\mathbf{p} \in Ass_f(A/dA) = Ass(A/dA)$, so $\mathbf{p} = (dA :_A b) = (A :_A y)$, where $y = \frac{b}{d}$, thus $\mathbf{p} \in Ass(K/A)$.
Suppose (ii) and let $\mathbf{p} \in Ass_f(A/dA)$; $\mathbf{p}$ is a minimal prime of $(dA :_A a) = (A :_A x)$ and $d \in \mathbf{p}$, since $x = \frac{a}{d}$; hence $\mathbf{p} \in Ass_f(K/A) = Ass(K/A)$, so $\mathbf{p} = (A :_A y) = (dA :_A dy) = (dA :_A c)$, and $c = dy$ is in A (since $d \in \mathbf{p}$), thus $\mathbf{p} \in Ass(A/dA)$ ■

Therefore we get, as N. DESSAGNES [7, theorem 1, p. 406] :

COROLLARY 1.4: *Let A be a Mori domain; then* $Ass_f(K/A) = Ass(K/A)$.

REMARK 1.5: Clearly, for any domain A, $Ass(K/A) = \bigcup\{Ass(A/I);\ I \text{ principal}\}$; if A is Mori, and I is a divisorial ideal, then Ass(A/I) consists of divisorial primes [Lemma 1.1] (and can be shown to be finite [11, theorem 2.1]); conversely divisorial primes of a Mori domain belong to Ass(K/A) [7, theorem 1, p. 406]; thus, if A is Mori,
$Ass(K/A) = \bigcup\{Ass(A/I);\ I \text{ principal}\} = \bigcup\{Ass(A/I);\ I \text{ invertible}\} = \bigcup\{Ass(A/I);\ I \text{ divisorial}\}$
$= \{\text{divisorial primes of } A\} = Ass_f(K/A)$.

It is immediate (and left to the reader) that $Ass_f(A/I) = Ass(A/I)$ for every ideal I if and only if $Ass_f(M) = Ass(M)$ for every module M, in conclusion we get the following set of implications:

a.c.c. on every ideal (A is noetherian)	⇒	$Ass_f(A/I) = Ass(A/I)$ for every ideal	⇔	$Ass_f(M) = Ass(M)$ for every module.
⇓		⇓		
a.c.c. on divisorial ideals (A is Mori)	⇒	$Ass_f(A/I) = Ass(A/I)$ for divisorial ideals		
⇓		⇓		
a.c.c. on invertible ideals		$Ass_f(A/I) = Ass(A/I)$ for invertible ideals		
⇓		⇓		
a.c.c. on principal ideals (A is a.c.c.p.)		$Ass_f(A/I) = Ass(A/I)$ for principal ideals	⇔	$Ass_f(K/A) = Ass(K/A)$

We give below an example to show that a.c.c. on principal ideals does not imply the equality $\mathrm{Ass}(A/I) = \mathrm{Ass}_f(A/I)$ for every principal ideal [example 3.1]. Also we may ask:

QUESTION 1.6: Does ascending chain condition on invertible ideals imply the equality $\mathrm{Ass}_f(A/I) = \mathrm{Ass}(A/I)$ for invertible or at least principal ideals?

2 MORI AND NON MORI DOMAINS, EXAMPLES

If A is a Mori domain, then $\mathrm{Ass}_f(A/I) = \mathrm{Ass}(A/I)$ for every divisorial ideal [theorem 1.2], but not necessarily for every ideal, even if A is Krull:

EXAMPLE 2.1: Let K be a field and $A = K[X_1,X_2,...,X_n,...]$ the ring of polynomials in infinitely many indeterminates indexed by $\mathbb{N}$; it is clear that A is Krull, hence Mori. Consider the ideals $I = ((X_1)^2,(X_2)^2,...,(X_n)^2,...)$ and $\mathfrak{m} = (X_1,X_2,...,X_n,...)$; $\mathfrak{m}$ is maximal, and if $f \in \mathfrak{m}$ (a polynomial in finitely many, say r indeterminates) there is an integer k such that $f^k \in I$ ($k = r+1$ will do) hence $\mathfrak{m}$ is the radical of I and the only element of $\mathrm{Ass}_f(A/I)$. Now if g is any polynomial not contained in I, and X_s an indeterminate which does not occur in g, then $X_s g \notin I$, thus $\mathfrak{m} \notin \mathrm{Ass}(A/I)$.

Conversely, the equality $\mathrm{Ass}(A/I) = \mathrm{Ass}_f(A/I)$ for every ideal I does not imply that A is noetherian nor even Mori. To give an example we first prove an easy lemma.

LEMMA 2.2: *Let A be a domain, if a maximal ideal $\mathfrak{m}$ of A is such that, for every nonzero ideal I of A, there exists an integer n and $\mathfrak{m}^n \subset I$, then*
i) $\mathfrak{m}$ is the only nonzero prime of A.
ii) For every ideal I of A, $\mathrm{Ass}_f(A/I) = \mathrm{Ass}(A/I)$.

Proof: i) $\mathfrak{m}$ is the radical of any nonzero ideal.
ii) If n is the smallest integer such that $\mathfrak{m}^n \subset I$, there is a in A, such that $a \in \mathfrak{m}^{n-1}$, but $a \notin I$, hence $\mathfrak{m} = (I :_A a)$ and $\mathfrak{m} \in \mathrm{Ass}(A/I)$ ∎

EXAMPLE 2.3: As S. GABELLI [9, Example p. 419], let $A = k + XD + X^2K[[X]]$, where k and K are fields and D a domain, but not a field, such that $k \subset D \subset K$. Then A is a quasi local domain (with maximal ideal $\mathfrak{m} = XD + X^2K[[X]]$) with a.c.c. on principal ideals hence on invertible ideals which is not Mori [3, Example 17]; if I is a nonzero ideal, a nonzero element x of A is a series $x = a_dX^d + a_{d+1}X^{d+1} + ...$and clearly $\mathfrak{m}^{d+2} \subset Ax \subset I$, hence $\mathrm{Ass}(A/I) = \mathrm{Ass}_f(A/I)$ for every ideal I [lemma 2.2].

3 PRINCIPAL IDEAL CONDITIONS

A.c.c. on principal ideals does not imply the equality $\mathrm{Ass}(A/I) = \mathrm{Ass}_f(A/I)$ for every principal ideal I, nor equivalently $\mathrm{Ass}(K/A) = \mathrm{Ass}_f(K/A)$ [proposition 1.3], according to next example (inspired by M. ROITMAN [15, Example 1.16]).

EXAMPLE 3.1: Let k be a field and $B = k[X,Y,(T_n)_{n\geq 1}]$ the ring of polynomials in the indeterminates X, Y and $(T_n)_{n\geq 1}$; it is clear that B is a Krull domain; we then let $A = k[X,Y,(XT_n)_{n\geq 1},(Y^nT_n)_{n\geq 1}]$ and $D = k[X,Y,(XT_n)_{n\geq 1}]$; thus $D \subset A \subset B$.
- D is Krull hence a.c.c.p., X is a prime element in D and $D \subset A \subset D[X^{-1}] = B[X^{-1}]$; X is also prime in B, which contains A, thus for every element a of A there is an integer n such that $aX^{-n} \notin A$; therefore A is a.c.c.p. [15, Lemma 1.17].
- X is prime in the Krull domain $B[Y^{-1}] = A[Y^{-1}]$, thus $\mathbf{p} = XA[Y^{-1}] \cap A$ is an height 1 prime of A, hence $\mathbf{p} \in \mathrm{Ass}_f(K/A)$ (where K is the field of fractions of A).
- Suppose $\mathbf{p} \subsetneq (A :_A x)$, then $xX \in A$ and $xXT_n \in A$, for every integer n; in particular $xXT_n \in B$, but, for n large xXT_n is not divisible by Y^n (in B), hence it is divisible by X and $xT_n \in B$; only finitely many indeterminates occur in x, thus $x \in B$ and there is n such that $Y^n x \in A$, therefore $(A :_A x)$ is strictly larger than $\mathbf{p}$, finally $\mathbf{p} \notin \mathrm{Ass}(K/A)$.

Conversely, the equality $\mathrm{Ass}_f(A/I) = \mathrm{Ass}(A/I)$ for every principal (resp. every invertible) ideal does not imply ascending chain condition on principal (resp. invertible) ideals:

EXAMPLE 3.2: Let p be a prime number and $A = \mathbb{Z}_p + X\mathbb{R}[[X]]$ (the important fact being here that the field of fractions of $\mathbb{Z}_p$ is strictly contained in $\mathbb{R}$). A is not a.c.c.p. since the ideals $(p^{-n}X)$ form a strictly increasing chain of principal ideals; however the only non trivial prime ideals of A are the principal ideal pA, and $\mathfrak{m} = X\mathbb{R}[[X]]$; clearly $pA = (A :_A p^{-1})$, and if $x \in \mathbb{R}$ but $x \notin \mathbb{Q}$, then $\mathfrak{m} = (A :_A x)$, indeed, $\mathfrak{m}x \subset A$ and conversely, if $(d + m)x \in A$, where $d \in \mathbb{Z}$, $m \in \mathfrak{m}$, then $dx \in \mathbb{Z}$, thus $d = 0$. Finally $\mathrm{Ass}_f(K/A) = \mathrm{Ass}(K/A) = \{pA, \mathfrak{m}\}$ and $\mathrm{Ass}_f(A/I) = \mathrm{Ass}(A/I)$ for every principal ideal [proposition 1.3]; now A is quasi local, hence every invertible ideal is principal; so a.c.c. does not hold on invertible ideals whereas $\mathrm{Ass}_f(A/I) = \mathrm{Ass}(A/I)$ for every invertible ideal.

We end up with a question:

QUESTION 3.3: Does the equality $\mathrm{Ass}_f(A/I) = \mathrm{Ass}(A/I)$ for *every* ideal imply ascending chain condition on invertible or at least principal ideals?

ACKNOWLEDGMENT: The author wishes to thank the referee for very useful corrections and remarks.

REFERENCES

1. BARUCCI V., On a class of Mori domains. *Comm in Algebra* **11**(1983), 1989-2001.
2. BARUCCI V., Strongly divisorial ideals and complete integral closure of an integral domain. *J. of Algebra* **99** (1986), 132-142.
3. BARUCCI V., DOBBS D.E. and FONTANA M., Conducive integral domains as pullbacks, *Manuscripta Math.* **54** (1986), 261-267.
4. BOURBAKI N., Algebre commutative. *Hermann*, Paris.
5. CAHEN P-J., Couples d'anneaux partageant un ideal. *Archiv der Math.* **51** (1988) 505-514.
6. CAHEN P-J., Construction B,I,D et anneaux localement ou résiduellement de Jaffard. *Archiv der Math.* **54** (19909 125-141.
7. DESSAGNES N.: Sur les anneaux de Mori. *CRAS Paris, Serie A* **286** (mars 1978), 405-407.
8. GABELLI S., On divisorial ideals in polynomial rings over Mori domains. *Comm in Algebra* **15** (1987), 2349-2370.
9. GABELLI S., On domains with ACC on invertible ideals. *Atti Acc. Lincei Rend. fis.* **82** (1988), 419-422.
10. GILMER R., Multiplicative ideal theory. *Marcel Dekker*, New York (1972).
11. HOUSTON E.G., T.G. LUCAS and VISWANATHAN T.M., Primary decomposition of divisorial ideals in Mori domains. *J. of Algebra* **117** (1988), 327-342.
12. QUERRE J., Sur une propriété des anneaux de Krull. *Bull. Sc. Math. 2-ème serie* **95** (1971) 341-354.
13. QUERRE J., Cours d'algèbre. *Masson*, Paris (1976).
14. ROITMAN M., On polynomial extensions of Mori domains over countable fields. *J. of pure and app. Alg.* **64** (1990) 315-328.
15. ROITMAN M., On the complete integral closure of a Mori domain. *J. of pure and app. Alg.* **66** (1990) 55-79.

5
Polynomials Whose Derivatives Are Integer-Valued in Number Fields

JEAN-LUC CHABERT Institut Supérieur des Sciences et Techniques. Université de Picardie. 48 rue Raspail, 02109 St Quentin, France

ABSTRACT: Carlitz proved that polynomials whose derivatives are integer-valued on Z are also those whose divided differences are integer-valued. We show that this result holds for the ring of integers A of some number fields. More precisely it holds if and only if each maximal ideal of A lying over a prime number p has an absolute ramification index strictly less than p.

1 INTRODUCTION

Carlitz's theorem

Everybody knows that:

1.1. The set **B** of integer-valued polynomials on $\mathbb{Z}$, i.e. $\mathbf{B} = \{P \in \mathbb{Q}[X] \mid P(\mathbb{Z}) \subset \mathbb{Z}\}$, is a free $\mathbb{Z}$-module generated by the binomial polynomials:

$$\binom{X}{n} = \frac{X(X-1)\ldots(X-n+1)}{n!}.$$

In fact this results from the Gregory-Newton formula. More recently, Straus [22] proved that:

1.2. The set **D** of polynomials which are integer-valued together with their derivatives of all orders is a free $\mathbb{Z}$-module generated by the polynomials:

$$d_n \binom{X}{n} \quad \text{where} \quad d_n = \prod_p p^{[n/p]}.$$

On the other hand, De Bruijn [12] characterized the set Δ^1 of polynomials which are integer-valued together with their first divided differences

$$\Delta P_m(X) = [P(X+m)-P(X)] / m \quad \text{where } m \in \mathbb{Z}\setminus\{0\}.$$

1.3. The set Δ^1 is a free $\mathbb{Z}$-module generated by the polynomials:

$$\delta_n^1 \binom{X}{n} \quad \text{where} \quad \delta_n^1 = \text{l.c.m.}\{1,2,\ldots,n\}.$$

Carlitz [8] extended this characterization to the set Δ^k of polynomials which are integer-valued together with their 1st, 2nd, ..., kth divided differences:

1.4. The set Δ^k is a free $\mathbb{Z}$-module generated by the polynomials:

$$\delta_n^k \binom{X}{n} \quad \text{where} \quad \delta_n^k = \text{l.c.m.}\ \{\ t_1.t_2....t_r \mid r \leq k\ ,\ t_1+t_2+...+t_r \leq n\ \}.$$

For k=1, this is De Bruijns's result and, for $k = \infty$, this shows that the set Δ of polynomials which are integer-valued together with all their divided differences coincides with the set **D** studied by Straus.

Extension to number fields

In 1919, Polya [18] and Ostrowski [17] introduced the notion of integer-valued polynomial on the ring of integers A of a number field K and they studied the set **B** of integer-valued polynomials on A, i.e. $\mathbf{B} = \{P \in K[X] \mid P(A) \subset A\}$. So the following question rises naturally: does the equality $\mathbf{D} = \Delta$, proved by Carlitz for $\mathbb{Z}$, hold for every ring of integers A ? In 1973 Barsky [1] answered affirmatively. But we have shown in a French written paper to appear in the *Acta arithmetica* that it is not always true and that it is possible to characterize the cases where the equality holds and where it does not.

For example let us consider the ring $A = \mathbb{Z}[i]$ of Gaussian integers and the polynomial:

$$P(X) = \frac{X^2(X-1)^2}{2}.$$

The polynomial P is integer-valued as well all its derivatives since P' belongs to A[X], while the first divided difference:

$$\Delta P_{1+i}(X) = \frac{P(1+i) - P(0)}{1+i} = -\frac{1+i}{2}$$

does not belong to A.

Here we are interested by the more general case where A is an order of a number field K, i.e. a subring of the ring of integers of K which is a free $\mathbb{Z}$-module with rank $[K:\mathbb{Q}]$. In the following text of the talk at Fès we explain how to compare the rings **D** and Δ and then how to get conditions on the rings A such that the equality $\mathbf{D} = \Delta$ holds.

In the second section we recall general results which are valid for any integral domain A and which concern inclusions and localizations. In the third section where we study the local case, A is a one-dimensional Noetherian local domain with finite residue field. Then, in the fourth section we restrict ourselves to the case where A is a discrete valuation domain; we determine conditions under which the equality $\mathbf{D} = \Delta$ holds and give some description of the spectrum of **D**. Finally, in the last section we globalize the previous results, obtain the desired characterization and give several examples.

2 NOTATIONS AND GENERAL RESULTS

Notations

Let A be any integral domain with quotient field K.

2.1. Let $\mathbf{B}_A$ be the ring of integer-valued polynomials on A:

$$\mathbf{B}_A = \{ P \in K[X] \mid P(A) \subset A \}.$$

2.2. For each integer k, let $\mathbf{D}_A^k$ be the ring of polynomials which are integer-valued on A together with their 1st, 2nd, ..., kth derivatives:

$$\mathbf{D}_A^k = \{ P \in \mathbf{B}_A \mid P' \in \mathbf{D}_A^{k-1} \}$$

and let Δ_A^k be the ring of polynomials which are integer-valued on A together with their 1st, 2nd, ..., kth divided differences:

$$\Delta_A^k = \{ P \in \mathbf{B}_A \mid \Delta_h P \in \Delta_A^{k-1}, h \in A \setminus \{o\} \}$$

where $\Delta_h P(X) = [P(X+h)-P(X)]/h$.

Of course, $\mathbf{D}_A^o = \Delta_A^o = \mathbf{B}_A$.

2.3. Let $\mathbf{D}_A$ be the set of polynomials which are integer-valued on A together with all their derivatives:

$$\mathbf{D}_A = \cap_k \mathbf{D}_A^k,$$

and let Δ_A be the set of polynomials which are integer-valued on A together with all their divided differences:

$$\Delta_A = \cap_k \Delta_A^k.$$

As for derivatives, finitely differences of order > n+1 of a polynomial of degree n are zero. If there is no confusion we will omit the reference to the ring A.

The rings **B** are studied in [3], [7], [9], $\mathbf{D}^k$ in [2, [4], [11], [19] and Δ^k in [14], [15], [16], [23].

Inclusions

2.4. PROPOSITION: *For each integer k, the ring Δ^k is contained in the ring $\mathbf{D}^k$.*

For A = $\mathbb{Z}$, this is a result of Carlitz [8] and, for any integral domain A, this is a result of Haouat and Grazzini [14]. Let us recall the proof.

2.5. LEMMA: *If P belongs to Δ^k (k>o), then P' belongs to Δ^{k-1}.*

Lemma's proof: For each polynomial P in K[X], let us write:

$$P(X+Y) = P(X) + Y\,P'(X) + Y^2 Q(X,Y) \quad \text{where } Q(X,Y) \in K[X,Y].$$

Let h be a non zero element of A such that hQ(X,Y) belongs to A[X,Y]. Then

$$P'(X) = [P(X+h)-P(X)]/h \ - \ h\,Q(X,h)\ ,$$

and

$$P'(X) = \Delta_h P(X) + R(X) \quad \text{where} \quad R(X) \in A[X].$$

Therefore, if P belongs to Δ^k (k>o), then $\Delta_h P$ belongs to Δ^{k-1} and P' also.

Proposition's proof: Let us prove by induction on k that Δ^k is contained in $\mathbf{D}^k$. For k=o, this results from the definitions. Let k be any positive integer. If P belongs to Δ^k, then, by lemma 2.5, P' belongs to Δ^{k-1}; by induction, Δ^{k-1} is contained in $\mathbf{D}^{k-1}$; thus P belongs to $\mathbf{D}^k$.

2.6. COROLLARY: *The domain Δ is contained in the domain* ***D***.

2.7. We have the following inclusions for any domain A:

$$\begin{array}{ccccccccccccccccc} A[X] & \subset & \Delta & \subset & \dots & \subset & \Delta^k & \subset & \Delta^{k-1} & \subset & \dots & \subset & \Delta^1 & & & & \\ & & \cap & & & & \cap & & \cap & & & & \cap & & & & \\ & & \mathbf{D} & \subset & \dots & \subset & \mathbf{D}^k & \subset & \mathbf{D}^{k-1} & \subset & \dots & \subset & \mathbf{D}^1 & \subset & \mathbf{B} & \subset & K[X] \end{array}$$

We will see that, as for the case A = $\mathbb{Z}$, each of these inclusions are strict except for the possible equality $\mathbf{D} = \Delta$.

Localizations

It is well-known that:

2.8. Every integer-valued polynomial on A is also an integer-valued polynomial on $S^{-1}A$ where S is any multiplicative subset of A [7].

This can be read as the following inclusion:

$$S^{-1}(\mathbf{B}_A) \subset \mathbf{B}_{S^{-1}A}$$

and can be extended with the following proposition.

2.9. PROPOSITION: *For each multiplicative subset S of A, one has the inclusion* $S^{-1}(\boldsymbol{R}_A) \subset \boldsymbol{R}_{S^{-1}A}$ *, where* ***R*** *is anyone of the rings* ***B***, $\boldsymbol{D}^k$, Δ^k, ***D*** *or* Δ.

Proof: The proof is immediate for the rings $\mathbf{D}^k$ and $\mathbf{D}$ because of the inclusion 2.8. Let us just recall the proof for Δ_A^1 (Haouat and Grazzini [16]).

Let P be in Δ_A^1, then, for each h∈ A\{o}, $\Delta_h P$ belongs to $\mathbf{B}_A$ and thus to $\mathbf{B}_{S^{-1}A}$. Let us fix x in $S^{-1}A$ and let us consider the polynomial:

$$Q_x(Y) = [P(x+Y)-P(x)]/Y = \Delta_Y P(x).$$

For each h in $A\backslash\{o\}$, $Q_x(h)$ belongs to $S^{-1}A$, then $Q_x(A)$ is contained in $S^{-1}A$ and $Q_x(S^{-1}A)$ is also contained in $S^{-1}A$, that is to say $\Delta_h P(x)$ belongs to $S^{-1}A$ for each x in $S^{-1}A$ and each h in $S^{-1}A\backslash\{o\}$.

In order to obtain equalities instead of inclusions we have to suppose that A is Noetherian.

2.10. PROPOSITION: *If A is a Noetherian domain, then, for each multiplicative subset S of A, the following equalities hold:*

$$S^{-1}(\boldsymbol{R}_A) = \boldsymbol{R}_{S^{-1}A}$$

where $\boldsymbol{R}$ *is anyone of the rings* $\boldsymbol{B}$, $\boldsymbol{D}^k$, Δ^k, $\boldsymbol{D}$ *or* Δ.

Proof: This equality is well-known for $\mathbf{R} = \mathbf{B}$ [7] and for $\mathbf{R} = \mathbf{D}^k$ or $\mathbf{R} = \mathbf{D}$ the proof is analogous to this special case: if P belongs to $\mathbf{D}^k_{S^{-1}A}$, i.e. P, P', ..., $P^{(k)}$ belong to $\mathbf{B}_{S^{-1}A}$, then there exists s in S such that sP, sP', ..., $sP^{(k)}$ belong to $\mathbf{B}_A$, i.e. sP belongs to $\mathbf{D}^k_A$. Indeed the A-module M generated by the values of P, P', ..., $P^{(k)}$ on A is contained in the A-module N generated by the coefficients of P, P', ..., $P^{(k)}$; as N is a finitely generated A-module, M is also a finitely generated A-module. This proof holds for $\mathbf{D}$ because a polynomial has only finitely many non-zero derivatives.

Let us give the proof for Δ^1. For each polynomial P with degree d, let us write

$$P(X+Y) = P(X) + Y\,P'(X) + Y^2\,P^{[2]}(X) + \ldots + Y^d\,P^{[d]}(X)$$

and let N be the A-module generated by the coefficients of P, P', $P^{[2]}$, ..., $P^{[d]}$. If P belongs to $\Delta^1_{S^{-1}A}$, then, for each h in $A\backslash\{o\}$ and each x in A, the element

$$\Delta_h P(x) = [P(x+h)-P(x)]/h = P'(x) + hP^{[2]}(x) + \ldots + h^{d-1}P^{[d]}(x)$$

belongs to the finitely generated A-module $N \cap S^{-1}A$. Let s be in S such that $s(N \cap S^{-1}A)$ is contained in A, then sP belongs to Δ^1_A.

2.11. COROLLARY: If A is a Noetherian domain, then the equality $\mathbf{D}_A = \Delta_A$ holds if and only if, for each maximal ideal $\mathfrak{m}$ of A, the equality $\mathbf{D}_{A_\mathfrak{m}} = \Delta_{A_\mathfrak{m}}$ holds.

3 THE LOCAL CASE

As we are interested by the case where A is the ring of integers of a number field or more generally where A is an order of a number field, the previous section allows us to restrict our study to the case where A is a local domain. Thus we may formulate the following hypothesis.

3.1. HYPOTHESES: Let A be a one-dimensional Noetherian domain with characteristic zero and finite residue field. Let us note K the quotient field of A, $\mathfrak{m}$ the maximal ideal of A and $q = p^f$ the cardinal of the residue field $A/\mathfrak{m}$.

3.2. PROPOSITION: *Under hypotheses 3.1,*
(i) the ring A[X] is strictly contained in the ring **D**,
(ii) for each k such that $1 \leq k < p$, Δ^{k-1} *is not contained in* $\mathbf{D}^k$, *thus* $\mathbf{D}^{k-1} \neq \mathbf{D}^k$ *and* $\Delta^{k-1} \neq \Delta^k$.

Proof: Since $\mathfrak{m}$ belongs to $Ass_A(K/A)$, there exists α in K\A such that $\mathfrak{m}\alpha \subset A$. Let us note $g_k(X) = \alpha(X^q-X)^k$. Obviously g_p does not belong to A[X], but it belongs to **D** since g_p' is in A[X]. Now we will show that, for $1 \leq k < p$, g_k belongs to Δ^{k-1} but does not belong to $\mathbf{D}^k$. Notice first that for any polynomials $P_1, ..., P_k$ in A[X] and any h in A\{0}:

$$\Delta_h(P_1...P_k)(X) = \Sigma_{1 \leq j \leq k} P_1(X)...P_{j-1}(X)\, \Delta_h P_j(X)\, P_{j+1}(X+h)...P_k(X+h).$$

Moreover, if at least s polynomials P_j take their values in $\mathfrak{m}$, then at least s-1 polynomials of each product in the previous sum take their values in $\mathfrak{m}$. Therefore if the k polynomials $P_1, ...,$ P_k take their values in $\mathfrak{m}$, the values of each divided differences of order $\leq$ k-1 of the product $P_1...P_k$ belong to $\mathfrak{m}$. So the divided differences of order $\leq$ k-1 of the polynomial $(X^q-X)^k$ take their values in $\mathfrak{m}$ and g_k belongs to Δ^{k-1}.

Suppose now that $1 \leq k < p$. One can check by induction on i that, for $o \leq i \leq k$, $g_k^{(i)}(X)$ may be written $\alpha(X^q-X)^{k-i}\, r_i(X)$ where $r_i(X) \in A[X]$ and $r_i(y) \notin \mathfrak{m}$ for each $y \in \mathfrak{m}$. Therefore $g_k^{(i)}(X)$ belongs to **B** for $o \leq i < k$, while $g_k^{(k)}(y)$ does not belong to A for each y in $\mathfrak{m}$. So g_k belongs to $\mathbf{D}^{k-1}$ (this results anyway from the inclusion $\Delta^{k-1} \subset \mathbf{D}^{k-1}$) but g_k does not belong to $\mathbf{D}^k$.

3.3. PROPOSITION: *Under hypotheses 3.1, if A is unibranch, that is to say if the integral closure A' of A is a local ring, then for each k the ring* Δ^k *is strictly contained in the ring* $\mathbf{D}^k$.

Proof: We know that the rings $\mathbf{D}^k$ and Δ^k are distinct because $\mathbf{D}^k$ has more prime ideals lying over $\mathfrak{m}$ than Δ_k. Indeed we recall there are exactly q such primes in Δ^k (assertion 3.4 below) whereas there is a one to one map from the set of primes of $\mathbf{D}^k$ lying over $\mathfrak{m}$ to the set of primes of **B** lying over $\mathfrak{m}$ (assertion 3.6), this latter set being infinite if A is unibranch (assertion 3.5).

3.4. For each k, the prime ideals of Δ^k lying over $\mathfrak{m}$ are the maximal ideals $\mathfrak{m}_a(\Delta^k) = \{P \in \Delta^k \mid P(a) \in \mathfrak{m}\}$ where a runs over some set of representatives of $A/\mathfrak{m}$ (Haouat and Grazzini [16]).

3.5. Let Â be the completion of A with respect to the $\mathfrak{m}$-adic topology. The prime ideals of **B** lying over $\mathfrak{m}$ are the maximal ideals $\mathfrak{m}_x = \{P \in \mathbf{B} \mid P(x) \in \mathfrak{m}\hat{A}\}$ where x is any element of Â [10]. There are infinitely many distinct such prime ideals $\mathfrak{m}_x$ if and only if A is unibranch (Cahen [5] and [6], Gilmer, Heinzer and Lantz [13]).

3.6. For each integer k and each polynomial P in **B**, there exists an integer s such that P^s belongs to $\mathbf{D}^k$ [11].

3.7. PROPOSITION: *Under hypotheses 3.1, if the ideal $\mathfrak{m}^{p-1}$ is contained in the ideal pA where p is the characteristic of $A/\mathfrak{m}$, then* **D** *and* Δ *are equal.*

To prove that **D** is contained in Δ it is enough to check the stability of **D** with respect to the operators Δ_h where $h \in A\setminus\{o\}$. This results from the following lemma in the case r=1.

3.8. LEMMA: *Under hypotheses 3.1, there exists an integer r such that $\mathfrak{m}^{r(p-1)}$ is contained in pA. Then for each polynomial P in* **D** *and each non zero element h of $\mathfrak{m}^r$, the polynomial $\Delta_h(P)$ belongs to* **D**.

Proof: Let P be in **D** and h be in $\mathfrak{m}^r$; then the polynomial

$$\Delta_h(P(X)) = [P(X+h)-P(X)]/h = \Sigma_{n\geq 1}\, (h^{n-1}/n!)\, P^{(n)}(X)$$

belongs to **D** because of the following remark.

For each $h \in \mathfrak{m}^r$ and each $n \in \mathbb{N}$, $h^{n-1}/n!$ belongs to A.

The element h^{n-1} belongs to $\mathfrak{m}^{r(n-1)}$, thus is belongs to $\mathfrak{m}^{kr(p-1)}$ where $k = [(n-1)/(p-1)]$ and then to p^kA. Every integer prime with p is invertible in the local ring A and it is enough to check that $k \geq v_p(n!)$ where $v_p(n!)$ is the greatest power of p which divides n!. It is equivalent to show that $(n-1)/(p-1) \geq v_p(n!)$, i.e. $n-1 \geq (p-1)\, v_p(n!)$ or $n > (p-1)\, v_p(n!)$. Indeed

$$v_p(n!) = [n/p] + [n/p^2] + [n/p^3] + ... < n/p + n/p^2 + n/p^3 + ... = n/(p-1).$$

3.9. REMARK: On the contrary if there exists some element h in $\mathfrak{m}$ such that h^{p-1} does not belong to pA while $\mathfrak{m}^p$ is contained in pA, then $\mathbf{D} \not\subset \Delta^1$ and $\mathbf{D} \neq \Delta$. For example $g(X) = (X^q-X)^p/p$ belongs to **B** and $g'(X)$ belongs to A[X] since $\mathfrak{m}^p \subset pA$, thus g(X) is in **D**. On the other hand $\Delta_h g(o) = (h^{q-1}-1)^p\, h^{p-1}/p$ belongs to A for h in $\mathfrak{m}$ if and only if $h^{p-1} \in pA$.

3.10. EXAMPLE: Let $A = \mathbb{Z}[\sqrt{d}]$ where d is a square-free integer and let $\mathfrak{m}$ be a prime ideal of A lying over 2. Then $A_\mathfrak{m}$ is a one-dimensional local Noetherian domain whose residue field is finite and has characteristic 2; $\mathfrak{m}^2 \subset 2A$, but $\mathfrak{m} \not\subset 2A$; thus $D_{A\mathfrak{m}} \neq \Delta_{A\mathfrak{m}}$.

3.11. COROLLARY: *Under hypotheses 3.1, let r be such that $\mathfrak{m}^{r(p-1)} \subset pA$. If $a \equiv b \pmod{\mathfrak{m}^r}$, then $\mathfrak{m}_a(\mathbf{D}) = \mathfrak{m}_b(\mathbf{D})$ where $\mathfrak{m}_a(\mathbf{D})$ is the maximal ideal $\{P \in \mathbf{D} \mid P(a) \in \mathfrak{m}\}$.*

Proof: If P belongs to **D** and if $a \equiv b \pmod{\mathfrak{m}^r}$, then $[P(b)-P(a)]/(b-a) = \Delta_{b-a}(P(a))$ belongs to A (lemma 3.8) and P(b)-P(a) belongs to $\mathfrak{m}^r$. Therefore $P \in \mathfrak{m}_a(\mathbf{D})$ is equivalent to $P \in \mathfrak{m}_b(\mathbf{D})$ and then $\mathfrak{m}_a(\mathbf{D}) = \mathfrak{m}_b(\mathbf{D})$.

4 THE CASE OF A DISCRETE VALUATION DOMAIN

The inclusions

4.1. PROPOSITION: *Let A be a discrete valuation domain. The rings **D** and Δ are equal if and only if the residue field of the valuation v is infinite or its characteristic p satisfies the inequality v(p) < p.*

Proof: If the residue field is infinite, then **B** = A[X] [7] and of course **D** = Δ. Let us suppose now that the residue field is finite and that its characteristic p satisfies the inequality v(p) < p. Then pA contains $\mathfrak{m}^{p-1}$, the characteristic of A is zero and proposition 3.7 shows that **D** = Δ. Conversely let t be such that v(t)=1 and let us consider the polynomial $h(X) = [(X^q - X) / t]^p$: h belongs to **B**, but $\Delta_t h(o) = [h(t)-h(o)] / t = (t^{q-1}-1)^p / t$ is not in A, thus h does not belong to Δ^1. If v(p) ≥ p, then $h'(x) = (X^q-X)^{p-1}.(qX^{q-1}-1).p / t^p$ belongs to A[X], h(X) belongs to **D** and **D** is not contained in Δ^1, a fortiori **D** and Δ are not equal.

4.2. REMARK: Let us suppose that A is a localization of the ring of integers of a number field, then A is a discrete valuation domain with characteristic zero and with finite residue field and, for each k, the domains $\mathbf{D}^k$ and Δ^k are distinct (proposition 3.3). Moreover, for each k, the domains $\mathbf{D}^k$ and $\mathbf{D}^{k+1}$ are also distinct, for, if $\mathbf{D}^{k+1}$ was equal to $\mathbf{D}^k$, **D** would also be equal to $\mathbf{D}^k$, but it is impossible because $\text{Spec}(\mathbf{D}^k/\mathfrak{m}\mathbf{D}^k)$ is infinite (propositions 3.5 and 3.6) and $\text{Spec}(\mathbf{D}/\mathfrak{m}\mathbf{D})$ is finite (corollary 3.11).

Spec(D/$\mathfrak{m}$D)

4.3. PROPOSITION: *Suppose that A is a discrete valuation domain with characteristic zero and with finite residue field. The following assertions are equivalent:*
(i) v(p) ≤ p-1,
*(ii) **D** = Δ,*
(iii) $\mathfrak{m}_a(\mathbf{D}) = \mathfrak{m}_b(\mathbf{D}) \Leftrightarrow v(a-b) \geq 1$.

Proof: (i) ⇔ (ii) results from proposition 4.1. The implication $\mathfrak{m}_a(\mathbf{D}) = \mathfrak{m}_b(\mathbf{D}) \Rightarrow v(a-b) \geq 1$ in (iii) is always true because X-a belongs to $\mathfrak{m}_a(\mathbf{D})$. Corollary 3.11 in the case r=1 shows that conversely, if v(p) ≤ p-1, $v(a-b) \geq 1 \Rightarrow \mathfrak{m}_a(\mathbf{D}) = \mathfrak{m}_b(\mathbf{D})$. Thus (i) implies (iii). Let us suppose that v(p) ≥ p. Then, as already seen in the proof of 4.1, the polynomial $h(X) = [(X^q - X) / t]^p$ belongs to **D**, it is in $\mathfrak{m}_0(\mathbf{D})$ but not in $\mathfrak{m}_t(\mathbf{D})$, so v(a-b)≥1 does not imply $\mathfrak{m}_a(\mathbf{D}) = \mathfrak{m}_b(\mathbf{D})$.

4.4. PROPOSITION: *Suppose that A is a discrete valuation domain with characteristic zero and with finite residue field. If p ≤ v(p) ≤ 2(p-1), then:*

$$\mathfrak{m}_a(D) = \mathfrak{m}_b(D) \Leftrightarrow v(a-b) \geq 2.$$

Proof: Since $v(p) \leq 2(p-1)$, corollary 3.11 in the case r=2 shows that $v(a-b)\geq 2$ implies $\mathfrak{m}_a(\mathbf{D}) = \mathfrak{m}_b(\mathbf{D})$. Conversely let us suppose that $v(a-b) < 2$. Let t be an element of A such that $v(t)=1$, let $a_1, a_2, ..., a_q$ be some set of representatives of $A/\mathfrak{m}$ which contains a but does not contain b and let g(X) be the polynomial $(X-a_1)(X-a_2)...(X-a_q) / t$. Then g belongs to **B**, g(a) = 0 and v(g(b)) = 0. Since $v(p) \geq p$, $(g^p)'$ belongs to A[X], then g^p belongs to **D** and to $\mathfrak{m}_a(\mathbf{D})$ but not to $\mathfrak{m}_b(\mathbf{D})$. Thus $\mathfrak{m}_a(\mathbf{D}) = \mathfrak{m}_b(\mathbf{D})$ implies $v(a-b)\geq 2$.

5 GLOBALIZATION

5.1. THEOREM: *Let A be a Dedekind domain. The rings* **D** *and* Δ *are equal if and only if, for each maximal ideal* $\mathfrak{m}$ *of A, either the residue field* $A/\mathfrak{m}$ *is infinite or its characteristic p satisfies* $\mathfrak{m}^{p-1} \subset pA_{\mathfrak{m}}$.

The theorem results from propositions 2.10 and 4.1.

The case of the ring of integers of a number field

5.2. PROPOSITION: *Let A be the ring of integers of a number field K. The rings* **D** *and* Δ *are equal if and only if for each maximal ideal* $\mathfrak{m}$ *of A the following equivalent conditions are satisfied (where p denotes the prime number lying under* $\mathfrak{m}$*):*
(i) the ramification index $e = e(\mathfrak{m}/p)$ *of* $\mathfrak{m}$ *in the extension* $K/\mathbb{Q}$ *is less than or equal to p-1,*
(ii) the exponent s of $\mathfrak{m}$ *in the different* $\mathfrak{D}_{A/\mathbf{Z}}$ *is less than or equal to p-2.*

Proof: Assertion (i) is the condition of proposition 5.1. Thus it remains to show that (i) and (ii) are equivalent. The exponent s of $\mathfrak{m}$ in the different $\mathfrak{D}_{A/\mathbf{Z}}$ is always greater than or equal to e-1 and the equality s = e-1 holds if and only if p does not divide e ([21], chap. III). In the present case, if $e \leq p-1$, then p does not divide e and $s = e-1 \leq p-2$; thus (i) implies (ii). Conversely if $s \leq p-2$, then $e \leq s+1 \leq p-1$, since $s \geq e-1$; thus (ii) implies (i).

Moreover:

5.3. THEOREM: *If A is the ring of integers of a number field, except the possible equality between* **D** *and* Δ*, all the following inclusions are strict:*

$$\begin{array}{ccccccccccccccc} A[X] & \subset & \Delta & \subset & \dots & \subset & \Delta^k & \subset & \Delta^{k-1} & \subset & \dots & \subset & \Delta^1 & & & & \\ & & \cap & & & & \cap & & \cap & & & & \cap & & & & \\ & & \mathbf{D} & \subset & \dots & \subset & \mathbf{D}^k & \subset & \mathbf{D}^{k-1} & \subset & \dots & \subset & \mathbf{D}^1 & \subset & \mathbf{B} & \subset & K[X] \end{array}$$

Proof: For each integer k the ring Δ^k is strictly contained in the ring $\mathbf{D}^k$ (propositions 2.10 and 3.3), the ring $\mathbf{D}^k$ is itself strictly contained in the ring $\mathbf{D}^{k-1}$ (proposition 2.10 and remark 4.2).

It remains to show that $\Delta^{k-1} \neq \Delta^k$ are distinct: for each k, there exists a prime number p strictly greater than k and a maximal ideal $\mathfrak{m}$ of A lying over p. With respect to the localization $A_\mathfrak{m}$, one has $\Delta^{p-1}_{A_\mathfrak{m}} \neq \Delta^{p-2}_{A_\mathfrak{m}}$ (proposition 3.2), therefore $\Delta^{p-1}_A \neq \Delta^{p-2}_A$ (proposition 2.10) and $\Delta^k_A \neq \Delta^{k-1}_A$.

Quadratic fields

5.4. PROPOSITION: *When A is the ring of integers of a quadratic field $K = \mathbb{Q}[\sqrt{d}]$ where d is a square-free integer, the equality $\mathbf{D} = \Delta$ holds if and only if $d \equiv 1 \pmod 4$.*

Proof: Since $[K/\mathbb{Q}] = 2$, $e = e(\mathfrak{m}/p) = 1$ or 2, the only case to reject is the case where e = 2 and p = 2. It is well-known that the prime number p is ramified if and only if $d \equiv 2$ or $3 \pmod 4$ ([20], chap. V); so the remaining case is $d \equiv 1 \pmod 4$ and $A = \mathbb{Z} + (1+\sqrt{d})/2\ \mathbb{Z}$.

Cyclotomic fields

5.5. PROPOSITION: *When A is the ring of integers of a cyclotomic field $K = \mathbb{Q}[\zeta]$ where ζ is a primitive n-th root of unity, the equality $\mathbf{D} = \Delta$ holds if and only if n is a product of distinct prime numbers.*

Proof: The prime numbers which are ramified in the extension $A/\mathbb{Z}$ are those which divide n and the corresponding ramification index is $p^{k-1}(p-1)$ where p^k is the greatest power of p which divides n ([20], chap. IV). So the condition of proposition 5.2 is satisfied if and only if each k is equal to 1.

Counterexamples ($\Delta \neq \mathbf{D}$)

(i) The ring of integers A of the number field $K = \mathbb{Q}[\sqrt[p]{p}]$ where p is any prime number.
Let ξ be a p-th root of unity. Since $\xi^p = p$ and $p = [K:\mathbb{Q}]$, $\mathfrak{m} = \xi A$ is the only prime ideal of A lying over p and $e(\mathfrak{m}/p) = p$.
(ii) The ring $\mathbb{Z}[i]$ of Gaussian integers.
In this case $i = \sqrt[2]{-1}$ and $-1 \equiv 3 \pmod 4$ (quadratic case) or $i = \sqrt[4]{1}$ and $4 = 2^2$ (cyclotomic case).
(iii) The ring $\mathbb{Z}[\sqrt{d}]$ where d is a square-free integer.
If $d \equiv 2$ or $3 \pmod 4$, $\mathbb{Z}[\sqrt{d}]$ is the ring of integers of $\mathbb{Q}[\sqrt{d}]$ (proposition 5.4).
If $d \equiv 1 \pmod 4$, the example 3.10 shows that $\mathbf{D}_{A_\mathfrak{m}} \neq \Delta_{A_\mathfrak{m}}$ where $\mathfrak{m}$ is a maximal ideal of A lying over 2.

REFERENCES

1. BARSKY (D.), Fonctions k-lipschitziennes sur un anneau local et polynômes à valeurs entières, *Bull. Soc. Math. France*, t. **101** (1973) p. 397-411.
2. BRIZOLIS (D.) and STRAUS (E.G.), A basis for the ring of doubly integer-valued polynomials, *J. reine angew. Math.*, t. **286/287** (1976) p. 187-195.
3. CAHEN (P-J.), Polynômes à valeurs entières, *Canad. J. Math.*, t. **24** (1972) p. 747-754.
4. CAHEN (P.-J.), Polynômes et dérivées à valeurs entières, *Ann. Sci. Univ. Clermont*, Ser. Math., t. **10** (1975) p. 25-43.
5. CAHEN (P.-J.), Polynômes à valeurs entières sur un anneau non analytiquement irréductible, *J. reine angew. Math.*, t. **418** (1991) p. 131-137.
6. CAHEN (P.-J.), Integer-valued polynomials on a subset, *Trans. Amer Math. Soc*, to appear.
7. CAHEN (P.-J.) and CHABERT (J.-L.), Coefficients et valeurs d'un polynôme, *Bull. Sc. Math.*, t. **95** (1971) p. 295-304.
8. CARLITZ (L.), A note on integral-valued polynomials, *Indagationes Math.*, Ser. A, t. **62** (1959) p. 294-299.
9. CHABERT (J.-L.), Anneaux de "polynômes à valeurs entières" et anneaux de Fatou, *Bull. Soc. math. France*, t. **99** (1971) p. 273-283.
10. CHABERT (J.-L.), Les idéaux premiers de l'anneau des polynômes à valeurs entières, *J. reine angew. Math.*, t. **293/294** (1977) p. 275-283.
11. CHABERT (J.-L.), Polynômes à valeurs entières ainsi que leurs dérivées, *Ann. Sci. Univ. Clermont*, Ser. Math., t.**18** (1979) p. 47-64.
12. DE BRUIJN (N.G.), Some classes of integer-valued functions, *Nederl. Akad. Wetensch. Proc.*, Ser. A, t. **58** (1955) p. 363-367.
13. GILMER (R.), HEINZER (W.) and LANTZ (D.), The Noetherian Property in Rings of Integer-Valued Polynomials, *Trans. Amer. Math. Soc.*, to appear.
14. HAOUAT (Y.) and GRAZZINI (F.), Polynômes et différences finies divisées, *C. R. Acad. Sc. Paris*, Ser. A, t. **284** (1977) p. 1171-1173.
15. HAOUAT (Y.) and GRAZZINI (F.), Différences finies divisées sur un anneau S(2), *C. R. Acad. Sc. Paris*, Ser. A, **286** (1978) p. 723-725.
16. HAOUAT (Y.) and GRAZZINI (F.), Polynômes de Barsky, *Ann. Sci. Univ. Clermont*, Ser. Math., t. **18** (1979) p. 65-81.
17. OSTROWSKI (A.), Über ganzwertige Polynome in algebraischen Zahlkörpern, *J. reine angew. Math.*, t. **149** (1919) p. 117-124.
18. POLYA (G.), Über ganzwertige Polynome in algebraischen Zahlkörpern, *J. reine angew. Math.*, t. **149** (1919) p. 97-116.
19. ROGERS (K.) and STRAUS (E.G.), Infinitely integer-valued polynomials over an algebraic number field *Pacific J. of Math.*, t. **118** (1985) p. 507-522.
20. SAMUEL (P.), *Théorie Algébrique des Nombres*, Paris, Hermann (1967).
21. SERRE (J.-P.), *Corps locaux*, Paris, Hermann (1962).
22. STRAUS (E.G.), On the polynomials whose derivatives have integral values at integers *Proc. Amer. Math. Soc.*, t. **2** (1951) p. 24-27.
23. WAGNER (C.G.), Polynomials over **GF**(q,x) with Integral-valued Differences, *Arch. Math.*, t. **27** (1976) p. 495-501.

6
Krull Dimension of Graded Algebras

Rachid CHIBLOUN Département de Mathématiques, Faculté des Sciences, Meknès, Morocco

Artibano MICALI Département des Mathématiques, Université de Montpellier II, 34095 Montpellier Cedex 05, France

We know (cf. [2], Theorem 3.3) that if A is a noetherian ring and M any finitely generated A-module, then :

$$\dim (S_A(M)) = \mathrm{Sup}_{P\in \mathrm{spec}(A)} (\dim (A/P) + \mu (M_P))$$

where dim (A/P) points out the Krull dimension of A/P, $S_A(M)$ the symmetric algebra of M and $\mu(M_P)$ the minimal number of generators of the localization of M at P.

This formula was first of all demonstrated in [6] in the local case, then in [4] under hypothesis of catenarity, and finally in [2] for finitely generated modules over noetherian rings. The goal of this paper is to prove that this formula is still valid to give away the Krull dimension of graded algebra.

1. PRELIMINARIES

All rings considered in this work are commutative with identity. For an integral domain A, Fr(A) denotes its quotient field. If A and B are integral domains such that $A \subset B$,we say that the couple (A, B) is an extension of A, and we denote by $\mathrm{tr.deg.}_A B$ the transcendence degree of Fr(B) over Fr(A). We say that an extension (A, B) satisfies the altitude inequality formula (resp. altitude formula) if for each prime ideal Q of B, $\mathrm{ht}\, Q + \mathrm{tr.deg.}_{A/(Q\cap A)} B/Q \leq \mathrm{ht}\, Q\cap A + \mathrm{tr.deg.}_A B$ (resp. $\mathrm{ht}\, Q + \mathrm{tr.deg.}_{A/(Q\cap A)} B/Q = \mathrm{ht}\, Q\cap A + \mathrm{tr.deg.}_A B$). We say that an integral domain A satisfies the altitude inequality formula [5] (resp. altitude formula) if each extension (A, B), where B is a finitely generated A-algebra, satisfies the altitude inequality formula (resp. altitude formula). It is known that arbitrary noetherian domains, arbitrary Prüfer domains, arbitrary stably strong S-domains [5] and arbitrary universally catenarian domains [1] satisfy the altitude inequality formula.

Afterwards we give some results that are already found in the literature (cf. [2] and [3]).

LEMMA 1.1. Let A be a ring and B a finitely generated A-algebra containing A. If Q is a prime ideal of B and $P = Q \cap A$, and if $B \otimes_A k(P)$ is an integral domain, where k(P) is the residue field of A_P, Then

$$\dim (B \otimes_A k(P)) = \mathrm{ht}\,(Q/PB) + \mathrm{tr.deg.}_{A/P}\, B/Q.$$

LEMMA 1.2. Let $A = \oplus_{n \geq o} A_n$ be an integral noetherian graded ring. Then

$$\dim A = \dim A_o + \mathrm{tr.deg.}_{A_o} A.$$

LEMMA 1.3. ([2, Proposition 1.4]). Let A be an integral noetherian domain and B an integral finitely generated A-algebra containing A. Then

$$\dim A + \mathrm{tr.deg.}_A B - 1 \leq \dim B \leq \dim A + \mathrm{tr.deg.}_A B$$

($\dim A < \infty$ if and only if $\dim B < \infty$).

These lemmas lead us to the following result :

THEOREM 1. 4. Let $A = \oplus_{n \geq o} A_n$ be a noetherian graded ring, Q a prime ideal not graded of A and $P = Q \cap A_o$. Then $\mathrm{tr.deg.}_{A_o/P} A/Q = \mathrm{tr.deg.}_{A_o/P} (A/Q^{gr}) - 1$, and

$$\dim (A/Q^{gr}) - 2 \leq \dim (A/Q) \leq \dim (A/Q^{gr}) - 1,$$

where $Q^{gr} = \sum_{n \geq o} (Q \cap A_n)$ is the biggest graded ideal of A contained in Q.

PROOF. Since The graded ring A is noetherian, its ideal $A_+ = \oplus_{n \geq 1} A_n$ is finitely generated, that is, A is a finitely generated A_o-algebra. There exists $m \in \mathbb{N}^*$ and I an ideal of $B = A[X_1, \ldots, X_m]$ such that $A = B/I$. The ideal I is not necessarily graded for natural graduation of B.

However, we know that some prime ideals, S and R, exist in B with $S \subset R$, $Q = R/I$, $Q^{gr} = S/I$ and $P = R \cap A_0 = S \cap A_0$. We have $A/Q = B/R$ and $A/Q^{gr} = B/S$, then taking into account Lemma 1.1, $\text{tr.deg.}_{A_0/P}\ A/Q = \text{tr.deg.}_{A_0/P}\ B/R = m - \text{ht}\ (R/PB)$, and $\text{tr.deg.}_{A_0/P}\ A/Q^{gr} = \text{tr.deg.}_{A_0/P} B/S = m - \text{ht}\ (S/PB)$. Then $\text{tr.deg.}_{A_0/P}\ A/Q^{gr} - \text{tr.deg.}_{A_0/P}\ A/Q = \text{ht}\ (R/PB) - \text{ht}(S/PB) = \text{ht}\ (R/S) = \text{ht}\ (Q/Q^{gr}) = 1$.

Concerning the second formula, Lemma 1.3 says that $\dim A_0/P + \text{tr.deg.}_{A_0/P} A/Q - 1 \leq \dim(A/Q) \leq \dim (A_0/P) + \text{tr.deg.}_{A_0/P} A/Q$, and taking account of the first part of Theorem we obtain $\dim (A_0/P) + \text{tr.deg.}_{A_0/P} A/Q^{gr} - 2 \leq \dim (A/Q) \leq \dim (A_0/P) + \text{tr.deg.}_{A_0/P} A/Q^{gr} - 1$. Lemma 1.2 tells us $\dim A/Q^{gr} = \dim (A_0/P) + \text{tr.deg.}_{A_0/P} A/Q^{gr}$. Then $\dim (A/Q^{gr}) - 2 \leq \dim(A/Q) \leq \dim (A/Q^{gr}) - 1$.

2. MAIN THEOREM

First of all, we give two lemmas in order to establish the main theorem :

LEMMA 2.1. Let A be a ring such that A/P satisfies the altitude inequality formula for each prime ideal P of A. Let B a finitely generated A-algebra containing A, then

$$\dim B \leq \text{Sup}_{P \in \text{spec}(A)} (\dim (A/P) + \dim (B \otimes_A k(P))).$$

PROOF. We know that $\dim (B) = \text{Sup}_{Q \in \text{spec}(B)} \dim (B/Q)$, and if Q is a prime ideal of B then $\dim(B/Q) \leq \dim (A/P) + \text{tr.deg.}_{A/P} B/Q$, where $P = Q \cap A$. However $B \otimes_A k(P) \cong S^{-1}(B/PB)$, where $S = A/P - \{0\}$. There exists a prime ideal Q' of $B \otimes_A k(P)$ which corresponds to $S^{-1}(Q/PB)$. Therefore $(B \otimes_A k(P)) / Q' \cong S^{-1}(B/PB) / S^{-1}(Q/PB) = S'^{-1}(B/Q)$, where S' is the homomorphic image of S by the canonical map B/PB --------> B/Q and so we have $\text{Fr}(B/Q) = \text{Fr}(S'^{-1}(B/Q)) \cong \text{Fr}((B \otimes_A k(P))/Q')$. Then $\text{tr.deg.}_{A/P} B/Q = \text{tr.deg.}_{A/P} (B \otimes_A k(P)) / Q'$. Since $(B \otimes_A k(P)) / Q'$ is an integral finitely generated k(P)-algebra, then $\dim (B \otimes_A k(P)) / Q' = \text{tr.deg.}_{k(P)} (B \otimes_A k(P)) / Q'$, which involves $\text{tr.deg.}_{A/P} B/Q = \dim (B \otimes_A k(P)) / Q' \leq \dim(B \otimes_A k(P))$. Therefore $\dim (B/Q) \leq \dim (A/P) + \dim (B \otimes_A k(P))$, consequently $\dim (B) \leq \text{Sup}_{P \in \text{spec}(A)} (\dim (A/P) + \dim (B \otimes_A k(P)))$.

LEMMA 2.2. Let A be a ring, P a prime ideal of A and $B = \oplus_{n\geq 0} B_n$ a graded finitely generated algebra over $B_0 = A$. Then

$$\dim (B\otimes_A k(P)) = \mathrm{ht}\,((P + B_+)/PB).$$

PROOF. First, we remind that $P + B_+ = PB + B_+$ is a graded ideal of B. The k(P)-algebra $B\otimes_A k(P)$ is finitely generated, therefore noetherian and $B\otimes_A k(P) = S^{-1}(B/PB)$, where $S = A/P - \{0\}$. As PB is a graded ideal of B, the ring B/PB is graded, each element of S is homogeneous, consequently $S^{-1}(B/PB)$ is a graded noetherian algebra. Therefore $\dim (B\otimes_A k(P)) = \dim S^{-1}(B/PB) = \sup_{\substack{Q \text{ graded } \in \operatorname{spec}(B/PB) \\ Q\cap S = \varnothing}} (\mathrm{ht}\, S^{-1}Q) = \sup_{\substack{Q' \in \operatorname{spec}(B),\ Q' \text{ contains } PB \\ Q'/PB \text{ graded and } Q'/PB \cap A/P = \{0\}}} (\mathrm{ht}\, Q'/PB)$.

However, $Q'/PB \cap A/P = \{0\}$ if and only if $Q'\cap A = P$. Therefore the graded ideal Q'/PB is contained in $(P + B_+)/PB$, which involves $\dim (B\otimes_A k(P)) = \mathrm{ht}\,((P + B_+)/PB)$.

THEOREM 2.3. Let A be a ring such that A/P satisfies the altitude inequality formula for each prime ideal P of A and $B = \oplus_{n\geq 0} B_n$ a graded finitely generated algebra over $B_0 = A$. Then

$$\dim B = \sup_{P\in \operatorname{spec}(A)} (\dim (A/P) + \dim (B\otimes_A k(P))).$$

PROOF. We note that if $P_1 \subsetneq P_2$ are two prime ideals of A then $P_1 + B_+ \subsetneq P_2 + B_+$ in B. This involves that for each prime ideal P of A, we have $\dim A/P + \mathrm{ht}\,((P + B_+)/PB) \leq \dim B$. Finally the demonstration of the theorem follows from Lemmas 2.1 and 2.2.

EXAMPLE 2.4.

Two important examples of graded algebras for which the formula of Theorem 2.3 is applied are the symmetric algebra $S_A(M)$ and the divided powers algebra $\Gamma_A(M)$, where M is a finitely generated A-module and A/P satisfies the altitude inequality formula for each prime ideal P of A.

REFERENCES

[1] Bouvier, A., Dobbs, D., Fontana, M. (1988). Universally catenarian integral domains, Advances in Math. 72 : 211-238.

[2] Chibloun, R., Micali, A., Olivier, J. P. (to appear). Sur la dimension des algèbres symétriques, Rend. Sem. Mat. Univ. Politec. Torino, Conference in Commutative Algebra and Algebraic Geometry in honor of P. Salmon, Torino, September 1990.

[3] Giral, J. M. (1981). Krull dimension, transcendence degree and subalgebras of finitely generated algebras, Arch. Math. 36 : 305-312.

[4] Huneke, C., Rossi, M. E. (1986). The dimension and components of symmetric algebras, J. Algebra 98 : 200-210.

[5] Kabbaj, S. (1986). La formule de la dimension pour les S-domaines forts universels, Boll. Un. Mat. Ita. Algebra e Geometria VI, Vol. V-D, (1) : 145-161.

[6] Micali, A., Olivier, J. P. (non published). Théorie de la dimension dans les algèbres symétriques, Montpellier 1971.

7
Radices in Commutative Rings

DOUGLAS L. COSTA and GORDON E. KELLER[1] Department of Mathematics, University of Virginia, Charlottesville, Virginia 22903-3199 USA

0 INTRODUCTION

Recent research by the authors [4,6] has shown that a structure, which they have named a *radix*, in commutative rings plays a fundamental role in the classification of normal subgroups of the special linear group. The purpose of this article is to introduce the notion of a radix to commutative algebraists, to explain how radices arose naturally, to show that they behave in some ways that commutative algebraists might consider "natural," and to suggest that it might prove fruitful for commutative ring theorists to examine structures more general than ideals.

1 WHAT IS A RADIX?

Let A be a commutative ring with identity. A *radix* in A is an additive subgroup ℞ satisfying the following two closure axioms:

For all $t \in A$ and $x \in$ ℞,

$$tx^3 \in ℞, \tag{1}$$

$$(t^3 - t)x^2 + t^2x \in ℞. \tag{2}$$

We remark that an equivalent definition is obtained by replacing (1) by the weaker closure axiom:

$$t^2x^3 \in ℞. \tag{1'}$$

This is proved in [5].

It is easy to see by standard algebraic manipulations (substitutions, linearizations, etc.) that axioms (1) and (2) entail many other closure properties of radices. Closure properties derivable from (1) and (2) include the following:

[1]Both authors' research supported by a grant from the NSA.

For all $t \in A$, $x \in R_X$, and for all units $u \in A$,

$$(t^2 - t)x^2 + 2tx \in R_X, \tag{3}$$

$$t^{2n}x \in R_X, \text{ for all } n \geq 2, \tag{4}$$

$$u^2 x \in R_X. \tag{5}$$

In fact, every radix R_X of A contains a relatively large ideal of A. In order to later state the appropriate result, we observe that there is a unique largest ideal of A contained in R_X, called the *core* of the radix and denoted by core(R_X). Note that the axiom (1) can be restated as

$$\sum_{x \in R_X} x^3 A \subseteq \text{core}(R_X). \tag{1}$$

2 HOW DO RADICES ARISE IN A NATURAL WAY?

To answer this question, we need to discuss a little of the theory of linear groups over commutative rings, and for this we are obliged to begin with a string of definitions.

Let $X = [a_{ij}]$ be an $n \times n$ matrix with entries in a commutative ring A. The *general linear group* is defined by $GL(n, A) = \{X \mid X \text{ is invertible}\}$. The *special linear group* is defined by $SL(n, A) = \{X \mid \det X = 1\}$. The *elementary group* $E(n, A)$ is the subgroup generated by the matrices of the form $E_{ij}(x) = I + xe_{ij}$, where I is the $n \times n$ identity matrix and e_{ij} is the the "matrix unit" having a 1 in the (i, j)-entry and zeroes elsewhere. Note that $E(n, A) \subseteq SL(n, A) \subseteq GL(n, A)$.

There are relative versions of these groups as well. Thus, let J be an ideal of A. Then $GL(n, A; J)$ is the *general congruence subgroup* defined by J: it is the set of matrices in $GL(n, A)$ which are congruent to scalars modulo J. Otherwise put, it is the inverse image of the center of $GL(n, A/J)$ under the obvious homomorphism from $GL(n, A)$ to $GL(n, A/J)$. The kernel of this homomorphism is another normal subgroup of $GL(n, A)$, denoted by $GL'(n, A; J)$. It consists of those invertible matrices which are congruent to the identity modulo J.

The *principal congruence subgroup* defined by J is $SL(2, A; J) = SL(2, A) \cap GL'(n, A; J)$. We also set $L(2, A; J) = SL(2, A) \cap GL(n, A; J)$. Finally, the relative elementary group $E(2, A; J)$ is defined to be the *normal* subgroup of $E(2, A)$ generated by $\{E_{ij}(x) \mid x \in J,\ 1 \leq i \neq j \leq n\}$.

Now the motivating question which underlies everything in this paper is: What are the normal subgroups of $SL(n, A)$ and $GL(n, A)$? For technical reasons, this question is usually transformed into the following slightly weaker problem: Characterize the subgroups of $SL(n, A)$ and $GL(n, A)$ which are normalized by $E(n, A)$. Since for many commutative rings (e.g., semi-local rings or algebraic number rings) it holds that $E(n, A) = SL(n, A)$, this latter problem is nearly equivalent to the former one.

For $n \geq 3$, the $E(n, A)$-normalized subgroups of $GL(n, A)$ for any commutative ring A can be completely characterized by the use of ideals in A. This characterization is the culmination of work of Bass, Milnor, and Serre in [1]. It was done for $n \geq 4$ by

Wilson in [11], for $n \geq 3$ by Borevich and Vavilov in [2,3], and generalized and finalized by Vaserstein in [10].

The main tool in this characterization is the notion of the *level ideal* of a matrix in $GL(n, A)$. If $X = [a_{ij}]$ is in $GL(n, A)$, the level ideal $\ell(X)$ of X is defined by

$$\ell(X) = \sum_{i \neq j} A a_{ij} + \sum_{i \neq j} A(a_{ii} - a_{jj}).$$

Clearly, for an ideal J of A, $X \in GL'(n, A; J)$ if and only if $\ell(X) \subseteq J$. For a subgroup N of $GL(n, A)$ we define the *level ideal* $\ell(N)$ of N by $\ell(N) = \sum_{x \in N} \ell(X)$. The main result for $n \geq 3$ is then (see [10]):

THEOREM 1. Let A be a commutative ring, and let $n \geq 3$. Then

(1) For each ideal J of A,

$$\begin{aligned} [GL(n, A), E(n, A; J)] &= [E(n, A), GL'(n, A; J)] \\ &= [E(n, A), E(n, A; J)] = E(n, A; J). \end{aligned}$$

(2) If N is an $E(n, A)$-normalized subgroup of $GL(n, A)$, then $E(n, A; \ell(N)) \subseteq N$.

In the statement of the theorem, [H,K] means the subgroup generated by the commutators $[h, k] = h^{-1}k^{-1}hk$, where $h \in H$ and $k \in K$.

The characterization of $E(n, A)$-normalized subgroups is immediate from Theorem 1.

COROLLARY. Let A be a commutative ring, let $n \geq 3$, and let N be a subgroup of $GL(n, A)$. Then the following are equivalent:

(i) N is $E(n, A)$-normalized.

(ii) $E(n, A; \ell(N) \subseteq N$.

(iii) For some ideal J of A, we have

$$E(n, A; J) \subseteq N \subseteq GL'(n, A; J).$$

Moreover, when these hold we have $[E(n, A), N] = E(n, A; J)$.

We now consider the case $n = 2$. The first thing to observe here is that the solution in terms of ideals given above doesn't work even if A is a field. Actually, Dickson's 1901 result [7] shows that for K a field, $PSL(n, K)$ is simple unless $n = 2$ and $|K| \leq 3$. Since $E(n, K) = SL(n, K)$ and $L(n, K; 0)$ is the center of $SL(n, K)$, this means that the corollary above holds for K even when $n = 2$, provided that $|K| > 3$. On the other hand, $SL(2, 2)$ and $SL(2, 3)$ have non-central, proper normal subgroups and these do not fit the characterization given in the corollary.

These two examples of bad behavior express themselves whenever a commutative ring A has a residue field of 2 or 3 elements. For instance, the work of Klingenberg [8] and Lacroix [9] showed that the corollary above holds for all local rings (A, m) except if $n = 2$ and either $|A/m| = 2$ or $|A/m| = 3$ and $\ell(N) = A$.

Can these difficulties in the case $n = 2$ be overcome? The answer is yes, provided that one is willing to use some new tools to attack the problem. The tools necessary are radices, von Neumannizing ideals, and C-groups. We have already defined radices in commutative rings. The results which follow come from [4].

The *von Neumannizing ideal of degree k*, $vn_k(A)$, in a commutative ring A is defined by

$$vn_k(A) = \sum_{x \in A} (x^k - x)A$$

for $k \geq 2$. If J is an ideal of A, we define

$$vn_k(J) = \sum_{x \in J} (x^k - x)A.$$

Obviously, $vn_k(A)$ is the smallest ideal of A such that $A/vn_k(A)$ satisfies the identity $x^k - x$ for all x. The following theorem, which is a nice exercise for commutative algebra students, gives the main properties of these ideals.

THEOREM 2. Let A be a commutative ring. Then for all integers $k \geq 2$:

(i) $vn_k(A) = \cap\{m \mid m$ is a maximal ideal of A, A/m is finite, and $|A/m| - 1$ divides $k-1\}$;

(ii) for each maximal ideal m of A, $m \supseteq vn_k(A)$ if and only if A/m is finite and $|A/m| - 1$ divides $k - 1$;

(iii) for each ideal J of A, $vn_k(A/J) = (vn_k(A) + J)/J$;

(iv) for each ideal J of A, $vn_k(J) = J \cap vn_k(A)$;

(v) for any multiplicative system S of A, $vn_k(S^{-1}A) = S^{-1}(vn_k(A))$.

In particular, we have $vn_2(A) = \cap\{m \mid m$ is a maximal ideal of A and $A/m \simeq GF(2)\}$, so that $vn_2(A) = A$ if and only if A has no residue class fields isomorphic to $GF(2)$.

To define the C-groups (the "self-reproducing" groups), we begin by defining certain matrices in $E(2, A)$. For each $x \in A$, let $C(x) = E_{21}(x)E_{12}(x) = \begin{bmatrix} 1 & x \\ x & 1+x^2 \end{bmatrix}$. For a subset S of the ring A, we then define $C(S)$ to be the *normal* subgroup of $E(2, A)$ generated by $\{C(x) \mid x \in S\}$.

The matrices $C(x)$ satisfy the following identities:

(a) $[E_{21}(1), C(x)]^{E_{12}(-1)} = C(x)$ for all $x \in A$,

(b) $[E_{21}(t), E_{12}(x)] = [E_{21}(t), C(x)]$ for all $x, t \in A$,

(c) $C(x)^{-1}C(y) = C(y - x)^{E_{12}(x)}$ for all $x, y \in A$.

The next result is an immediate consequence of these identities.

THEOREM 3. Let S be a subset of a commutative ring A. Then

(1) $[E(2, A), C(S)] = C(S)$,

(2) $[E(2, A), E(2, A; S)] = C(S)$,

(3) $C(S) = C(\langle S \rangle)$, where $\langle S \rangle$ is the additive subgroup of A generated by S.

(It is in the sense of (1) that we refer to the C-groups as "self-reproducing.")

For a matrix $X = \begin{bmatrix} a & b \\ c & d \end{bmatrix}$ in $SL(2, A)$, we define $\rho(X) = a^2 - 1 + ab$. This map from $SL(2, A)$ to A assumes some importance because of the next result, the "commutator factorization theorem."

THEOREM 4. Let $X = \begin{bmatrix} a & b \\ c & d \end{bmatrix}$ be a matrix in $SL(2, A)$, and let $x = \rho(X)$. Assume that a is a unit in A. Then

(1) $[E_{21}(1), X] = C(x)^{\begin{bmatrix} a^{-1} & 0 \\ 0 & a \end{bmatrix}^{E_{12}(1)}} \begin{bmatrix} a^2 & -(a - a^{-1})^2 \\ 0 & a^{-2} \end{bmatrix}$, and

(2) If $X \in N$ and N is an $E(2, A)$-normalized subgroup of $SL(2, A)$, then $C(\rho(X)) \in [E(2, A), N]$.

Thus, for a subgroup N of $SL(2, A)$ we define $\rho(N)$ to be the additive subgroup of A generated by $\{\rho(X) | \, X \in N\}$. At this point it is natural to ask what properties $\rho(N)$ has as a sub-object of A.

THEOREM 5. If A is a commutative ring and N is any $E(2, A)$-normalized subgroup of $SL(2, A)$, then $\rho(N)$ is a radix of A.

Without giving a formal proof of this theorem we can give some indication of why it holds via the following identities which together show that if $E_{12}(x) \in N$, so that $x \in \rho(N)$, and N is $E(2, A)$-normalized, then the radix closure axioms (1′), (2), (3) hold for x with respect to $\rho(N)$:

$$\begin{aligned} \rho([E_{21}(t), E_{12}(x)]) &= (t^2 - t)x^2 - 2tx + t^2x^3, \\ \rho\left([E_{12}(x)^{E_{21}(-t)}\right) &= (t^2 - t)x^2 - 2tx + x, \\ \rho({}^*[E_{21}(t), E_{12}(x)]) &= (t^2 - t^3)x^2 + (t^2 - 2t)x. \end{aligned}$$

The asterisk here denotes the transpose of the matrix.

We may now ask: Does every radix of A occur as $\rho(N)$ for some $E(2, A)$-normalized subgroup of $SL(2, A)$?

THEOREM 6. Let R_X be a radix in a commutative ring A. Set

$$\begin{aligned} G(R_X) &= \{X \in SL(2, A) | \, \rho(X), \rho(X^{-1}), \rho({}^*X), \rho({}^*X^{-1}) \in R_X \\ &\quad \text{and } (a^2 - 1)vn_2(A) + (d^2 - 1)vn_2(A) \subseteq R_X\}. \end{aligned}$$

Then

(1) $G(R_X)$ is normal in $SL(2, A)$, and

(2) $\rho(G(R_X)) = \rho(E(2, A; R_X)) = \rho(C(R_X)) = R_X$.

Moreover, if N is any $E(2, A)$-normalized subgroup of $SL(2, A)$ with $\rho(N) = R_X$, then $N \subseteq G(R_X)$.

This last result should make it clear that radices arise "naturally." It also indicates something of their significance for characterizing $E(2, A)$- normalized subgroups of

$SL(2,A)$. In order to make this more precise, we need to know how large an ideal is contained in a radix. The next theorem, alluded to earlier, gives a lower bound for the core. (See [4].)

THEOREM 7. Let $\mathrm{R\!\!\!/}$ be a radix of a commutative ring A and let $(\mathrm{R\!\!\!/})$ be the ideal generated by $\mathrm{R\!\!\!/}$. Then for any $x \in (\mathrm{R\!\!\!/})$, the ideal $4xA + 2x^2A + x^3A + x^2(vn_2(A))^2 + 2x(vn_2(A))^3 + (x^2 + 2x)vn_2(A)$ is contained in core($\mathrm{R\!\!\!/}$).

Notice that it follows that if 2 is invertible in A, then every radix is an ideal of A, so we don't have anything new in this case. Also, notice that if A has no residue fields of 2 elements, then $vn_2(A) = A$, so that $x^2A \subseteq \mathrm{R\!\!\!/}$ for every $x \in \mathrm{R\!\!\!/}$. With radix axiom (2) this implies that $t^2x \in \mathrm{R\!\!\!/}$ for $x \in \mathrm{R\!\!\!/}$ and $t \in A$. These two facts make $\mathrm{R\!\!\!/}$ a *Jordan ideal* of A, i.e., $x^2t \in \mathrm{R\!\!\!/}$ and $t^2x \in \mathrm{R\!\!\!/}$ for all $x \in \mathrm{R\!\!\!/}$ and $t \in A$. (Vaserstein calls these *quasi-ideals.*) So radices really constitute new structures only in the case that A has residue class fields of 2 elements.

3 HOW CAN NORMAL SUBGROUPS OF $SL(2,A)$ BE CHARACTERIZED USING RADICES?

We begin this section by describing what we call *generic normal subgroups* of $SL(2,A)$, for A an arbitrary commutative ring. For this purpose, let J be an ideal of A, and let U be a group of units in A/J.

Define a subgroup $G(J,U)$ of $SL(2,A)$ by

$$G(J,U) = \{X \in SL(2,A) \mid \ell(X) \subseteq J \text{ and } X \equiv uI (\text{mod } J) \text{ for some } u \in U\}.$$

If $\phi : SL(2,A) \to SL(2,A/J)$ is the canonical homomorphism and $\tilde{U} = \{uI \mid u \in U\}$, then $\tilde{U}$ is normal in $SL(2,A/J)$ and $G(J,U) = \phi^{-1}(\tilde{U})$. This shows that $G(J,U)$ is a normal subgroup of $SL(2,A)$.

If in addition to J and U, we also consider a radix $\mathrm{R\!\!\!/}$, we set $G(J,U,\mathrm{R\!\!\!/}) = G(J,U) \cap G(\mathrm{R\!\!\!/})$. By Theorem 6 and the previous paragraph, $G(J,U,\mathrm{R\!\!\!/})$ is a normal subgroup of $SL(2,A)$. We call $G(J,U,\mathrm{R\!\!\!/})$ the *generic normal subgroup* defined by J, U, and $\mathrm{R\!\!\!/}$.

Next, let N be any $E(2,A)$-normalized subgroup of $SL(2,A)$. We associate to N its level ideal $J = \ell(N)$ and its radix $\rho(N)$. Notice that taken modulo $J = \ell(N)$, N is a group of scalar matrices and so defines a group of units $U = u(N)$ in A/J. With these objects defined by N we may now assign to N its associated generic normal subgroup $G(N) = G(\ell(N), u(N), \rho(N))$. Observe that $N \subseteq G(N)$. In fact, we can decompose $G(N)$ as follows:

THEOREM 8. Let A be a commutative ring, and let N be an $E(2,A)$-normalized subgroup of $SL(2,A)$. Then

$$G(N) = SL(2,A;\text{core}(\rho(N))) \cdot E(2,A;\rho(N)) \cdot N.$$

For the rest of this section we assume that A is either an algebraic number ring having infinitely many units or a ring satisfying Bass' first stable range condition (an SR_2-ring). The latter include the semi-local rings, for example. For all these rings $E(2,A) = SL(2,A)$, so that the $E(2,A)$-normalized subgroups of $SL(2,A)$ are the same as the normal subgroups. Also, for any ideal J of A, $SL(2, A; J) = E(2, A; J)$ for all these rings except the totally imaginary number rings. (The theorem below still holds true in that case, however.) Thus, if N is a normal subgroup of $SL(2, A)$, then Theorem 8 displays the generic group of N as $G(N) = E(2, A; \rho(N)) \cdot N$.

By what is essentially a localization argument, one can show that $C(\rho(N)) \subseteq N$, using Theorem 4. Taking commutators on both sides of this containment and using the self-reproducing property of C-groups, we get

$$C(\rho(N)) \subseteq [E(2, A), N].$$

Then

$$\begin{aligned}[E(2,A), G(N)] &= [E(2,A), E(2,A); \rho(N)] \cdot [E(2,A), N] \\ &= C(\rho(N))[E(2,A), N] = [E(2,A), N].\end{aligned}$$

Thus, the commutators of the generic groups give all possible commutator groups of normal subgroups with $E(2, A)$. Moreover, this calculation proves:

THEOREM 9. Let A be an algebraic number ring with infinitely many units or an SR_2-ring, and let N be a subgroup of $SL(2, A)$. Then N is normal in $SL(2, A)$ if and only if

(i) $\rho(N)$ is a radix, and

(ii) $[E(2, A), G(N)] \subseteq N \subseteq G(N)$.

In other words, the triples $(J, U,$ ℞$)$ parameterize the normal subgroups of $SL(2, A)$ in the way that ideals did for $SL(n, A)$, $n \geq 3$.

4 WHAT CAN BE SAID ABOUT RADICES AS SUB-OBJECTS OF COMMUTATIVE RINGS?

Having seen that radices arise more or less naturally in the study of normal subgroups of $SL(2, A)$, we now explore some purely ring-theoretic results on the structure of radices in commutative rings and especially in algebraic number rings. Throughout, A will be a commutative ring. We begin with a simple observation.

LEMMA. An additive subgroup P of a commutative ring A is a radix if and only if P/core(P) is a radix in A/core(P).

This simple fact means that in practice one may often assume that core(℞) $= 0$ when working with a radix ℞ of A. A somewhat deeper observation is the following:

THEOREM 10. If ℞ and ℞$'$ are radices of a commutative ring A, then so is ℞ + ℞$'$.

A proof of this theorem will appear in the forthcoming article [6] along with the remaining theorems in this section.

When A decomposes as a direct sum, then so do its radices:

THEOREM 11. If e is an idempotent in a commutative ring A and ℞ is a radix, then e℞ is a radix in eA, $(1-e)$℞ is a radix in $(1-e)A$ and ℞ $= e$℞ $+ (1-e)$℞. Indeed, the radices of A are all of the form ℞$_1$ + ℞$_2$, where ℞$_1$ is a radix of eA and ℞$_2$ is a radix of $(1-e)A$.

Proof. Observe that for every $x \in ℞$, $ex = (e^3 - e)x^2 + e^2x \in ℞$. With this fact in hand the rest of the proof is straightforward.

THEOREM 12. Let ℞ be a radix of a commutative ring A, and let I and J be ideals of A such that $I + J = A$. Then $(℞ + I) \cap (℞ + J) = ℞ + (I \cap J)$.

Proof. One containment is obvious, so let $z \in (℞ + I) \cap (℞ + J)$. Since the ideal $I \cap J$ is contained in all three of the radices in question we may work modulo $I \cap J$, and so we assume that $I \cap J = 0$. Now write $z = x + i = y + j$ with $x, y \in ℞$, $i \in I$, and $j \in J$. Then $i - j = y - x \in ℞$. Writing $1 = e + (1 - e)$, where $e \in I$ and $1 - e \in J$, we have $e^2 = e$. Then $i = ei = e(i - j) = (e^3 - e)(i - j)^2 + e^2(i - j) \in ℞$ and hence $z = x + i \in ℞$. This proves the theorem.

We may now apply these general facts to radices in Dedekind domains. For this purpose it is convenient to define a *primary radix* to be a radix which contains a power of a maximal ideal.

THEOREM 13. Let A be a Dedekind domain, and let ℞ be a non-zero radix of A. Then ℞ can be expressed uniquely as a finite intersection of primary radices of A.

Proof. Since $℞ \neq 0$, $\text{core}(℞) \neq 0$ by radix axiom (1). Let $\text{core}(℞) = p_1^{n_1} \cdots p_t^{n_t} = p_1^{n_1} \cap \cdots \cap p_t^{n_t}$, where $p_1, \ldots, p_t$ are distinct maximal ideals of A. Then by Theorem 12 we have $℞ = ℞ + p_1^{n_1} \cap \cdots p_t^{n_t} = \bigcap_{i=1}^{t}(℞ + p_i^{n_i})$. By Theorem 10, each $℞ + p_i^{n_i}$ is a radix of A which is evidently primary. The Chinese remainder theorem and Theorem 11 then show the uniqueness of this decomposition.

Because of Theorem 9 it is of some interest to determine all the radices in an algebraic number ring A. According to Theorem 13 it is sufficient to determine the primary ones. So suppose ℞ is a radix of the algebraic number ring A and that ℞ contains a power of the maximal ideal p. Then it follows that $\text{core}(℞) = p^n$ for some integer n and by the lemma it suffices to determine all radices with trivial core in rings of the form A/p^n. Now these rings are finite special PIR's, and the radices in such rings are completely determined in the paper [6].

As a simple example, let $k = GF(2)$, and let $A = k[[X]]/(X^3)$. Then $℞ = \{0, X\}$ is a radix of A but is not even a Jordan ideal of A.

5 WHAT REMAINS TO BE DONE?

The theory of radices is still in its infancy and much undoubtedly remains to be discovered. In particular, can one give some structure theorem for radices in general local rings? And then, is there any sense in which radices "localize" well? What properties, other than those already listed, do radices have?

One may think of radices in the following more general context. Let t and x be indeterminates (or finite sets of indeterminates) over $\mathbf{Z}$, and let $\mathcal{F} \subseteq \mathbf{Z}[t, x]$ be a set of polynomials. In [4] we defined an $\mathcal{F}$-ideal in a commutative ring A to be an additive subgroup Λ such that for all $t \in A$, for all $x \in \Lambda$, and for all $f \in \mathcal{F}$,

$$f(t, x) \in \Lambda.$$

Clearly, radices are a particular kind of $\mathcal{F}$-ideal. (Ideals are too, of course, as they are defined by the polynomial tx.)

In this context, it seems natural to ask: Do other $\mathcal{F}$-ideals arise naturally? And in the abstract it seems interesting to ask: Given a set $\mathcal{F} \subseteq \mathbf{Z}[t, x]$ does there always exist a finite set $\mathcal{F}_0 \subseteq \mathbf{Z}[t, x]$ such that $\mathcal{F}$-ideals and $\mathcal{F}_0$-ideals are one and the same thing?

References

[1] H. Bass, J. Milnor, and J.-P. Serre, Solution of the congruence subgroup problem for $SL_n(n \geq 3)$ and $Sp_{2n}(n \geq 2)$, Publ. I.H.E.S. 33 (1967), 59-137.

[2] Z. I. Borevich and N. A. Vavilov, On the subgroups of the full linear group over a commutative ring, Soviet Math. Dokl. 26(1982), 679-691.

[3] Z. I. Borevich and N. A. Vavilov, The distribution of subgroups in the full linear group over a commutative ring, Proc. Steklov Inst. Math. 3(1985), 27-46.

[4] D. Costa and G. Keller, The $E(2, A)$ sections of $SL(2, A)$, Ann. Math. 134(1991), 159-188.

[5] D. Costa and G. Keller, Radix redux: Normal subgroups of symplectic groups, J. reine angew. Math. 427(1992), 51-105.

[6] D. Costa and G. Keller, Power residue symbols and the central sections of $SL(2, A)$, in preparation.

[7] L. E. Dickson, Theory of linear groups in arbitrary fields, Trans. A.M.S. 2(1901), 363-394.

[8] W. Klingenberg, Lineare Gruppen über lokalen Ringen, Amer. J. Math. 83(1961), 137-153.

[9] N. H. J. Lacroix, Two-dimensional linear groups over local rings, Canad. J. Math. 21(1969), 106-135.

[10] L. N. Vaserstein, On the normal subgroups of GL_n over a ring, L. N. S. 854, Springer (1981), 456-465.

[11] J. S. Wilson, The normal and subnormal structure of general linear groups, Proc. Camb. Phil. Soc. 71(1972), 163-177.

8

The Generalized Samuel Numbers

H. DICHI Département de Mathématiques, Université d'Abidjan,
Abidjan, Côte d'Ivoire

1. INTRODUCTION

Throughout this note, A will denote a commutative ring with unit element. For two filtrations $f = (I_n)$ and $g = (J_n)$ on A, we put $\bar{v}_f(g) = \lim_{n \to \infty} \frac{v_f(J_n)}{n}$ and $\bar{w}_f(g) = \lim_{n \to \infty} \frac{w_f(J_n)}{n}$, where for an ideal J on A, $v_f(J) = \sup\{ n ; J \subseteq I_n \}$ and $w_f(J) = \inf\{ n ; I_n \subseteq J \}$. A filtration $f = (I_n)$ on A is a decreasing sequence of ideals I_n such that $I_0 = A$ and $I_n I_m \subseteq I_{n+m}$ for all integers n,m. The set $\mathbb{F}(A)$ of filtrations on the ring A is ordered by $f = (I_n) \leq g = (J_n)$ if $I_n \subseteq J_n$ for all n. The numbers $\bar{v}_f(g)$ and $\bar{w}_f(g)$ are called the generalized Samuel numbers of the filtrations f and g. They have been studied in [10] by Samuel in the particular case when $f = (I^n)$ and $g = (J^n)$ are adic filtrations on a noetherian ring A, and by Ayégnon and Sangaré in [1] in case when f and g are AP filtrations on A . A filtration $f = (I_n)$ is said to be an AP filtration (resp. a strongly AP filtration) if there exists a sequence (k_n) of positive integers such that $\lim_{n \to \infty} \frac{k_n}{n} = 1$ and $I_{k_n m} \subseteq I_n^m$ for all n,m (resp. if $I_n = I_r^n$ for some $r \geq 1$, $\forall$ $n \geq 0$). Clearly, any

Correspondence to: Henri Dichi, Université d'Evry Val d'Essonne Boulevard des Coquibus, 91025 Evry Cedex, France.

adic filtration is strongly AP and any strongly AP filtration is an AP filtration. The existence in $\overline{\mathbb{R}}_+ = \mathbb{R}_+ \cup \{\infty\}$ of the generalized Samuel number $\bar{v}_f(g)$ has been proved for any AP filtration g on A and the existence of $\bar{w}_f(g)$ has been proved for any AP filtrations f and g on a noetherian ring A in ([1], Proposition 2.6 and Corollary 2.8). Unfortunately, we know that the generalized Samuel numbers don't exist for all filtrations on A , even if A is a noetherian ring, as seen in ([1], 2.9).

We begin the section 2 of this note by giving several properties of the generalized Samuel number $\bar{w}_f(g)$ and by proving the relation $\bar{w}_f(g) = \lim_{n \to \infty} n\, \bar{w}_{I_n}(g)$ for any AP filtration f (Theorem 1) where $\bar{w}_I(g)$ denotes the number $\bar{w}_f(g)$ when $f = (I^n)$ is the I-adic filtration on A. We also prove that $\bar{w}_f(g) \geq \bar{v}_f(g)$ for any AP filtrations f and g if f is a separated and nonnilpotent filtration (Theorem 3). A filtration $f = (I_n)$ is said to be separated and nonnilpotent if $\bigcap_{n \geq 0} I_n = (0)$ and $I_n \neq (0)$ for all n.

In section 3, we define for two filtrations f and g on A the numbers $\bar{a}_f(g) = \lim_{n \to \infty} \frac{a_f(g^{(n)})}{n}$ and $\bar{b}_f(g) = \lim_{n \to \infty} \frac{b_f(g^{(n)})}{n}$ where $a_f(g) = \sup \{r \in \mathbb{N}^*;\ g \leq f^{(r)}\}$, $b_f(g) = \inf \{r \in \mathbb{N}^*;\ f^{(r)} \leq g\}$ and where $f^{(r)}$ is the filtration $(I_{nr})_{n \geq 0}$ if $f = (I_n)$. We prove that the numbers $\bar{a}_f(g)$ and $\bar{b}_f(g)$ exist in $\overline{\mathbb{R}}_+$ for all filtrations on the ring A and that these numbers coincide respectively with $\bar{v}_f(g)$ if g is AP and with $\bar{w}_f(g)$ if f is AP (Theorem 5). If f and g are two filtrations on the noetherian ring A, the relationships between the number $\bar{a}_f(g)$, the asymptotic closure of the filtration f and the notion of integral dependence between f and g are established. For $f = (I_n) \in \mathbb{F}(A)$, the asymptotic closure of f is the filtration $\bar{f} = (\bar{I}_n)$ where $\bar{I}_n = \{ x \in A;\ \bar{v}_f(x) \geq n \}$. An element $x \in A$ is said to be integral over the filtration $f = (I_n)$ if there exists an integer $m \geq 1$ and $a_j \in I_j$ for $j = 1,\dots,m$ such that $x^m + a_1x^{m-1} + a_2x^{m-2} + \dots + a_m = 0$. The prüferian closure of f is the filtration $P(f) = (P_n(f))$ where $P_n(f)$ is the ideal of elements of A which are integral over the filtration $f^{(n)}$, and following [3] , a filtration g is said to be integral over f if $g \leq P(f)$. We prove that, if g is an AP filtration on the noetherian ring A, then $\bar{v}_f(g) \geq 1$ iff $g \leq \bar{f}$ (Theorem 7); in case where f is strongly AP, we obtain that $\bar{v}_f(g) \geq 1$ iff g is integral over f. We close this section by giving a generalization to filtrations of Theorems 2 and 4 of [10] (Corollary 3).

2. GENERALIZED SAMUEL NUMBERS ASSOCIATED WITH FILTRATIONS

We use in the proof of Theorem 1 the following relations which have been proved in [1] : For any integer $n \geq 1$, we have $\bar{w}_f(g) = n\,\bar{w}_{f^{(n)}}(g) = \frac{\bar{w}_f(g^{(n)})}{n}$

THEOREM 1.- Let $f = (I_n)$ and $g = (J_n)$ be two filtrations on the ring A. Then the generalized Samuel number $\bar{w}_f(g)$ exists in $\bar{\mathbb{R}}_+$ for any AP filtration f and $\bar{w}_f(g) = \lim_{n\to\infty} n\,\bar{w}_{I_n}(g)$.

Proof.- We prove first the existence of $\bar{w}_f(g)$ for any AP filtration $f = (I_n)$. Let us write $g = (J_n)$; we can suppose that $\liminf \frac{w_f(J_n)}{n} = a \in \mathbb{R}_+$ and that $\lim_{n\to\infty} w_f(J_n) = \infty$. As f is AP , there exists a sequence of integers $(k_j) \geq 1$ such that (*) $I_{nk_j} \subseteq I_j^n$ for all $j,n \geq 1$ with $\lim_{j\to\infty} \frac{k_j}{j} = 1$. Let ε be a

real number > 0 and $u_n = w_f(J_n)$; then there exists an integer $\ell \geq 1$ and an integer $m \geq \ell$ such that $\frac{u_m}{m} \leq a+\varepsilon$, $a-\varepsilon \leq \inf_{n\geq \ell} \frac{u_n}{n} \leq a$ and $\forall\, n \geq \ell$, $\frac{k_{u_n}}{u_n} \leq 1+\varepsilon$. Let s, n be integers ≥ 1, we obtain from (*) that $u_{ns} \leq n\, k_{u_s}$ and if we write $n = q_n m + r_n$ where $0 \leq r_n \leq m-1$, we have $u_n \leq u_{m(q_n+1)} \leq (q_n+1)k_{u_m}$; then $\forall\, n \geq m$, $a-\varepsilon \leq \limsup \frac{u_n}{n} \leq \frac{k_{u_m}}{m} \leq (1+\varepsilon)(a+\varepsilon)$. By taking limits as $\varepsilon \dashrightarrow 0$, we obtain $a = \limsup \frac{u_n}{n}$, so $\bar{w}_f(g)$ exists. To prove that $\bar{w}_f(g) = \lim_{n\to\infty} n\, \bar{w}_{I_n}(g)$ for any AP filtration $f = (I_n)$ and any filtration $g = (J_n)$, we can suppose that $k_n \geq n$ for all n. Then, from the inequalities $f_{I_{k_n}} \leq f^{(k_n)} \leq f_{I_n}$ we have $\bar{w}_{I_n}(g) \leq \bar{w}_{f^{(k_n)}}(g) = \frac{1}{k_n} \bar{w}_f(g) \leq \bar{w}_{I_n}(g)$ and when $n \to \infty$, we obtain $\liminf k_n \bar{w}_{I_n}(g) \leq \bar{w}_f(g) \leq \liminf n\, \bar{w}_{I_n}(g) \leq \liminf k_n \bar{w}_{I_{k_n}}(g)$. Hence $\bar{w}_f(g) = \liminf n\, \bar{w}_{I_n}(g)$. Now, it is enough to prove that $\limsup n\bar{w}_{I_n}(g) \leq \bar{w}_f(g)$. We can suppose that $\bar{w}_f(g) \in \mathbb{R}_+$. Let p, q $\in \mathbb{N}$ such that $\bar{w}_f(g) < \frac{p}{q}$, then $\bar{w}_f(g^{(q)}) < p$ and there exists an integer N such that for all integer $m \geq N$, $w_f(J_{mq}) \leq mp$. Hence, we have $I_{mp} \subseteq J_{mq}$ for all integer $m \geq N$. Let n be an integer ≥ 1; by the division algorithm we have $m = nq_m + r_m$ where $0 \leq r_m \leq n$. So, $I_n^{(q_m+1)p} \subseteq I_{n(q_m+1)p} \subseteq I_{mp} \subseteq J_{mq}$, and we obtain $\frac{w_{I_n}(J_{mq})}{mq} \leq \frac{p(q_m+1)}{mq}$.
Taking limits when $m \to \infty$, we obtain $\bar{w}_{I_n}(g) \leq \frac{1}{n}\frac{p}{q}$ and $\bar{w}_f(g) \geq \limsup n\, \bar{w}_{I_n}(g)$.□

COROLLARY 1.- Let f, g ,h be filtrations on the ring A such that f is AP . Then $\bar{w}_f(gh) \le \bar{w}_f(g) + \bar{w}_f(h)$.

Proof.- It is easy to see that if I, J, K are ideals on A, we have $w_I(JK) \le w_I(J) + w_I(K)$. The remainder of the proof follows from Theorem 1.□

COROLLARY 2.- Let f,g,h be filtrations on the ring A such that f and g are AP . Then $\frac{1}{\bar{w}_{fg}(h)} \ge \frac{1}{\bar{w}_f(h)} + \frac{1}{\bar{w}_g(h)}$.

Proof.- From Theorem 1, it is enough to prove the relation $\frac{1}{\bar{w}_{IJ}(h)} \ge \frac{1}{\bar{w}_I(h)} + \frac{1}{\bar{w}_J(h)}$ if I, J are ideals and if h is a filtration on A. As $\bar{w}_{IJ}(h) \le \inf(\bar{w}_I(h), \bar{w}_J(h))$, we can suppose that $\alpha = \bar{w}_I(h)$ and $\beta = \bar{w}_J(h)$ are positive real numbers. Put $h = (H_n)$ and let p, p′, q, q′ be integers ≥ 1 such that $\alpha < \frac{p}{q} = r$ and $\beta < \frac{p'}{q'} = r'$. Then $\bar{w}_I(h^{(q')}) < p$ and $\bar{w}_J(h^{(q')}) < p'$, and there exists an integer m such that for all $n \ge m$, we have $w_I(H_{nq}) < np$ and $w_J(H_{nq'}) < np'$, that is $I^{np} \subseteq H_{nq}$ and $J^{np'} \subseteq H_{nq'}$. In particular, $I^{npp'} \subseteq H^{p'}_{nq} \subseteq H_{nqp'}$ and $J^{npp'} \subseteq H^{p}_{nq'} \subseteq H_{npq'}$, then $(IJ)^{npp'} \subseteq H_{n(p'q+pq')}$ and we have $w_{IJ}(H_{n(p'q+pq')}) \le npp'$. Dividing by $n(p'q+pq')$ and taking limits as $n \to \infty$, we obtain $\bar{w}_{IJ}(h) \le \frac{rr'}{r+r'}$. Hence $\bar{w}_{IJ}(h) \le \frac{\alpha\beta}{\alpha+\beta}$ and $\frac{1}{\bar{w}_{IJ}(h)} \ge \frac{\alpha+\beta}{\alpha\beta} = \frac{1}{\bar{w}_I(h)} + \frac{1}{\bar{w}_J(h)}$. □

THEOREM 2.- Let $f = (I_n)$ and $g = (J_n)$ be filtrations on A. We have :

(i) If f is AP, then the sequence $\left(\frac{\bar{w}_f(J_n)}{n}\right)_{n\ge1}$ is convergent ;

(ii) If f and g are AP filtrations, then $\bar{w}_f(g) = \lim\limits_{n\to\infty} \frac{\bar{w}_f(J_n)}{n}$.

Proof.- The proof of the convergence of the sequence $\left(\frac{\bar{w}_f(J_n)}{n}\right)_{n\ge1}$ is similar to ([1], Prop 2.6(i)). Suppose now that f and g =

(J_n) are AP filtrations and let (k_j) be a sequence of integers ≥ 1 where $\lim_{j \to \infty} \frac{k_j}{j} = 1$ and $J_{k_j n} \subseteq J_j^n$ for all j, n $\geq$ 1. The relations $J_{k_j}^n \subseteq J_{k_j n} \subseteq J_j^n$ give $\bar{w}_f(J_j) \leq k_j \bar{w}_f(g) \leq \bar{w}_f(J_{k_j})$. Taking limits when $j \to \infty$ we obtain (ii).□

THEOREM 3.- Let f and g be AP filtrations on A such that the filtration f is separated and nonnilpotent. Then $\bar{w}_f(g) \geq \bar{v}_f(g)$.

Proof.- Let f = (I_n), g = (J_n) be AP filtrations. We obtain from ([1],Prop 3.6) and from theorem 1 that $\bar{v}_f(g) = \lim_{n \to \infty} n\, \bar{v}_{I_n}(g)$ and $\bar{w}_f(g) = \lim_{n \to \infty} n\, \bar{w}_{I_n}(g)$. As f is an AP separated nonnilpotent filtration, the filtration f_{I_n} has the same properties for all integer n $\geq$ 1 and we obtain $w_{I_n}(J_p) \geq v_{I_n}(J_p)$ for all n,p $\geq$ 1 ; this completes the proof . □

THEOREM 4.- Let f and g be filtrations on the noetherian ring A such that f is separated, g is nonnilpotent and $\sqrt{g} \subseteq \sqrt{f}$. Then if $\bar{w}_g(f) \neq 0$, we have $\bar{v}_f(g) \in \mathbb{R}_+^*$ and $\bar{v}_f(g)\, \bar{w}_g(f) = 1$.

Proof.- The proof is similar to that ([1],Prop 4.3).□

3. THE NUMBERS $\bar{a}_f(g)$ AND $\bar{b}_f(g)$ ASSOCIATED WITH TWO FILTRATIONS

We know that the real number $\bar{v}_f(g)$ exists if g is AP [1] and that $\bar{w}_f(g)$ exists for any AP filtration f on the ring A (Theorem 1). Further, we have examples of filtrations such that $\bar{v}_f(g)$ does not exist ([1],2.9). So, it would be interesting to define on the set of all filtrations on A two real numbers which coincide respectively with $\bar{v}_f(g)$ if g is AP and with $\bar{w}_f(g)$ if f is AP. It is the goal of this section.

DEFINITIONS.- Let f, g be filtrations on the ring A.

(1) We put $a_f(g) = \sup\{r \in \mathbb{N}^*;\ g \le f^{(r)}\}$ if $\{r \in \mathbb{N}^*;\ g \le f^{(r)}\}$ is not empty and $a_f(g) = 0$ otherwise.

(2) We put $b_f(g) = \inf\{r \in \mathbb{N}^*;\ f^{(r)} \le g\}$ if $\{r \in \mathbb{N}^*;\ f^{(r)} \le g\}$ is not empty and $b_f(g) = \infty$ otherwise.

(3) We put $\bar{a}_f(g) = \lim_{n \to \infty} \frac{a_f(g^{(n)})}{n}$ and $\bar{b}_f(g) = \lim_{n \to \infty} \frac{b_f(g^{(n)})}{n}$ if these limits exist in $\bar{\mathbb{R}}_+$.

Notations : If I, J are filtrations on A and if $f_I = (I^n)$ and $f_J = (J^n)$, we put : $a_{f_I}(g) = a_I(g)$, $b_{f_I}(g) = b_I(g)$, $a_f(f_J) = a_f(J)$ and $b_f(f_J) = b_f(J)$. Similarly, we adopt the notations $\bar{a}_I(g)$, $\bar{b}_I(g)$, $\bar{a}_f(J)$ and $\bar{b}_f(J)$.

REMARKS 1.- Let $f = (I_n)$, $g = (J_n)$, $h = (H_n)$ be filtrations on the ring A and let I, J be ideals on A. The following properties hold :

(1) For any integer $k \ge 1$, we have : $v_f(J_k) \ge ka_f(g)$, $w_f(J_k) \le kb_f(g)$.

(2) $a_f(J) = v_f(J)$ et $b_I(g) = w_I(J_1)$.

(3) If $g \le h$, then $a_f(g) \ge a_f(h)$ and $b_f(g) \ge b_f(h)$.

(4) For any integer $k \ge 1$, we have $ka_f(g) \le a_f(g^{(k)})$ and $b_f(g^{(k)}) \le kb_f(g)$.

(5) For any real number $\lambda > 0$, we have $\bar{a}_f(g) = \lambda\, \bar{a}_{f^{(\lambda)}}(g) = \frac{1}{\lambda}\, \bar{a}_f(g^{(\lambda)})$ and $\bar{b}_f(g) = \lambda \bar{b}_{f^{(\lambda)}}(g) = \frac{1}{\lambda}\, \bar{b}_f(g^{(\lambda)})$.

(6) $\bar{a}_f(g+h) \le \min(\bar{a}_f(g), \bar{a}_f(h))$.

(7) $\bar{a}_f(gh) \ge \bar{a}_f(g) + \bar{a}_f(h)$.

THEOREM 5.- The numbers $\bar{a}_f(g)$ and $\bar{b}_f(g)$ exist in $\bar{\mathbb{R}}_+$ for any filtrations f and g on the ring A. Further, if g is AP, then $\bar{v}_f(g) = \bar{a}_f(g)$ and if f is AP, then $\bar{w}_f(g) = \bar{b}_f(g)$.

Proof.- The proof of the existence of $\bar{a}_f(g)$ and of $\bar{b}_f(g)$ is similar to that ([6],Lemma 0.2.1). Now, suppose that g is AP and show that $\bar{a}_f(g) = \bar{v}_f(g)$. Let us write $g = (J_n)$ and let (k_j) be a sequence of integers ≥ 1 such that $\lim_{j\to\infty} \frac{k_j}{j} = 1$ and such that $f_{J_k} \leq g^{(k_j)} \leq f_{J_j}$ for all $j \geq 1$. As $\bar{a}_f(J) = \bar{v}_f(J)$ for any ideal J on A , the relations $\frac{a_f(J_j^n)}{n} \leq \frac{a_f(g^{(nk_j)})}{n} \leq \frac{a_f(J_{k_j}^n)}{n}$ give us the following inequalities when $n \to \infty$: $\bar{v}_f(J_j) \leq \bar{a}_f(g^{(k_j)}) = k_j\, \bar{a}_f(g) \leq \bar{v}_f(J_{k_j})$. Therefore, we have from ([1],Prop 2.6) that $\bar{v}_f(g) = \lim_{j\to\infty} \frac{\bar{v}_f(J_j)}{j}$; this completes the proof. □

THEOREM 6.- Let f, g be filtrations on A. If $\bar{a}_f(f) = 1$, then $\bar{b}_f(g) \geq \bar{a}_f(g)$.

Proof.- Suppose that $\bar{a}_f(f) = 1$ and let g be a filtration on A . Then, we have $a_f(f^{(n)}) \in \mathbb{N}$ for all integer $n \geq 1$ and we can suppose that $b_f(g^{(n)}) \in \mathbb{N}$ for all $n \geq 1$. Let us write $v_n = b_f(g^{(n)})$; then from the relation $f^{(v_n)} \leq g^{(n)}$ and from Remark 1 (3) and (5) we obtain the inequality $v_n\, \bar{a}_f(f) \geq n\, \bar{a}_f(g)$. It follows that $\frac{v_n}{n} \geq \bar{a}_f(g)$ and $\bar{b}_f(g) \geq \bar{a}_f(g)$. □

Let $f = (I_n)$ be a filtration on A. We put $\bar{I}_n = \{ x \in A ; \bar{v}_f(x) \geq n \}$, where $\bar{v}_f$ is the homogeneous pseudovaluation associated with f. $\bar{I}_n$ is an ideal on A and $\bar{f} = (\bar{I}_n)$ is a filtration on A called the **asymptotic closure of f.** We said that f is asymptotically closed if $f = \bar{f}$.

LEMMA 1.- Let f, g be filtrations on A. Then :

(i) If $g \leq f$, then $\bar{a}_f(g) \geq 1$;

(ii) If $\bar{a}_f(g) \geq 1$, then $g \leq \bar{f}$.

Proof.- (ii) Suppose that $\bar{a}_f(g) \geq 1$ and let us write $f = (I_n)$ and $g = (J_n)$. Then we have for any integer $n \geq 1$, $f_{J_n} \leq g^{(n)}$, that is $\bar{a}_f(J_n) \geq \bar{a}_f(g^{(n)}) \geq n$ from the hypothesis $\bar{a}_f(g) \geq 1$ and from Remark 1.(5). Since $a_f(J_n^k) = v_f(J_n^k)$ from Remark 1.(2), we obtain $\bar{a}_f(J_n) = \lim_{k \to \infty} \frac{a_f(J_n^k)}{k} = \bar{v}_f(J_n)$, then $\bar{v}_f(J_n) \geq n$ and $J_n \subset \bar{I}_n = \{x \in A \mid \bar{v}_f(x) \geq n\}$; so $g \leq \bar{f}$. □

LEMMA 2.- Let f, g be filtrations on the noetherian ring A. Then if g is an AP filtration, we have the following equalities :

$\bar{a}_f(g) = \bar{a}_{\bar{f}}(g) = \bar{v}_f(g) = \bar{v}_{\bar{f}}(g)$.

Proof. We know from Theorem 5 that $\bar{a}_f(g) = \bar{v}_f(g)$ and $\bar{a}_{\bar{f}}(g) = \bar{v}_{\bar{f}}(g)$. So, it is enough to prove that $\bar{v}_f(g) = \bar{v}_{\bar{f}}(g)$. Let us write $g = (J_n)$. Since $\bar{v}_f(x) \leq v_{\bar{f}}(x) + 1 \leq \bar{v}_f(x) + 1$ for all $x \in A$, we obtain from the relations $\bar{v}_f(J_n) = \inf_{x \in J_n} \bar{v}_f(x)$ and $v_{\bar{f}}(J_n) = \inf_{x \in J_n} v_{\bar{f}}(x)$ that $\frac{\bar{v}_f(J_n)}{n} \leq \frac{v_{\bar{f}}(J_n)}{n} \leq \frac{\bar{v}_f(J_n)}{n} + \frac{1}{n}$.

As g is an AP filtration, $\bar{v}_f(g)$ is the limit of the sequence $\left(\frac{\bar{v}_f(J_n)}{n}\right)$ when $n \to \infty$ ([1],Prop 2.6) and we obtain $\bar{v}_f(g) = \bar{v}_{\bar{f}}(g)$. □

The following theorem is a generalization of ([1], Theorem 5.6). Indeed, we know from ([3],Prop 4.7) that any strongly AP filtration satisfies the relation $\bar{f} = P(f)$ where P(f) is the **prüferian closure** of f.

THEOREM 7.- Let f, g be filtrations on the noetherian ring A such that g is AP . Then $\bar{v}_f(g) \geq 1$ iff $g \leq \bar{f}$.

Proof.- We know from the above lemmas that $g \leq \bar{f}$ if $\bar{v}_f(g) \geq 1$.

Conversely if $g \le \bar{f}$, we obtain $\bar{a}_{\bar{f}}(g) \ge 1$ and the proof follows from Lemma 2 . □

THEOREM 8.- Let f, g , h be filtrations on the ring A such that f is asymptotically closed. Then $\bar{a}_f(g+h) = \min(\bar{a}_f(g), \bar{a}_f(h))$.

Proof.- Let us write $\lambda = \min(\bar{a}_f(g), \bar{a}_f(h))$; we can assume that $0 < \lambda < \infty$. Let $r = \frac{p}{q}$ be a rational number such that $0 < r < \lambda$. Then $\bar{a}_{f^{(r)}}(g) \ge 1$ and $\bar{a}_{f^{(r)}}(h) \ge 1$, and we obtain from the lemma 1 that $g + h \le \overline{f^{(r)}}$ hence $\bar{a}_{f^{(r)}}(g + h) \ge 1$ (*). Let us write $f = (I_n)$, $\overline{f^{(r)}} = (J_n)$, $\bar{f} = (\bar{I}_n)$ and $\bar{f}^{(r)} = (K_n)$. Then $J_n = \{x \in A \ / \ \bar{v}_f(x) \ge nr\}$ and $K_n = \{x \in A \ / \ \bar{v}_f(x) \ge \{nr\}\}$ where $\{\beta\}$ is the greatest integer less than or equal to β, if $\beta \in \mathbb{R}_+^*$. By taking $n = sq$ with s large enough, we obtain $J_n = K_n$ and $\overline{(f^{(r)})^{(q)}} = (\bar{f}^{(r)})^{(q)} = \bar{f}^{(rq)}$. Then, it follows from (*) that $\bar{a}_f(g + h) \ge r$, so $\bar{a}_f(g + h) \ge \lambda$. If $\lambda = \infty$, it follows from the above proof that $\bar{a}_f(g + h) \ge t$ for any integer t, hence $\bar{a}_f(g + h) = \infty$. The remainder of the proof follows from Remark 1.(6). □

COROLLARY 3.- Let f, g , h be filtrations on the noetherian ring A such that g and h are AP. Then $\bar{v}_f(g+h) = \min(\bar{v}_f(g), \bar{v}_f(h))$.

Proof.- This follows from Lemma 2 and from Theorem 8 . □

The above corollary generalizes to filtrations the result ([10], Theorem 2) given by P.Samuel for ideals on a noetherian local ring .

Corollary 4.([10],Theorem 2)- Let I,J,K be ideals on the noetherian ring A. Then $\bar{v}_K(I+J) = \min(\bar{v}_K(I), \bar{v}_K(J))$.

Proof.- Take $f = f_K$, $g = f_I$, $h = f_J$ in Corollary 3 , then $f_I + f_J = f_{I+J}$.□

REFERENCES

1. P.Ayégnon and D.Sangaré (1990). Generalized Samuel numbers and AP filtrations, J. Pure Appl. Algebra. 65, 1-13.

2. W.Bishop (1975). A theory of multiplicity for multiplicative filtrations, J.Reine Angew. Math. 277, 8-26.

3. H.Dichi (1989). Integral dependence over a filtration, J. Pure Appl. Algebra. 58, 7-18.

4. H.Dichi (1993). Strongly AP filtrations, integral dependence and prüferian equivalence in Dedekind domains, J. Pure Appl. Algebra (to appear)

5. H.Dichi, D.Sangaré (1991). Filtrations, asymptotic and prüferian closures. Cancellation laws. Proc. Amer. Math. Soc. Vol 113, n°3, 617-624.

6. M.Lejeune and B.Teissier (1974). Clôture intégrale des idéaux et équisingularité, Séminaire de Mathématiques. Ecole Polytéchnique .

7. J.W.Petro (1961). Some results in the theory of pseudovaluations. Ph.D.dissertation. State University of Iowa City, Iowa.

8. J.W.Petro (1975). Concerning filtrations on commutative rings, Manuscripta Math. 15, 261-270.

9. D.Rees (1988). Lectures on the asymptotic theory of ideals. London Math.Society. Lectures Notes Series 113.

10. P.Samuel (1952). Some asymptotic properties of powers of ideals, Ann of Math. 56 , 11-21.

11. D.Sangaré. Sur diverses généralisations de la formule $\bar{v}_{I^n} = \frac{1}{n}\bar{v}_I$ aux pseudo-valuations associés à une filtration. Département de mathématiques. Université d'Abidjan. Côte d'Ivoire.

9
Some Locally Trivial Star-Theoretic Properties of Integral Domains

David E. Dobbs*
Department of Mathematics
University of Tennessee
Knoxville, Tennessee 37996-1300
U.S.A.

Muhammad Zafrullah
Department of Mathematics
Winthrop College
Rock Hill, South Carolina 29733
U.S.A.

ABSTRACT. Using the t- and the v-operations, we introduce two properties, dubbed (T) and (V), that hold for all quasi-local domains and for some special non-quasi-local domains. We show that a non-quasi-local domain D satisfies (T) (resp., (V)) if and only if D is a t-linkative (resp., a t-linkative H-) domain. Upshots include new characterizations of non-quasi-local Prüfer domains and non-quasi-local Dedekind domains.

Throughout this note, D denotes a (commutative integral) domain; Max(D), the set of maximal ideals of D; U(D), the set of

* Supported in part by a University of Tennessee Faculty Research Award.

units of D; and F(D) (resp., f(D)), the set of all (resp., all finitely generated) fractional ideals of D. It is often the case that if a property $\mathcal{P}$ of (some) domains holds locally for D (in the sense that D_M satisfies $\mathcal{P}$ for all $M \in \mathrm{Max}(D)$), then either D also satisfies $\mathcal{P}$ or at least some significant aspect of $\mathcal{P}$ is acquired by D. For instance, the former situation arises with $\mathcal{P}$ as "integrally closed"; and the latter is exemplified with $\mathcal{P}$ as "Krull domain," for if D is locally a Krull domain, then D is at least completely integrally closed. The principal aim of this note is to give two examples of properties $\mathcal{P}$ which hold locally for each domain D, although only very special non-quasi-local domains can satisfy either of these two $\mathcal{P}$. What makes these properties $\mathcal{P}$ especially interesting is that they are connected with some domains of current interest in multiplicative ideal theory: see the characterizations in Corollaries 2, 3 and 5.

Our examples $\mathcal{P}$ involve star-operations, in the sense of Sections 32 and 34 of Gilmer (1972), with which we assume familiarity; in particular, they involve the v- and the t-operations. (Recall that if $A \in F(D)$, then we define $A_v = (A^{-1})^{-1}$; and $A_t = \cup F_v$, where F ranges over the finitely generated (non-zero) fractional ideals contained in A. It is easy to see that $A_t = A_v$ if $A \in f(D)$.) Additional background on these operations can be found in Jaffard (1960); for background on the related topic of Prüfer v-multiplication domains (for short,

PVMDs), see Mott and Zafrullah (1981).

Our focus is on the following two properties, (V) and (T). We say that D satisfies (V) if, for each $A \in F(D)$ and $M \in Max(D)$, the conditions $A \subset M$ and $A_v \neq D$ jointly entail that $A_v \subset M$. (Note that our use of the inclusion symbol, $\subset$, does not preclude the possibility of equality.) Similarly, we say that D satisfies (T) if, for each $A \in F(D)$ and $M \in Max(D)$, the conditions $A \subset M$ and $A_t \neq D$ jointly entail that $A_t \subset M$. We claim that (V) and (T) each hold locally for each domain D. It is enough to show that each quasi-local domain (D,M) satisfies (V) and (T). We give the proof for (V); the argument for (T) is similar. Suppose that $A \in F(D)$ satisfies $A \subset M$ and $A_v \neq D$. Since $A_v \subset D_v = D$, we have that $A_v \subset D \backslash U(D) = M$, proving the claim.

In Corollaries 2 and 3, we shall see the effect of supposing that a non-quasi-local domain D satisfies (V) or (T). Notice, first, for motivation that each principal ideal domain satisfies both (V) and (T); indeed, this follows from the fact that $A_v = A$ for each principal fractional ideal A.

It is convenient to approach Corollaries 2 and 3 by means of a result couched in terms of an arbitrary star-operation, $*$. To do this, Proposition 1 determines when a maximal ideal M is a $*$-ideal, in the sense that $M^* = M$.

PROPOSITION 1. Suppose that the domain D is non-quasi-local, let $M \in Max(D)$, and let $*$ be a star-operation on $F(D)$. Then the following conditions are equivalent:

(1) The conditions $A \subset M$ and $A^* \neq D$ jointly entail that $A^* \subset M$;

(2) M is a $*$-ideal.

Proof: (2) $\Rightarrow$ (1): Assume (2) and suppose that $A \in F(D)$ satisfies $A \subset M$. Then $A^* \subset M^* = M$, yielding (1).

(1) $\Rightarrow$ (2): We have that $M \subset M^* \subset D^* = D$. If the assertion fails, then maximality of M forces $M^* = D$. Since D is not quasi-local, we may choose a nonunit $d \in D\backslash M$. Consider $A = dM \in F(D)$. We have that $A \subset M$ and, since d is a nonunit, that $A^* = dM^* = dD \neq D$. By (1), it follows that $dD \subset M$, whence $d \in M$, the desired contradiction.

In order to develop Corollaries 2 and 3, we need to recall some material. As usual, an <u>overring</u> of D is a ring contained betwen D and its quotient field. An overring R of D is called <u>t-linked</u> (<u>over</u> D) if $A \in f(D)$, $A^{-1} = D$ entails $(AR)^{-1} = R$. The notion of t-linked overring was introduced by Dobbs et al. (1989) in order to develop a t-theoretic analogue of a known characterization of Prüfer domains and a t-theoretic characterization of PVMDs; this notion has been studied further by Dobbs et al. (1990) and Dobbs et al. (1992). As in Dobbs et al. (1992), we say that D is <u>t-linkative</u> if each overring of D is t-linked over D. It is known (see Theorem 2.6 and Corollary 2.7 of Dobbs et al. (1989)) that D is t-linkative if and only if each M

in Max(D) is a t-ideal; and that many domains of classical interest, such as arbitrary Prüfer domain, are t-linkative. By combining the first of these facts with the case $* = t$ of Proposition 1, we immediately infer the following result.

COROLLARY 2. Suppose that the domain D is non-quasi-local. Then D satisfies (T) if and only if D is t-linkative.

Next, recall from Glaz and Vasconcelos (1977) that a domain D is called an <u>H-domain</u> in case $A \in F(D)$, $A^{-1} = D$ implies that there exists $B \in f(D)$ such that $B \subset A$ and $B^{-1} = D$. It was shown in Proposition 2.4 of Houston and Zafrullah (1988) that a domain D is an H-domain if and only if each maximal t-ideal of D is divisorial (that is, is a v-ideal). We are now ready to characterize the non-quasi-local domains that satisfy (V).

COROLLARY 3. Suppose that the domain D is non-quasi-local. Then the following conditions are equivalent:

(1) D is a t-linkative H-domain;

(2) D satisfies (T) and D is an H-domain;

(3) D satisfies (V).

Proof: Of course, Corollary 2 yields that (1) $\Leftrightarrow$ (2). Also, by Proposition 1, D satisfies (V) (resp., (T)) if and only if each $M \in$ Max(D) is divisorial (resp., a t-ideal). As each divisorial ideal is a t-ideal, it now follows from the above characterization of H-domains (Proposition 2.4 of Houston and Zafrullah (1988)) that (2) $\Leftrightarrow$ (3).

Recall that $A \in F(D)$ is called v-invertible (resp., t-invertible) in case $(AA^{-1})_v = D$ (resp., $(AA^{-1})_t = D$). It is clear that invertible ⇒ t-invertible ⇒ v-invertible. It was shown in Theorem 2.6 of Dobbs et al. (1989) that a domain D is t-linkative if and only if each t-invertible ideal of D is invertible. We next give a useful case in which the second implication is reversible.

PROPOSITION 4. If D is an H-domain and $A \in F(D)$ is v-invertible, then A is t-invertible.

Proof: If the assertion fails, then the t-ideal $B = (AA^{-1})_t$ is a proper ideal of D, and so is contained in some maximal t-ideal M of D. By the characterization of H-domains in Proposition 2.4 of Houston and Zafrullah (1988)], M is divisorial. Thus

$$D = (AA^{-1})_v = B_v \subset M_v = M,$$

the desired contradiction.

In Corollary 5, we shall collect several applications to domains of classical interest. First, for motivation, we observe via Corollary 2 and the remarks preceding it (and the fact that (T) holds locally for each domain) that each Prüfer domain satisfies (T). Similarly, since each Dedekind domain is trivially an H-domain, we see via Corollary 3 (and the fact that (V) holds locally for each domain) that each Dedekind domain satisfies (V). For non-quasi-local domains, partial converses of these observations are given in Corollary 5 (b), (c). Next, it is convenient to recall here that a domain D is called a v-domain in case each $A \in f(D)$ is v-invertible.

COROLLARY 5: Suppose that the domain D is non-quasi-local. Then:

(a) If D satisfies (V) and $A \in F(D)$, then the following conditions are equivalent:

(1) A is v-invertible;

(2) A is t-invertible;

(3) A is invertible.

(b) D is a Dedekind domain if and only if D satisfies (V) and D is completely integrally closed.

(c) D is a Prüfer domain if and only if D satisfies (T) and D is a PVMD.

(d) If D is a v-domain, then the following conditions are equivalent:

(i) D satisfies (V);

(ii) D is a PVMD and each maximal ideal of D is invertible;

(iii) D is a Prüfer domain and each maximal ideal of D is invertible;

(iv) D is a Prüfer domain and each maximal ideal of D is divisorial.

Proof: (a) By the remarks preceding Proposition 4, it is enough to show that if A is v-invertible, then A is invertible. Since Corollary 3 ensures that D is an H-domain, it follows from Proposition 4 that A is t-invertible. However, since Corollary 3 also ensures that D is t-linkative, it now follows from Theorem

2.6 of Dobbs et al. (1989) that A is invertible.

(b) We noted above that each Dedekind domain satisfies (V); and it is standard that each Dedekind domain is completely integrally closed. Conversely, suppose that the non-quasi-local domain D satisfies (V) and is completely integrally closed. As is well known (see Gilmer (1972)), the latter condition ensures (in fact, is equivalent to the condition) that each $A \in F(D)$ is v-invertible. Hence, by (a), each $A \in F(D)$ is invertible, and so D is a Dedekind domain.

(c) We noted above that each Prüfer domain satisfies (T); and it is standard that each Prüfer domain is a PVMD. Conversely, suppose that the non-quasi-local domain D satisfies (T) and is a PVMD. By Corollary 2, the (T) property ensures that D is t-linkative, and so each maximal ideal of D is a t-ideal. Thus, by a result of Griffin (cf. Proposition 0.1 of Houston (1986)), the PVMD condition yields that D is locally a valuation domain; that is, D is a Prüfer domain.

(d) (i) $\Rightarrow$ (ii): Assume (i). If $A \in f(D)$, then A is v-invertible (since D is a v-domain), hence t-invertible by (a), and so D is a PVMD. Next, consider $M \in \mathrm{Max}(D)$. As noted in the proof of Corollary 3, it follows from Proposition 1 that M is divisorial. As Corollary 3 also ensures that M is a maximal t-ideal, it follows from the proof of Proposition 1.6 of Houston and Zafrullah (1988) that $M = (a) : (b)$ for suitable nonzero elements $a, b \in D$. Accordingly, since D is a PVMD, it follows (cf. Mott and Zafrullah (1981)) that M is t-invertible. Hence, by (a), M is invertible, as desired.

(ii) ⇒ (iii): Assume (ii). By (c), it is enough to show that D satisfies (T), that is, by Corollary 2, that each maximal ideal of D is a t-ideal. This follows easily from (ii), since each invertible ideal is a t-ideal.

(iii) ⇒ (iv): This is evident, since each invertible ideal is divisorial.

(iv) ⇒ (i): Assume (iv). By (c), D satisfies (T), and so each maximal ideal of D is a t-ideal. Hence, by (iv), each maximal t-ideal of D is divisorial. By the characterization in Proposition 2.4 of Houston and Zafrullah (1988), it follows that D is an H-domain. An application of Corollary 3 now yields (i).

REMARK 6. (a) It is of some interest to observe that (T) is equivalent to the following property: for each $A \in f(D)$ and $M \in \mathrm{Max}(D)$, the conditions $A \subset M$ and $A_t \neq D$ jointly entail that $A_t \subset M$. (The proof that this property is equivalent to (T) follows easily from the definition of the t-operation.)

(b) In contrast with (a), we next record a variant of (T) which holds for all (not necessarily quasi-local) domains D. Indeed, using the definition of the t-operation, one sees easily that this holds for the following property: for each ideal $A \subset D\backslash U(D)$ such that $A_t \neq D$, it follows that $A_t \subset D\backslash U(D)$. By letting the v-operation replace the t-operation in the description of this property, we similarly obtain a variant of (V) which holds for all (not necessarily quasi-local) domains. Of course, these observations would not lead to anything beyond the above, at least in the spirit of the definitions of (V) and (T), because $D\setminus U(D)$ is an ideal of D if and only if D is quasi-local.

(c) Another property related to (V) and (T) is given by the following: Pic(D) = 0; that is, that each invertible ideal A of D (necessarily, $A \in f(D)$) is principal. In particular, this property holds locally for each domain D. Indeed, this property holds for each quasi-semi-local domain: cf. Proposition 5, page 113 of Bourbaki (1972).

REFERENCES

Bourbaki, N. (1972). Commutative Algebra, Addison-Wesley, Reading.

Dobbs, D. E., Houston, E. G., Lucas, T. G. and Zafrullah,M. (1989). t-linked overrings and Prüfer v-multiplication domains, Comm. Algebra, 17: 2835-2852.

Dobbs, D. E., Houston, E. G., Lucas, T. G. and Zafrullah,M. (1990). t-linked overrings as intersections of localizations, Proc. Amer. Math. Soc., 109: 637-646.

Dobbs, D. E., Houston, E. G., Lucas, T. G., Roitman, M. and Zafrullah, M., (1992). On t-linked overrings, Comm. Algebra, 20: 1463-1488.

Gilmer, R. (1972). Multiplicative Ideal Theory, Dekker, New York.

Glaz, S. and Vasconcelos, W. V. (1977). Flat ideals II, Manuscripta Math., 22: 325-341.

Houston, E. G. (1986). On divisorial prime ideals in Prüfer v-multiplication domains, J. Pure Appl. Algebra, 42: 55-62.

Houston, E. G. and Zafrullah, M. (1988). Integral domains in which each t-ideal is divisorial, Michigan Math. J., 35: 291-300.

Jaffard, P. (1960). Les Systèmes d'Idéaux, Dunod, Paris.

Mott, J. L. and Zafrullah, M. (1981). On Prüfer v-multiplication domains, Manuscripta Math., 35: 1-26.

10
The Altitude Formula

OTHMAN ECHI Département de Mathématiques, Faculté des Sciences de Sfax, Route de Soukra 3038, Sfax, Tunisia

All rings considered in this article are commutative with identity, integral and finite-dimensional. We say that (A,B) satisfies *the altitude formula* if : $A \subset B$ and

$$\text{ht } P = \text{ht } p + \text{t.d.}[\,B : A\,] - \text{t.d.}[\,B/P : A/p].$$

(A,B) satisfies *the altitude inequality formula* if :

$$\text{ht } P \leq \text{ht } p + \text{t.d.}[\,B : A\,] - \text{t.d.}[\,B/P : A/p]$$

for each prime ideal P of B and $p = P \cap A$ (t.d.[B:A] is the transcendental degree of the quotient field of B over the quotient field of A). We say that a domain A satisfies the altitude formula if (A,B) satisfies the altitude formula whenever B is a finite type A-algebra . We prove here that A stisfies the altitude formula if and only if for each nonnegative integer n and each pair $P \subset Q$ of prime ideals of the polynomial ring $A[X_1,..., X_n]$ such that $P \cap A = (0)$,

$$\text{we have : ht } (Q/P) = \text{ht } Q - \text{ht } P.$$

We set $A[X_1,...,X_n] = A\,[n]$ and $p\,[X_1,...,X_n] = p\,[n]$, for each prime ideal p of A.
It will be convenient to say that a domain A satisfies *residually the altitude formula* if A/P satisfies the altitude formula, for each prime ideal P of A.
It is proved here that A satisfies residually the altitude formula if and only if, for each nonnegative integer n and each pair $P \subset Q$ of prime ideals of A[n], we have the equality :

$$(*)\ \ \text{ht } (Q/P) - [\text{ht } Q - \text{ht } P] = \text{ht } (q/p) - [\text{ht } q - \text{ht } p].$$

Where $q = Q \cap A$ and $p = P \cap A$.
Let $A \subset B$ be an integral extension of domains such that B satisfies the altitude formula, It is proved here that if ht q = ht $(q \cap A)$, for each prime ideal q of B, then A satisfies also the altitude formula.
A ring A is said to be *catenarian* in case for each pair $P \subset Q$ of prime ideals of A, we have ht (Q/P) = ht Q - ht P. A is *universally catenarian* if the polynomial rings A[n] are catenarian for each positive integer n.
In this article we reproduce the results [2, Corollary (4.8)] and [5, Theorem (2.3)] concerning universally catenarian domains.

1. THE ALTITUDE FORMULA

a) Jaffard domains and altitude inequality formula

Let A be a domain, following [1] and [4], we say that:
- A is a *Jaffard* domain if dim (A[n]) = n + dim A, for each nonnegative integer n.
- A is a *locally Jaffard* domain if A_p is a Jaffard domain for each prime ideal p of A.
- A is a *residually Jaffard* domain if A/p is a Jaffard domain for each prime ideal p of A.
- A is a*totally Jaffard* domain if A_p is a residually Jaffard for each prime ideal p of A.

PROPOSITION(1.0) [5, Lemma (1.4)]
Let A be a domain, then the following conditions are equivalent :
i) A is a locally Jaffard domain ;
ii) A satisfies the altitude inequality formula ;
iii) (A, A[n]) satisfies the altitude formula, for each nonnegative integer n ;
iv) ht (p[n]) = ht p, for each nonnegative integer n and each prime ideal p of A.

PROPOSITION(1.1)
Let A be a domain, then the following conditions are equivalent :
i) A is a totally Jaffard domain ;
ii) A/p satisfies the altitude inequality formula, for each prime ideal p of A;
iii) ht(q/p) = ht(q[n]/p[n]), for each p⊂q prime ideals of A, and for each nonnegativeinteger n
iv) For each pair p ⊂ q of prime ideals of A, for each nonnegative integer n and each prime ideal Q of A [n], such that q = Q ∩ A, we have ht (Q/p[n]) - ht Q = ht (q/p) - ht q.

Proof :
It follows from (1.0) that i)⇔ ii)⇔ iii)
iii) ⇒ iv)
Following [3, Theorem 1], we have:

$$\text{ht }(Q/p[n]) = \text{ht}(Q/q[n]) + \text{ht }(q[n]/p[n])$$

$$\text{and ht } Q = \text{ht }(Q/q[n]) + \text{ht } q[n]$$

This leads to ht (Q/p[n]) - ht Q = ht (q[n]/p[n]) - ht q[n] = ht (q/p) - ht q.
iv) ⇒ iii)
From iv) we get : ht (Q/q[n]) - ht Q = -htq, then ht (q[n]) = ht q,
and ht (q[n]/p[n]) - ht (q[n]) = ht (q/p) - ht q (By the condition iv).
wich gives, ht (q[n]/p[n]) = ht (q/p) completing the proof.

b/ Characterization

THEOREM(1.2)
Let A be a domain, then the following conditions are equivalent:
i) A satisfies the altitude formula ;
ii) For each nonnegative integer n and for each pair P ⊂ Q of prime ideals of A [n], such that P∩ A = (0), we have ht (Q/P) = htQ - ht P.

Proof :

i) $\Rightarrow$ ii)

B = A[n]/P is a finite type A-algebra contaning A then

$$\begin{aligned} \text{ht}\,(Q/P) &= \text{ht}\,q + \text{t.d.}[B:A] - \text{t.d.}[B/(Q/P) : (A/q)] \\ &= \text{ht}\,q + \text{t.d.}[(A[n]/P) : A] - \text{t.d.}[(A[n]/Q) : (A/q)] \qquad (1) \end{aligned}$$

A[n] is also a finite type A-algebra, then :

$$\text{ht}\,Q = \text{ht}\,q + \text{t.d.}[A[n] :A] - \text{t.d.}[(A[n]/Q) : (A/q)] \qquad (2)$$

$$\text{ht}\,P = \text{t.d.}[A[n] : A] - \text{t.d.}\ [(A[n]/P) : A] \qquad (3)$$

By combining (1),(2) and (3), we obtain the equality ht Q/P = ht Q - ht P.

ii) $\Rightarrow$ i)

First we prove that ht(p[n]) = ht p, for each nonnegative integer n and each prime ideal p ofA. Set $P = (X_1,...,X_n) = X_1\ A[n] +...+ X_n\ A\ [n]$ and $Q = (p, X_1,...,X_n) = p + P$.

Since $P \cap A = (0)$, we get ht (Q/P) = htQ - ht P which gives

$$\text{ht}\,p = \text{ht}\,Q - \text{ht}\,P = \text{ht}\,Q - n \qquad (4)$$

By [3, Theorem 1], we have ht Q = ht p[n] + ht (Q /p[n]) = ht p[n] + n (5)

Combining (4) and (5), we obtain : ht p[n] = ht p, which proves that (A, A[n]) satisfies the altitude formula [5, Lemma (1.4)], the equalities (2) and (3) are then satisfied,
if in addition ht (Q/P) = ht Q - ht P for each pair $P \subset Q$ of prime ideals of A[n], such that $P \cap A = (0)$ then the equality (1) is satisfied (with B = A[n]/P) which completes the proof.

COROLLARY(1.3) [2, Corollary (4.8)]
If A is a universally catenarian domain, then A satisfies the altitude formula.

Proof : It follows from condition (ii) of theorem (1.2).

2. THE CLASS OF DOMAINS SATISFYING RESIDUALLY THE ALTITUDE FORMULA

THEOREM(2.1)
Let A be a domain, then the following conditions are equivalent :
i) A satisfies residually the altitude formula ;
ii) For each nonnegative integer n, and each pair $P \subset Q$ of prime ideals of A[n], set $p = P \cap A$ and $q = Q \cap A$, we have :
(*) ht (Q/P) - [ht Q - ht P] = ht (q/p) - [ht q - ht p].

Proof :

i) $\Rightarrow$ ii)

A/p satisfies the altitude formula, then following theorem (1.2) :

ht(Q/P) = ht (Q/p[n]) - ht (P/[n]) (1), and following Proposition (1.1) :
ht(Q/p[n]) = h Q + ht(q/p) - ht q (2)
ht(P/p[n]) = ht P + 0 - ht p (3)

Combining (1), (2) and (3), we obtain :

(*) ht(Q/P) - [ht Q - ht P] = ht(q/p) - [ht q - ht p]

To prove that ii) $\Rightarrow$ i), we establish a technical result. It will be convenient to say that a domain A satisfies the condition (*) in case the condition ii) of theorem (2.2) holds.

LEMMA (2.2)
If a domain A satisfies the condition (*), then for each prime ideal t of A, A/t satisfies the condition (*).

Proof of (2.2):
A pair of prime ideals of (A/t)[n], is obtained by a pair $P \subset Q$ of prime ideals of A[n] containing t[n].
Applying the condition (*) to the pairs $P \subset t[n]$ and $t[n] \subset P$, we obtain :

$$\text{ht}(Q/P) - [\text{ht } Q - \text{ht } P] = \text{ht}(q/p) - [\text{ht } q - \text{ht } p \qquad (4)$$

$$\text{ht}(Q/t[n]) - [\text{ht } Q - \text{ht}(t[n])] = \text{ht}(q/t) - [\text{ht } q - \text{ht } t] \qquad (5)$$

$$\text{ht}(P/t[n]) - [\text{ht } P - \text{ht}(t[n])] = \text{ht}(p/t) - [\text{ht } p - \text{ht } t] \qquad (6)$$

Combining (4), (5) and (6) we obtain :

$$\text{ht}(Q/P) - [\text{ht}(Q/t[n]) - \text{ht}(P/t[n])] = \text{ht}(q/p) - [\text{ht}(q/t) - \text{ht}(p/t)]$$

(which is the condition (*) for the pair $P/t[n] \subset Q/t[n]$ of prime ideals of (A/t) [n]).
ii) $\Rightarrow$ i)
By Lemma (2.2) it suffices to prove that A satisfies the altitude formula.
If $P \cap A = (0)$, the condition (*) is

$$\text{ht}(Q/P) - [\text{ht } Q - \text{ht } P] = \text{ht } q - [\text{ht } q - 0] = 0$$

which gives $\text{ht}(Q/P) = \text{ht } Q - \text{ht } P$, the conclusion follows easily from Theorem (1.2).

COROLLARY (2.3) [5, Theorem (2.3)]
Let A be a domain, then the following conditions are equivalent:
i) A is universally catenarian ;
ii) A is catenarian and satisfies residually the altitude formula.

Proof :
i) $\Rightarrow$ ii) [2, corollary 4.8]
ii) $\Rightarrow$ i)
For each pair $P \subset Q$ of prime ideals of A[n], we have :

$$(*) \ \text{ht}(Q/P) - [\text{ht } Q - \text{ht } P] = \text{ht}(q/p) - [\text{ht } q - \text{ht } p].$$

Since A is catenarian, $\text{ht}(q/p) = \text{ht } q - \text{ht } p$, and so $\text{ht}(Q/P) = \text{ht } Q - \text{ht } P$, which proves that A is universally catenarian.

We next establish a result analogous to [2, Theorem (6.1)].

PROPOSITION (2.4)
Let $A \subset B$ an integral extension of domains such that B satisfies the altitude formula. Suppose that ht q = ht $(q \cap A)$ for each prime ideal q of B. Then A satisfies the altitude formula.

Proof :
1) if $A \subset B$ is an integral extension of domains such that ht q = ht $(q \cap A)$, for each prime ideal q of B, suppose that B is locally Jaffard, then we prove that for each nonnegative

integer n, and each prime ideal Q of B[n], we have ht Q = ht (Q ∩ (A [n])).
By induction it suffices to prove the result for B[X].

Let Q be a prime ideal of B[X]. Set P = Q ∩(A[X]), q = Q∩B and p = q∩A = P∩A.

- If Q = q[X], then ht Q = ht q and ht P = ht p (A and B are locally Jaffard [1])and then ht Q = ht P, (Since ht q = ht(q∩A) = ht p).

- If q[X] ⊂ Q , then by the INC propriety, we obtain p[X] ⊂ P, in this case we have, ht Q = ht q + 1= ht (q∩A) + 1 = ht p + 1 = ht P.

2) Now we prove that A satisfies the altitude formula. Let P ⊂ Q a pair of prime ideals of A[n] such that P∩A = (0). By the lying-over, going-up properties there exist P'⊂ Q', pair of prime ideals of B[n] such that ; P'∩(A[n]) = P, Q'∩(A[n]) = Q

Since B satisfies the altitude formula and P'∩B = (0), then htQ'/P') = htQ'- ht P' (1).
In the other hand we have :

$$\text{ht} (Q'/ P') \leq \text{ht} (Q /P) \leq \text{ht}Q - \text{ht } P \qquad (2)$$

$$\text{ht } Q' = \text{ht } Q, \quad \text{ht } P' = \text{ht } P \qquad (3)$$ (B is locally Jaffard).

Combining (1), (2) and (3) we get : htQ - ht P ≤ ht (Q/P) ≤ ht Q - ht P, which gives ht (Q/P) = ht Q - ht P, this proves that A satisfies the altitude formula [Theorem (1.2)].

REFERENCES

[1] D.F. ANDERSON, A. BOUVIER, D.E. DOBBS, M. FONTANA, S. KABBAJ. On Jaffard domains. Expo. math,**5**, (1988), 145-175.

[2] A. BOUVIER, D.E. DOBBS, M. FONTANA.Universally catenarian integral domains. Advances in Math,**72**, (1988), 211-238.

[3] J. W.BREWER, P.R. MONTGOMERY, E. A RUTTER, W.J.HEINZER.Krulldimension of polynomial rings. Lecture notes in Mathematics springer-verlag , **311**, (1972),26-45

[4] P.J. CAHEN.Construction B.I.D. et anneaux localement ou residuellement de Jaffard. Archiv. der Math, **54**,125-141,(1990).

[5] S.KABBAJ.Formule de la dimension pour les S-domaines forts universels. Bolletino U.M.I Algebra e Geometria ,**5**,145-161, (1986).

11
Absolutely Pure Modules and Locally Injective Modules

ALBERTO FACCHINI Dipartimento di Matematica, Università di Udine, Via Zanon 6, 33100 Udine, Italy

In his paper [1] Goro Azumaya asked the following question: if every absolutely pure left module over a ring R is locally injective, is R left Noetherian? In this paper we give a negative answer to that question.

The motivation of Azumaya's question was the following. It had been proved [11, pp. 60-61] that *(a)* every flat left R-module is locally projective if and only if R is left perfect, and *(b)* every locally projective left R-module is projective if and only if R is left perfect. These theorems can be "dualized formally" as: *(a')* every absolutely pure left R-module is locally injective if and only if R is left Noetherian, and *(b')* every locally injective left R-module is injective if and only if R is left Noetherian. Now *(b')* was proved to be a Theorem in [8, Th. 3], while here we prove that *(a')* is false.

The technique we make use of is to study the notions of absolutely pure module and locally injective module over almost maximal valuation domains. Over these rings absolutely pure modules turn out to be divisible modules, and locally injective modules turn out to be h-divisible modules. It is then easy to construct an example of an almost maximal valuation domain which is not

The author is a member of GNSAGA of CNR. This research was supported by Ministero dell'Università e della Ricerca Scientifica e Tecnologica (Fondi 40% e 60%), Italy.

Noetherian and over which there exists a divisible not h-divisible module. This example provides a negative answer to Azumaya's question.

Some of the results we make use of to answer Azumaya's question have already appeared in the mathematical literature (Theorem 1 appeared in [8], Proposition 2 in [9, Lemma 1] and Theorem 4 in [7]), but we provide some of the proofs in order to make the paper as self-contained as possible.

Any ring considered will be associative with an identity element which acts as the identity operator on any module over the ring. Let R be a ring, B a left R-module and A a submodule of B. The submodule A is said to be a *locally split* submodule of B [1] (or *strongly pure* in B [8]) if for each $a \in A$ there exists a homomorphism $\alpha: B \to A$ such that $\alpha(a) = a$. It is easy to prove that if A is a locally split submodule of B, then for every finitely generated submodule A' of A there exists a homomorphism $\alpha: B \to A$ whose restriction to A' is the identity. From this fact the following theorem follows easily.

THEOREM 1. *Let R be a ring and M be a left R-module. The following conditions are equivalent:*

(a) Every homomorphism $A \to M$ extends to a homomorphism $B \to M$ whenever A is a finitely generated submodule of an arbitrary left R-module B.

(b) The module M is locally split in every left R-module containing M as a submodule.

(c) For every finite subset X of M there exists an injective R-submodule E of M such that $X \subseteq E$.

Proof: [8, Th. 3.1 and Prop. 3.3]. □

A left R-module M is said to be *locally injective* [1] (or *finitely injective*, or *strongly absolutely pure* [8]) if it satisfies the equivalent conditions of Theorem 1.

Recall that a *valuation domain* is a commutative integral domain whose ideals are totally ordered by inclusion. A valuation domain R is said to be *almost maximal* if every finitely solvable system of congruences of the form

$$x \equiv a_i \pmod{L_i}, \quad i \in I,$$

where I is an arbitrary index set, $a_i \in R$, L_i are ideals of R and $\bigcap_{i \in I} L_i \neq 0$, has a global solution in R. Equivalently, a valuation domain R with field of fractions Q is almost maximal if and only if Q/R is an injective R-module [5].

PROPOSITION 2 [9]. *Let R be a valuation domain with field of fractions Q. Then R is almost maximal if and only if for each integer $n \geq 1$ every homomorphic image of Q^n is an injective R-module.*

Proof: If Q/R is injective, then R is almost maximal by [5, Th. 4].

Conversely, assume that R is almost maximal. We shall prove that every homomorphic image E of Q^n is injective by induction on n. For $n = 1$ the module E is either zero or isomorphic to Q/J for some ideal J of R, so that E is injective by [5, Th. 4] again. Suppose that the result is true for $n-1$ and let $\varphi: Q^n \to E$ be an epimorphism. Fix a one-dimensional vector subspace V of the n-dimensional vector space Q^n and set $F = \varphi(V)$. Then F is an injective R-module by the case $n = 1$, and $\varphi: Q^n \to E$ induces an epimorphism $\varphi': Q^n/V \to E/F$. Since Q^n/V is a vector space over Q of dimension $n-1$, the inductive hypothesis yields that E/F is an injective R-module. But since F and E/F are both injective modules, the module E must be injective too. This concludes the proof of the proposition. □

A module M over a commutative integral domain R is said to be *h-divisible* [6] if it is an homomorphic image of an injective R-module. Since every injective R-module is a homomorphic image of a direct sum of copies of the field of fractions Q of R, a module over a domain R is h-divisible if and only if it is a homomorphic image of a direct sum of copies of Q.

THEOREM 3. *Let R be a commutative integral domain. Then every locally injective module is h-divisible. The converse holds if the domain R is an almost maximal valuation domain.*

Proof: Let M be a locally injective module over an integral domain R. By Theorem 1 for every element $m \in M$ there exists an injective submodule E_m of M with $m \in E_m$. For each $m \in M$ let $\varepsilon_m: E_m \to M$ be the canonical embedding, and let $\varepsilon: \bigoplus_{m \in M} E_m \to M$ be the morphism induced by the ε_m's. Then $\varepsilon: \bigoplus_{m \in M} E_m \to M$ is onto. Now the conclusion follows immediately, because the injective modules E_m are h-divisible, so that their direct sum $\bigoplus_{m \in M} E_m$ is h-divisible, and therefore its homomorphic image M is h-divisible too.

Conversely let M be an h-divisible module over an almost maximal valuation domain R. By Theorem 1 in order to prove that M is locally injective it is sufficient to show that for every finite subset X of M there exists an injective R-submodule E of M such that $X \subseteq E$. Let X be a finite subset of M. Since M is h-divisible, there exist a vector space V over the field of fractions Q of R and an epimorphism $\varphi: V \to M$. But X is a finite subset of M, so that there is finite dimensional Q-vector subspace W of V such that $\varphi(W) \supseteq X$. By Proposition 2 the R-module $\varphi(W)$ is injective because R is an almost maximal valuation domain. Therefore $E = \varphi(W)$ is the required injective submodule of M containing X. □

Recall that a left R-module M is called *absolutely pure* [7] (or *fp-injective*) if it satisfies the following equivalent conditions: *(a)* M is a pure submodule of every left R-module containing M as a submodule; *(b)* $\mathrm{Ext}^1_R(F, M) = 0$ for every finitely presented R-module F, i.e., every exact sequence $0 \to M \to B \to F \to 0$ with F finitely presented splits.

THEOREM 4 [7]. *Let R be a Prüfer domain. An R-module M is absolutely pure if and only if it is divisible, i.e., $rM = M$ for every non-zero $r \in R$.*

Proof: If M is an absolutely pure R-module and r is a non-zero element of R, let $E(M)$ denote the injective envelope of M. Then $E(M)$ is divisible and M is pure in $E(M)$, so that $rM = M \cap rE(M) = M \cap E(M) = M$, i.e., M is divisible.

Conversely, if M is a divisible module and N is any left module containing it, in order to show that M is pure in N it is sufficient to show that $M \cap rN = rM$ for every $r \in R$ [10]. This is trivial for $r = 0$, and for $r \neq 0$ one has $M \cap rN \subseteq M = rM$ because M is divisible. Since the other inclusion $M \cap rN \supseteq rM$ also is trivial, this concludes the proof the Theorem. □

In particular Theorem 4 holds for modules over valuation domains.

THEOREM 5. *Let R be an almost maximal valuation domain with field of fractions $Q \neq R$. The following statements are equivalent:*

(1) *the projective dimension of the R-module Q is 1;*

(2) *every absolutely pure R-module is locally injective.*

Proof: Over an almost maximal valuation domain absolutely pure modules are exactly divisible modules (Theorem 4) and locally injective modules are exactly h-divisible modules (Theorem 3). Therefore Condition (2) is equivalent to *"every divisible R-module is h-divisible"*. By [3] every divisible R-module is h-divisible if and only if the projective dimension of Q is 1. □

Theorem 5 allows us to give a negative answer to the question asked by Azuamaya in the last two lines of his paper [1], i.e., whether it is true that if every absolutely pure left R-module is locally injective then the ring R must be left Noetherian.

Let Γ be any non trivial countable totally ordered abelian group non isomorphic to the group $\mathbf{Z}$ of integers (e.g., the group $\mathbf{Q}$ of rationals), and let k be any field. If $R = k[[\Gamma]]$ is the formal power series ring of Γ over k, then R is an (almost) maximal valuation domain with value group Γ [2, Th. I.5.11]. In particular R is not Noetherian because Γ is non trivial and is not isomorphic to $\mathbf{Z}$. Since Γ is countable, the field of fractions Q of R is countably generated as an R-module. In particular Q has projective dimension 1 as an R-module [4]. By Theorem 5 every absolutely pure R-module is locally injective. Therefore R is an example that provides a negative answer to Azumaya's question.

REFERENCES

1. G. Azumaya, Finite splitness and finite projectivity, *J. Algebra*, *106*: 114–134 (1987).
2. L. Fuchs and L. Salce, "Modules over valuation rings," Marcel Dekker, New York and Basel, 1985.
3. R. M. Hamsher, On the structure of a one-dimensional quotient field, *J. Algebra*, *19*: 416–425 (1971).

4. I. Kaplansky, The homological dimension of a quotient field, *Nagoya Math. J.*, *27*: 139–142 (1966).
5. E. Matlis, Injective modules over Prufer rings, *Nagoya Math. J.*, *15*: 57–69 (1959).
6. E. Matlis, "Torsion-free modules," The University of Chicago Press, Chicago and London, 1972.
7. C. Megibben, Absolutely pure modules, *Proc. Amer. Math. Soc.*, *26*: 561–566 (1970).
8. V. S. Ramamurthi and K. M. Rangaswamy, On finitely injective modules, *J. Austral. Math. Soc.*, *16*: 239–248 (1973).
9. L. Salce and P. Zanardo, On a paper of I. Fleischer, *Abelian Group Theory*, (R. Göbel and E. Walker, eds.), Lecture Notes in Math. 874, Springer-Verlag, Berlin-Heidelberg-New York, 1981, pp. 76–86.
10. R. B. Warfield, Jr., Purity and algebraic compactness for modules, *Pacific J. Math.*, *28*: 699–719 (1969).
11. B. Zimmermann-Huisgen, "Direct products of modules and algebraic compactness," Habilitationsschrift, Tech. Univ. München, 1980.

12
Krull and Valuative Dimensions of the $A + XB[X]$ Rings

M. FONTANA Dipartimento di Matematica, Terza Università degli Studi di Roma, 00146 Roma, Italy

L. IZELGUE and S. KABBAJ Département de Mathématiques et Informatique, Faculté des Sciences "Dhar El-Mehraz", Université de Fès, Fès, Morocco

0. INTRODUCTION

All the rings considered in this paper are integral domains, i.e. commutative rings with identity and non zero-divisors. Given a finite dimensional ring A, we say that A is a *Jaffard domain* if $\dim A = \dim_v A$ [2]. The previous property is not a local property and thus we say that A is a *locally Jaffard domain* if $A_{\mathfrak{p}}$ is a Jaffard domain, for each prime ideal $\mathfrak{p}$ of A. Noetherian domains and, in the locally finite dimensional case, Prüfer domains, stably strong S-domains and universally catenarian domains are examples of locally Jaffard domains. As a matter of fact, the locally Jaffard domains coincide with the *rings satisfying the inequality formula* [3], [4, Théorème 1.5], [16, Lemme 1.4]. Besides the locally Jaffard domains, further examples of Jaffard domains are given by the polynomial rings with the coefficients on a Jaffard domain and by some class of rings arising from the pullback diagrams of a special type (cf. [2], [3], [7], [9], [13] and [16]).

In [10] D. Costa, J. L. Mott and M. Zafrullah introduced the rings of the type $D^{(S)} := D + XD_S[X]$, where D is an integral domain and S is a multiplicatively closed subset of D. If $S := D \setminus \{0\}$ then $D^{(S)} = D + XK[X]$, where $K := \mathrm{qf}(D)$ (= quotient

These authors were supported in part by a *NATO Collaborative Research Grant* CRG N. 900113

field of D). Some properties of the prime spectrum of $D^{(S)}$ were investigated in [10], even if the problem of an exact determination of $\dim D^{(S)}$ was not settled in the general case. In [13], the authors were interested in a more general situation concerning the rings $D^{(S,\, r)} := D + (X_1,..., X_r)D_S[X_1,..., X_r]$. After proving a formula for the Krull dimension of $D^{(S,\, r)}$ [13, Theorem 3.2], they showed that $D^{(S,\, r)}$ is a Jaffard domain with $\dim D^{(S,\, r)} = r + \dim D$ if and only if D is a Jaffard domain [13, Theorem 3.5].

In the present work, we investigate an even more general construction, considering the ring $R := A + XB[X] = \{f \in B[X] \mid f(0) \in A\}$, where $A \subset B$ is a ring extension, X is an indeterminant over B. The ring R is a particular case of the constructions B, I, D introduced by J.-P.Cahen in [9] (cf. also [12]).

The ring $\mathrm{Int}(B, A) := \{f \in B[X] \mid f(A) \subset A\}$ is a subring of $A + XB[X]$ and a deeper knowledge of the properties of the rings of the type $A + XB[X]$ may have some interesting consequences for the theory of the rings of integer-valued polynomials [1].

In Section 1, we study the structure of the prime spectrum of $R = A + XB[X]$, clarifying the relation among $\mathrm{Spec}(R)$ and the spectra of A and $B[X]$. Furthermore, we will provide upper and lower bounds to $\mathrm{ht}_R XB[X]$ by means of $\mathrm{tr.deg.}_A B$.

In the second section, we will take care of the theory of the dimension and of the transfer of the related properties in the constructions $A + XB[X]$. We will generalize some results previously established for the domains of the type $D + XK[X]$ and $D + XD_S[X]$, where K is a field containing D and S is a multiplicative subset of D. We will prove, among other facts, that $R = A + XB[X]$ is a Jaffard domain and $\dim R = 1 + \dim A$ if and only if A is a Jaffard domain and $\mathrm{deg.tr.}_A B = 0$.

Section 3 is devoted to the investigation of several examples showing the limits of some of the results previously established. We will show also that some of the results, holding for the constructions $D + XD_S[X]$ and for $D + XK[X]$, can not be extended in their classical form to the general construction $A + XB[X]$. We will take this opportunity to describe a new class of Jaffard domains, different from all the known classes.

1. THE PRIME SPECTRUM

We start by establishing some links between the prime ideals of $R = A + XB[X]$ and those of A and $B[X]$. The following lemma is a consequence of some general theorems concerning the pullback constructions [12].

LEMMA 1.1. *Let $A \subset B$ be an extension of rings, X is an indeterminant over B and $R = A + XB[X]$.*

(a) *The ideal $XB[X]$ is a prime ideal of R and $R/XB[X]$ is canonically isomorphic to A. From a topological point of view, the map ${}^a g$: $\mathrm{Spec}(A) \to \mathrm{Spec}(R)$, corresponding to the canonical projection $g : R \to A$, is a closed embedding and it induces an order-isomorphism of $\mathrm{Spec}(A)$ onto $\mathfrak{X} := \{\mathfrak{p} \in \mathrm{Spec}(R) \mid XB[X] \subset \mathfrak{p}\}$, $\mathfrak{p} \mapsto \mathfrak{p} + XB[X]$. In particular, $\mathfrak{X}$ is a subspace of $\mathrm{Spec}(R)$ stable under specialization.*

(b) *The set $S := \{X^n \mid n \geq 0\}$ is a multiplicatively closed subset of R and of $B[X]$ such that $S^{-1}R = S^{-1}B[X] = B[X, X^{-1}]$. Moreover, by contraction, we obtain an order-isomorphism $\{\mathfrak{Q} \in \mathrm{Spec}(B[X]) \mid X \notin \mathfrak{Q}\} \to \mathfrak{Y} := \{\mathfrak{P} \in \mathrm{Spec}(R) \mid X \notin \mathfrak{P}\}$, and thus $\mathfrak{Y}$ is a subspace of $\mathrm{Spec}(R)$ stable under generalization.*

(c) *The spectral space $\mathrm{Spec}(R)$ is canonically homeomorphic to the amalgamated sum of $\mathrm{Spec}(A)$ and $\mathrm{Spec}(B[X])$ with respect to $\mathrm{Spec}(B)$.*

PROOF. **(a)** The map $g : R \to A$, $X \mapsto 0$, is a surjective homomorphism with $\mathrm{Ker}(g) = XB[X]$. Therefore $R/XB[X]$ is isomorphic to A and $XB[X]$ is a prime ideal of R. It is clear that the continuous map ${}^a g : \mathrm{Spec}(A) \to \mathrm{Spec}(R)$ is closed and injective.

(b) Since S is a multiplicatively closed subset of R then $S^{-1}R = A[X^{-1}] + S^{-1}(XB[X]) = S^{-1}(B[X]) = B[X, X^{-1}] = B[\mathbb{Z}]$. We deduce easily that $\mathfrak{Y}$, $\mathrm{Spec}(B[X, X^{-1}])$ and $\{\mathfrak{P} \in \mathrm{Spec}(B[X]) \mid X \notin \mathfrak{P}\}$ are bijectively equivalent.

(c) is a consequence of the general properties of the pullback constructions, cf. [12].

∎

For the constructions $D + XK[X]$ and $D + XD_S[X]$, it is known that ht $XK[X]$ = ht $XD_S[X] = 1$ ([10] and [13] ; see also Step 1 of the proof of the following Lemma 1.3). Nevertheless, the previous result does not hold in the general case $R = A + XB[X]$. More precisely, we will see later (Example1.5) that, for each $n \geq 1$, there exist $A \subset B$ such that $\mathrm{ht}_R\, XB[X] = n$. Next goal is an approximation of the height of $XB[X]$ inside R .

THEOREM 1.2. *Let* $R = A + XB[X]$, $N := A \setminus \{0\}$ *and* $k := \mathrm{qf}(A)$.

(a) $\mathrm{ht}_R\, XB[X] = \dim N^{-1}B[X] = \dim (B[X] \otimes_A k)$.

(b) $1 \leq \mathrm{ht}_R\, XB[X] \leq 1 + \mathrm{tr.deg.}_A\, B$.

In order to prove this theorem, we need the following lemma:

LEMMA 1.3. *In the same situation of* Theorem 1.2,

$$\mathrm{ht}_R\, XB[X] = 1 + \mathrm{Sup}\{\mathrm{ht}_{B[X]}\, \mathfrak{q}[X] \mid \mathfrak{q} \in \mathrm{Spec}(B) \text{ and } \mathfrak{q} \cap A = (0)\} .$$

PROOF. ***Step 1*** *:* If for each $\mathfrak{q} \in \mathrm{Spec(B)} \setminus \{(0)\}$ we have that $\mathfrak{q} \cap A \neq (0)$, then $\mathrm{ht}_R\, XB[X] = 1$.

As a matter of fact, in this situation, we have $N^{-1}B = \mathrm{qf}(B) =: L$. We deduce that $N^{-1}R = N^{-1}A + XN^{-1}B[X] = k + XL[X]$. Therefore $\mathrm{ht}_R\, XB[X] = \mathrm{ht}_{N^{-1}R}\, XN^{-1}B[X] = 1$ since $\dim N^{-1}R = 1$ [2, Proposition 2.15].

Step 2 *:* There exists a non zero prime ideal $\mathfrak{q} \in \mathrm{Spec}(B)$ such that $\mathfrak{q} \cap A = (0)$.

First at all, for each $\mathfrak{q} \in \mathrm{Spec}(B)$ such that $\mathfrak{q} \cap A = (0)$ we have $\mathfrak{q}[X] \cap R \subset XB[X]$. As a matter of fact, $((\mathfrak{q}[X] + XB[X])/XB[X]) \cap A = (0)$, hence $(\mathfrak{q}[X] + XB[X]) \cap R = XB[X] \cap R = XB[X]$ thus $\mathfrak{q}[X] \cap R \subset XB[X]$.

If $\mathfrak{q} \in \mathrm{Spec}(B)$ is such that $\mathfrak{q} \cap A = (0)$, then we obtain $\mathrm{ht}_{B[X]}\, \mathfrak{q}[X] = \mathrm{ht}_R\, (\mathfrak{q}[X] \cap R)$ (Lemma 1.1). Therefore, $\mathrm{ht}_R\, XB[X] \geq 1 + \mathrm{ht}_{B[X]}\, \mathfrak{q}[X]$.

Let $(0) \subset \mathfrak{P}_1 \subset ... \subset \mathfrak{P}_n \subset XB[X]$ be a chain of prime ideals realizing the height of $XB[X]$ inside R . We claim that, for each $i \in \{1, ..., n\}$, $X \notin \mathfrak{P}_i$. If not, we would have $XR = XA + X^2 B[X] \subset \mathfrak{P}_i \subset XB[X]$, hence $X^2B[X] \subset \mathfrak{P}_i \subset XB[X]$ and thus $\sqrt{(X^2B[X])} = \mathfrak{P}_i = XB[X]$ inside R : a contradiction. By Lemma 1.1, the previous chain lifts to a chain $(0) \subset \mathfrak{Q}_1 \subset ... \subset \mathfrak{Q}_n$ of the same length inside $B[X]$, hence ht $\mathfrak{P}_n$ = ht $\mathfrak{Q}_n$ = n . By [15], $\mathrm{ht}_{B[X]}\,\mathfrak{Q}_n$ can be realized by a special chain $(0) \subset \subset \mathfrak{Q}'_1 \subset ... \subset \mathfrak{Q}'_{n-1} \subset \mathfrak{Q}_n$ of prime ideals of $B[X]$. Let $\mathfrak{q} := \mathfrak{Q}_n \cap B$, then either $\mathfrak{q}[X] = \mathfrak{Q}'_{n-1}$ or $\mathfrak{q}[X] = \mathfrak{Q}_n$. In any case, we have $(0) \subset \mathfrak{q} \cap A = \mathfrak{Q}_n \cap B \cap A = \mathfrak{Q}_n \cap R \cap A = \mathfrak{P}_n \cap A \subset XB[X] \cap A = (0)$.

If $\mathfrak{q}$ is not maximal among the prime ideals $\mathfrak{p}$ of B such that $\mathfrak{p} \cap A = (0)$, then let $\mathfrak{q}' \in \mathrm{Spec}(B)$ such that $\mathfrak{q} \subsetneq \mathfrak{q}'$ and $\mathfrak{q}' \cap A = (0)$. We deduce that $\mathfrak{q}[X] \cap R \subset \mathfrak{q}'[X] \cap R \subset XB[X]$, $n - 1 = \mathrm{ht}_{B[X]}\,\mathfrak{q}[X] = \mathrm{ht}_R\,(\mathfrak{q}[X] \cap R)$ and $n \leq \mathrm{ht}_{B[X]}\,\mathfrak{q}'[X] = = \mathrm{ht}_R\,(\mathfrak{q}'[X] \cap R) < \mathrm{ht}_R\,XB[X] = n + 1$. Therefore, $\mathrm{ht}_{B[X]}\,\mathfrak{q}'[X] = n$ and $\mathrm{ht}_R\,XB[X] = 2 + \mathrm{ht}_{B[X]}\,\mathfrak{q}[X] = 1 + \mathrm{ht}_{B[X]}\,\mathfrak{q}'[X]$.

Let $\mathfrak{q}$ be maximal among the prime ideals $\mathfrak{p}$ of B such that $\mathfrak{p} \cap A = (0)$. Necessarily we must have $\mathfrak{q}[X] = \mathfrak{Q}_n$: otherwise, $\mathfrak{q}[X] \subsetneq \mathfrak{Q}_n$ implies the existence of the following chain of prime ideals $\mathfrak{q}[X] \cap R \subset \mathfrak{Q}_n \cap R \subset XB[X]$. Therefore, we obtain the chain of prime ideals $(0) \subset (\mathfrak{Q}_n \cap R)/(\mathfrak{q}[X] \cap R) \subset (XB[X])/(\mathfrak{q}[X] \cap R)$ inside the integral domain $R/(\mathfrak{q}[X] \cap R)$ (isomorphic to $A + X(B/\mathfrak{q})[X]$). Because of the maximality of $\mathfrak{q}$, for each $\mathfrak{q}' \in \mathrm{Spec}(B/\mathfrak{q}) \setminus \{0\}$, we get that $\mathfrak{q}' \cap A \neq (0)$, hence we are in the situation of Step 1. We deduce that $\mathrm{ht}_{R'}\,X(B/\mathfrak{q})[X] = 1$, hence $R' := A + X(B/\mathfrak{q})[X]$. We reach a contradiction, since $(XB[X])/(\mathfrak{q}[X] \cap R)$ is isomorphic to $X(B/\mathfrak{q})[X]$ and $\mathrm{ht}(\,(XB[X])/(\mathfrak{q}[X] \cap R)\,) \geq 2$. Therefore we have that $\mathfrak{q}[X] = \mathfrak{Q}_n$ and thus $\mathrm{ht}_R\,XB[X] = 1 + n = 1 + \mathrm{ht}_{B[X]}\,\mathfrak{Q}_n = 1 + \mathrm{ht}_{B[X]}\,\mathfrak{q}[X]$.

We proved that $\mathrm{ht}_R\,XB[X] \leq 1 + \mathrm{Sup}\{\, \mathrm{ht}_{B[X]}\,\mathfrak{q}[X] \mid \mathfrak{q} \in \mathrm{Spec}(B) \text{ et } \mathfrak{q} \cap A = (0)\}$ from which the conclusion follows easily. ∎

PROOF OF THEOREM 1.2. (a) We consider the ring $N^{-1}R = N^{-1}A + XN^{-1}B[X] = k + XN^{-1}B[X]$. It is obvious that, for each $\mathfrak{q} \in \mathrm{Spec}(N^{-1}B)$, we have that $\mathfrak{q} \cap k = (0)$. By Lemma 1.1 $\mathrm{ht}_R\,XB[X] = \mathrm{ht}_{N^{-1}R}\,XN^{-1}B[X]$ and, by Lemma 1.3, it follows that $\mathrm{ht}_{N^{-1}R}\,XN^{-1}B[X] = 1 + \mathrm{Sup}\{\, \mathrm{ht}_{N^{-1}B[X]}\,\mathfrak{q}[X] \mid \mathfrak{q} \in \mathrm{Spec}(N^{-1}B)\} = \dim N^{-1}B[X]$.

(b) We have that $k \subset N^{-1}B[X] \subset L(X) = \mathrm{qf}(B[X])$. By [14, Theorem 20.9], $\dim N^{-1}B[X] \leq \mathrm{tr.deg.}_k\,L(X) = 1 + \mathrm{tr.deg.}_A\,B$, from the statement **(a)** the conclusion follows. ∎

We notice that Theorem 1.2 recovers, as a particular case, some of the known results concerning the domains of the type $D + XK[X]$ et $D + XD_S[X]$, where K is a field containing D and S is a multiplicatively closed subset of D . In the following Example 1.5 we will apply the full strength of Theorem 1.2 for computing the height of $XB[X]$ inside $A + XB[X]$, where B is not a field nor a localisation of A .

COROLLARY 1.4. *Let* $A \subset B$, $L := \mathrm{qf}(B)$ *and* S *be a multiplicatively closed subset of* A . *Set* $R := A + XB[X]$, $T := A + XL[X]$ *and* $A^{(S)} := A + XA_S[X]$.

(a1) *If* $\mathrm{qf}(A) \subset B$, *then* $\mathrm{ht}_R\, XB[X] = \dim B[X]$.

(a2) $\mathrm{ht}_T\, XL[X] = 1$.

(b1) *If* $A \subset B$ *is an algebraic extension of integral domains, then* $\mathrm{ht}_R\, XB[X] = 1$.

(b2) $\mathrm{ht}_{A^{(S)}}\, XA_S[X] = 1$.

PROOF. In order to prove **(a1)** it is sufficient to notice that $\mathrm{qf}(A) \subset B$ implies that $N^{-1}B = B$; **(b1)** follows by Theorem 1.2 (b). ∎

EXAMPLE 1.5. For each integer $n \geq 1$, there exist two integral domains $A \subset B$ such that $\mathrm{ht}_R\, XB[X] = n$, where $R := A + XB[X]$.

Let $A := \mathbb{Z}$, $B := \mathbb{Q}[X_1, ..., X_{n-1}]$. By Corollary 1.4 (a1), we obtain that $\mathrm{ht}_R\, XB[X] = \dim \mathbb{Q}[X_1, ..., X_{n-1}][X] = n$.

The following example shows that, for a domain of the type $R = A + XB[X]$, $\mathrm{ht}_R\, XB[X]$ can describe all the integer values between 1 and $1+ \mathrm{tr.deg.}_A B$. In particular, the boundaries established in Theorem 1.2 may not be improved.

EXAMPLE 1.6. Let $d \in \mathbb{N}$ and $t \in \{1, ..., 1+d\}$, then there exists an extension of integral domains $A \subset B$ such that $\mathrm{tr.deg.}_A B = d$ and $\mathrm{ht}_R\, XB[X] = t$, where $R := A + XB[X]$.

As a matter of fact, let k be a field and let $X, X_1, ..., X_{d+1}, Y_1, ..., Y_d$ be indeterminants over k . Set $A := k$ and $B := k(X_1, ..., X_{d-t+1})[Y_1, ..., Y_{t-1}]$. Then,

$\text{tr.deg.}_A\, B = \text{tr.deg.}_k\, k(X_1, ..., X_{d-t+1}, Y_1, ..., Y_{t-1}) = d$,

$\text{ht}_R\, XB[X] = \dim B[X] = \dim k(X_1, ..., X_{d-t+1})[Y_1, ..., Y_{t-1}]\, [X] = t$ (Corollary 1.4 (a1)).

2. Krull and valuative dimension of $A+XB[X]$

In this section we establish two of the main results of the present work. If $A \subset B$ is a given extension of integral domains, then the first one gives an approximation of the Krull dimension of $A+XB[X]$. In the second result, we determine the valuative dimension of $R = A+XB[X]$ by means of $\dim_v A$ and $\text{tr.deg.}_A\, B$. As a consequence, we will be able to study the transfer to R of the Jaffard and locally Jaffard properties.

THEOREM 2.1. *Let* $R := A+XB[X]$, $N := A \setminus \{0\}$ *and* $k := \text{qf}(A)$.

(a) $\text{Max}\{\dim N^{-1} B[X] + \dim A\ ;\ \dim B[X]\} \leq \dim R \leq \dim A + \dim B[X]$;

(b) *If* $k \subset B$ *then* $\dim R = \dim A + \dim B[X]$.

PROOF. **(a)** We know that $\text{ht}_R\, XB[X] + \dim R/XB[X] \leq \dim R$. Since $R/XB[X] \simeq A$ (Lemma 1.1) and $\text{ht}_R\, XB[X] = \dim N^{-1}B[X]$ (Theorem 1.2), then $\dim N^{-1}B[X] + \dim A \leq \dim R$. Furthermore, $S^{-1}R = B[X, X^{-1}]$ (Lemma 1.1) where $S := \{X^n \mid n \geq 0\}$. By [2, Proposition 1.14], $\dim B[X, X^{-1}] = \dim B[\mathbb{Z}] = \dim B[X]$. We deduce that $\dim B[X] \leq \dim R$, which implies the first inequality.

Let $\mathfrak{P}_0 = (0) \subset \mathfrak{P}_1 \subset ... \subset \mathfrak{P}_n$ be a chain of prime ideals of R which realizes the dimension of R . Let r be the maximum integer of $\{1, ..., n\}$ such that X does not belong to $\mathfrak{P}_r$, hence for each $m \leq r$, X does not belong to $\mathfrak{P}_m$. By using the order-isomorphisms $\{\mathfrak{Q} \in \text{Spec}(B[X]) \mid X \notin \mathfrak{Q}\} \to \mathfrak{Y} = \{\mathfrak{P} \in \text{Spec}(R) \mid X \notin \mathfrak{P}\}$ and $\text{Spec}(A) \to \mathfrak{X} = \{\mathfrak{P} \in \text{Spec}(R) \mid XB[X] \subset \mathfrak{P}\}$ (Lemma 1.10), we deduce that $n - r \leq \dim A$ and $r \leq \dim B[X]$. Therefore, $n = \dim R \leq \dim A + \dim B[X]$.

(b) If $k \subset B$, it is clear that, for each $\mathfrak{q} \in \text{Spec}(B)$, $\mathfrak{q} \cap A = (0)$. By Lemma 1.3 we deduce that $\text{ht}_R\, XB[X] = \dim B[X]$. The conclusion follows easily from Theorem 1.2 and from the point (a) . ∎

For the constructions $D^{(S)} := D+XD_S[X]$, it has been proved in [13, Proposition 3.1] and in [10, Theorem 2.6, Corollary 2.9] that $\dim D^{(S)} \leq \mathrm{Min}\{\dim D[X]\ ;\ \dim D + \dim D_S[X]\}$. Example 3.1 shows that an inequality of the same type does not hold for the general constructions of the type $A+XB[X]$. More precisely, we will construct a domain $R = A+XB[X]$ such that $\dim R > \dim A[X] > \dim B[X]$, with $\mathrm{qf}(A) = \mathrm{qf}(B)$ (hence, $\dim R = \dim A + \dim B[X]$, Theorem 2.1. (b)). Furthermore, Example 3.1 shows that the double inequality of Theorem 2.1 (a) may be strict, with $\dim A[X] < \dim R$ and $\mathrm{tr.deg.}_A B = 0$.

COROLLARY 2.2. *Let D be an integral domain, K its field of fractions and S a multiplicatively closed subset of D.*

(a) $\dim D+XK[X] = 1 + \dim D$;

(b) $\mathrm{Max}\{1+ \dim D\ ;\ \dim D_S[X]\} \leq \dim D+XD_S[X] \leq \dim D + \dim D_S[X]$.

PROOF. (a) (respectively, **(b)**) follows from Theorem 2.1 (b) (respectively from Theorem 2.1 (a)). ∎

THEOREM 2.3. *Let $A \subset B$ be two integral domains and $R := A + XB[X]$.*

(a) $\dim_v R = \dim_v A + \mathrm{tr.deg.}_A B + 1$.

(b) *The following statements are equivalent:*

(i) *A is a Jaffard domain and* $\mathrm{qf}(B)$ *is an algebraic extension of* $\mathrm{qf}(A)$;

(ii) *R is a Jaffard domain and* $\dim R = \dim A + 1$.

PROOF. (a) We use induction on $d := \mathrm{tr.deg.}_A B$.

Set $L := \mathrm{qf}(B)$ and $k := \mathrm{qf}(A)$.

Step 1 : $d = 0$, i.e. L is an algebraic extension of k. We have that $A[X] \subset R \subset B[X]$ and $L(X) = \mathrm{qf}(B[X]) = \mathrm{qf}(R)$ is an algebraic extension of $k(X) = \mathrm{qf}(A[X])$. By [2, Definition-Theorem 0.1], $1 + \dim_v A = \dim_v A[X] = \mathrm{Sup}\{\dim V \mid V$ is a $L(X)$-valuation overring of $A[X]\} \geq \dim_v R = \mathrm{Sup}\{\dim V \mid V$ is a valuation overring of $R\}$. Let V be an L-valuation overring of A such that $\dim V = \dim_v A$ and set $T := V + XL[X]$. The integral domain T is clearly an overring of R. We deduce that $\dim_v R \geq$

$\dim T = 1 + \dim V$ [2, Proposition 2.15] , [10, Corollary 2.10], and thus $\dim_v R = 1 + \dim_v A$.

Step 2 : $d = \text{tr.deg.}_A B = 1$. Let $y \in B$ be a transcendental element over k (its existence follows from the fact that $L = \text{qf}(B)$). We consider the integral domains $R[y] = A[y] + XB[X]$ and $R[y^{-1}] = A[y^{-1}] + XB[y^{-1}][X]$. In this situation, $\text{tr.deg.}_{A[y]} B = 0$ and also $\text{tr.deg.}_{A[y^{-1}]} B[y^{-1}] = 0$ since $\text{tr.deg}_A A[y^{-1}] = 1$. By Step 1, $\dim_v R[y] = \dim_v A[y] + 1 = \dim_v A + 2$ and $\dim_v R[y^{-1}] = \dim_v A[y^{-1}] + 1 = \dim_v A + 2$. Moreover, every overring of R is a valuation overring of $R[y]$ or $R[y^{-1}]$. We deduce that $\dim_v R = \text{Max}\{\dim_v R[y] \ ; \ \dim_v R[y^{-1}]\} = \dim_v A + 2$.

Step 3 : $d = \text{deg.tr.}_A B \geq 1$. We suppose that, for $A' \subset B'$ such that $r = \text{deg.tr.}_{A'} B' \leq d - 1$ then $\dim_v R' = \dim_v A' + r + 1$, where $R' := A' + XB'[X]$. Let $y \in B$ be a transcendental element over k , then $\text{tr.deg.}_{A[y]} B = \text{tr.deg.}_{A[y^{-1}]} B[y^{-1}] = d - 1$. With the same notation of Step 2, the inductive hypothesis implies that $\dim_v R[y] = \dim_v A[y] + (d - 1) + 1 = \dim_v A + d + 1$ and $\dim_v R[y^{-1}] = \dim_v A[y^{-1}] + (d - 1) + 1 = \dim_v A + d + 1$. As in Step 2, we have $\dim_v R = \text{Max}\{\dim_v R[y] \ ; \ \dim_v R[y^{-1}]\} = \dim_v A + d + 1$. This completes the proof of **(a)** .

(b) (i) $\Longrightarrow$ (ii)
$$\begin{aligned} \dim_v R &= \dim_v A + \text{tr.deg.}_A B + 1 \\ &= \dim A + 1, \quad \text{by hypothesis} \\ &= \dim A + \text{ht}_R XB[X], \quad \text{by Corollary 1.4 (b)} \\ &\leq \dim R, \quad \text{by Theorem 2.1 (a)} \\ &\leq \dim_v R, \end{aligned}$$

thus R is a Jaffard domain and $\dim R = 1 + \dim A$.

(ii) $\Longrightarrow$ (i) $\dim_v R = \dim_v A + d + 1 = \dim R = 1 + \dim A$, hence we deduce that $\dim_v A = \dim A$ and $d = 0$. Therefore, A is a Jaffard domain and $\text{qf}(B)$ is an algebraic extension of $\text{qf}(A)$. ▮

Among the applications of Theorem 2.3, we recover some "classical" result concerning the integral domains $D + XD_S[X]$ and $D + XK[X]$ (cf. [10]) proved in [2, Proposition 2.15 and 2.16], [7, Proposition 2.11 and Corollary 2.12] and [13, Proposition 3.4 and Theorem 3.5 (a)]:

COROLLARY 2.4. *Let D be an integral domain, S a multiplicatively closed subset of D and $K := \mathrm{qf}(D)$.*

(a) $\dim_v D + XD_S[X] = \dim_v D + XK[X] = \dim_v D + 1$;

(b) *D is a Jaffard domain if and only if $D + XK[X]$ is a Jaffard domain* ;

(c) *D is a Jaffard domain if and only if $D + XD_S[X]$ is a Jaffard domain and* $\dim D + XD_S[X] = \dim D + 1$. ∎

COROLLARY 2.5. *Let A be an integral domain having k as its field of fractions and let L be a field extension of k. Set $R = A + XL[X]$.*

(a) $\dim_v R = \dim_v A + \mathrm{tr.deg.}_k L + 1$,

(b) *R is a Jaffard domain if and only if A is a Jaffard domain and L is an algebraic extension of k.* ∎

It was proved in [2, Proposition 2.16 (b)] that if $A \subset B$ and if $\mathrm{qf}(A) = \mathrm{qf}(B)$, when A is a Jaffard domain, then $R = A + XB[X]$ is also a Jaffard domain and $\dim R = \dim A + 1$. The following corollary establishes, among other facts, that the converse holds as well (cf. also [2, Remark 2.17]):

COROLLARY 2.6. *Let $A \subset B$ be an extension of integral domains with the same field of fractions. Set $R := A + XB[X]$.*

(a) $\dim_v R = \dim_v A + 1$,

(b) *A is a Jaffard domain if and only if R is a Jaffard domain and* $\dim R = \dim A + 1$. ∎

Lastly we have:

COROLLARY 2.7. *Let $A \subset B$ be an extension of integral domains such that* $\mathrm{qf}(A) \subset B$. *Set $R := A + XB[X]$. Then, R is a Jaffard domain if and only if A is a Jaffard domain and* $\dim B[X] = 1 + \mathrm{tr.deg.}_A B$.

PROOF. By Theorem 1.2 (b), Lemma 1.3 and Corollary 1.4 (a1), we deduce that $\mathrm{ht}_R\, XB[X] = 1 + \mathrm{Sup}\{\mathrm{ht}\, \mathfrak{q}[X] \mid \mathfrak{q} \in \mathrm{Spec}(B) \text{ and } \mathfrak{q} \cap A = (0)\} = \dim B[X] \leq 1 + \mathrm{tr.deg.}_A B$. Theorem 2.3 (a) and Theorem 2.1 (b) lead to the conclusion, since $\dim_v R - \dim R = \dim_v A - \dim A + (1 + d - \dim B[X])$. ∎

In Section 3, we will give several examples showing the limits of the previous results.

THEOREM 2.8. *Let* $A \subset B$ *be an extension of integral domains and let* $R := A + XB[X]$. *We suppose that* A *is a locally Jaffard domain. The following statements are equivalent :*

(i) $B[X]$ *is locally Jaffard and* $\mathrm{ht}_R\, XB[X] = 1 + \mathrm{tr.deg.}_A B$;

(ii) R *is locally Jaffard.*

In order to prove this theorem we need the following lemma (cf. also [2, Corollary 1.16]):

LEMMA 2.9. *Let* B *be an integral domain, then* $B[X]$ *is locally Jaffard if and only if* $B[X, X^{-1}]$ *is locally Jaffard.*

PROOF. We suppose that $B[X, X^{-1}]$ is a locally Jaffard domain. Since $B[X, X^{-1}]$ is integral over $B[X + X^{-1}]$, for each $\mathfrak{q} \in \mathrm{Spec}(B[X + X^{-1}])$, if $T := (B[X + X^{-1}] \setminus \mathfrak{q})$, then $T^{-1}B[X, X^{-1}]$ is integral over $T^{-1}B[X + X^{-1}] = B[X + X^{-1}]_{\mathfrak{q}}$. Now $B[X, X^{-1}]$ is locally Jaffard, then $B[X + X^{-1}]_{\mathfrak{q}}$ is the same [2, Proposition 1.1 and Proposition 1.5 (a)]. Since $B[X + X^{-1}]$ is isomorphic to $B[X]$, then $B[X]$ is also locally Jaffard and the lemma is proved. ∎

PROOF OF THEOREM 2.8.

(i) $\Longrightarrow$ (ii). Let A and $B[X]$ be locally Jaffard domains and $\mathrm{ht}_R\, XB[X] = 1 + \mathrm{tr.deg.}_A B$, we want to prove that, for each $\mathfrak{P} \in \mathrm{Spec}(R)$, $R_{\mathfrak{P}}$ is a Jaffard domain. For such a prime $\mathfrak{P}$ two cases are possible :

Case 1 : $X \in \mathfrak{P}$. There exists $\mathfrak{p} \in \text{Spec}(A)$ such that $\mathfrak{P} = \mathfrak{p} + XB[X]$ (Lemma 1.1 (a)). It is easy to verify that $R_{\mathfrak{P}} = A_{\mathfrak{p}} + XT^{-1}B[X]$, where $T^{-1} := R \setminus (\mathfrak{p} + XB[X])$. We have that $A_{\mathfrak{p}} + XB_{\mathfrak{p}}[X] \subset R_{\mathfrak{P}} \subset A_{\mathfrak{p}} + XL[X]_{(X)}$, with $B_{\mathfrak{p}} := (A \setminus \mathfrak{p})^{-1}B$ and $L = \text{qf}(B)$, and we notice that all these rings have the same field of fractions. By the previous remarks, we deduce that :

- $\dim_v (A_{\mathfrak{p}} + XB_{\mathfrak{p}}[X]) = \dim_v A_{\mathfrak{p}} + \text{tr.deg.}_A B + 1$ (Theorem 2.3 (a));
- $A_{\mathfrak{p}} + XL[X]_{(X)}$ is the pullback of the inclusion $A_{\mathfrak{p}} \to L$ with respect to the canonical projection $L[X]_{(X)} \to L$.

By [2, Theorem 2.6], $\dim_v (A_{\mathfrak{p}} + XL[X]_{(X)}) = \dim_v A_{\mathfrak{p}} + \dim_v L[X]_{(X)} + \text{tr.deg.}_A L = \dim_v A_{\mathfrak{p}} + 1 + \text{tr.deg.}_A B$. On the other hand, $\dim R_{\mathfrak{P}} = \text{ht}_R \mathfrak{P} = \text{ht}_R(\mathfrak{p} + XB[X]) \geq \text{ht}_A \mathfrak{p} + \text{ht}_R XB[X] = \dim A_{\mathfrak{p}} + \text{tr.deg.}_A B + 1 = \dim_v A_{\mathfrak{p}} + \text{tr.deg.}_A B + 1 = \dim_v R_{\mathfrak{P}}$, thus $R_{\mathfrak{P}}$ is a Jaffard domain.

Case 2 : $X \notin \mathfrak{P}$. Set $S := \{X^n \mid n \geq 0\}$. Then $S \cap \mathfrak{P} = \emptyset$ and $R_{\mathfrak{P}} = (S^{-1}R)_{S^{-1}\mathfrak{P}}$ [14, Corollary 5.3]. On the other hand, $S^{-1}R = S^{-1}(B[X]) = B[X, X^{-1}]$ (Lemma 1.1 (b)), and $B[X]$ is locally Jaffard. It follows that $R_{\mathfrak{P}}$ is a Jaffard domain.

(ii) $\Longrightarrow$ (i). If R is locally Jaffard and $S := \{X^n \mid n \geq 0\}$, then $S^{-1}R = S^{-1}(B[X]) = B[X, X^{-1}]$ is the same. Lemma 2.9 allows to conclude that $B[X]$ is a locally Jaffard domain. Since $R_{XB[X]}$ is a Jaffard domain, then $\text{ht}_R XB[X] = \dim R_{XB[X]} = \dim_v R_{XB[X]}$. By replacing $\mathfrak{p}$ with (0) in (i) $\Longrightarrow$ (ii) of Case 1, we obtain that $\dim_v R_{XB[X]} = \dim_v \text{qf}(A) + \text{tr.deg.}_A B + 1 = \text{tr.deg.}_A B + 1$ and the proof of the theorem is complete. ∎

We notice that the hypothesis that A is a Jaffard domain (instead of a locally Jaffard domain) in Theorem 2.8 is not sufficient for the conclusion. As a matter of fact, it is sufficient to consider a Jaffard non locally Jaffard domain A [2, Example 3.2], and $B := \text{qf}(A)$. In this situation $B[X]$ is locally Jaffard and $\text{ht}_R XB[X] = 1 = 1 + \text{tr.deg.}_A B$ but R is not locally Jaffard [2, Corollary 2.12 (a)]. However, the hypothesis that A is locally Jaffard is not necessary in order that R is also locally Jaffard (cf. Example 3.1 (d)).

COROLLARY 2.10. *Let D be a domain and $K := \mathrm{qf}(D)$. The following are equivalent:*

(i) *D is locally Jaffard;*

(ii) *$D + XD_S[X]$ is locally Jaffard, for each multiplicatively closed subset S of D;*

(iii) *$D + XK[X]$ is locally Jaffard.*

PROOF. (i) $\Longrightarrow$ (ii) is a consequence of Theorem 2. 8, (i) $\Longrightarrow$ (ii) (cf. also [9, Proposition 1 (i)] ; (ii) $\Longrightarrow$ (iii) holds trivially and (iii) $\Longrightarrow$ (i) is a particular case of [2, Corollary 2.12 (a)] . ∎

Theorem 2.8 will give us the possibility to construct new examples of locally Jaffard domains (cf. Examples 3.1 and 3.5).

We notice that, if $X_1, \dots, X_n$ are indeterminants over $R = A + XB[X]$, then $R[X_1, \dots, X_n] = A[X_1, \dots, X_n] + XB[X_1, \dots, X_n][X]$ is a ring of the same type (i.e. $R[X_1, \dots, X_n] = A' + XB'[X]$ with $A' := A[X_1, \dots, X_n]$ and $B' := B[X_1, \dots, X_n]$). From the previous remark, we are led to studying the ring $R[X_1, \dots, X_n]$ by means of the techniques introduced above.

As a consequence of Theorem 2.1 (a) and 2.3 (a), and of the fact that $\dim R[X_1, \dots, X_n] \leq \dim_v R[X_1, \dots, X_n]$, we deduce :

(*) $\mathrm{Max}\{\dim A[X_1,\dots, X_n] + \mathrm{ht}_{R[X_1, \dots, X_n]} XB[X_1,\dots, X_n][X] \; ; \; \dim B[X_1, \dots, X_n][X]\} \leq$

$$\leq \dim R[X_1, \dots, X_n] \leq$$

$\leq \mathrm{Min}\{\dim_v A[X_1,\dots, X_n] + \text{tr.deg.}_A B + 1 \; ; \; \dim A[X_1,\dots, X_n] + \dim B[X_1,\dots, X_n][X]\}.$

Therefore, for n big enough, it is possible to evaluate the dimension of $R[X_1, \dots, X_n]$:

COROLLARY 2.11. *Let $A \subset B$ be an extension of integral domains, $R := A + XB[X]$, n an integer and let $X_1, \dots, X_n$ be indeterminants over R.*

(a) *If* $n \geq \mathrm{Max}\{\text{tr.deg.}_A B \; ; \; \dim_v A - 1\}$, *then* $\dim R[X_1, \dots, X_n] = \dim A[X_1, \dots, X_n] + \text{tr.deg.}_A B + 1$ *and $R[X_1, \dots, X_n]$ is a Jaffard domain.*

(b) *If* $n < \text{tr.deg.}_A B$ *and if B is a field, then* $\dim R[X_1, \dots, X_n] = \dim A[X_1, \dots, X_n] + n + 1$ *and $R[X_1, \dots, X_n]$ is not a Jaffard domain.*

PROOF. **(a)**. By Theorem 1.2 (b), we deduce that $ht_{R[X_1, \dots, X_n]}XB[X_1, \dots, X_n][X] \leq \text{tr.deg.}_A B + 1$. The integral domains R and $B[X]$ have in common the ideal $XB[X]$, then by [9, Lemma 3] we obtain :
$ht_{R[X_1,\dots,X_n]} XB[X_1, \dots, X_n][X] \geq ht_{B[X_1,\dots,X_n][X]}XB[X_1, \dots, X_n][X] + \text{Min}\{n, \text{tr.deg.}_A B\}$.
Therefore, $ht_{R[X_1,\dots,X_n]} XB[X_1, \dots, X_n][X] \geq \text{tr.deg.}_A B + 1$, thus we deduce the equality.
On the other hand, by (*) we get $\dim A[X_1, \dots, X_n] + \text{tr.deg.}_A B + 1 \leq$ $\leq \dim R[X_1, \dots, X_n] \leq \dim_v R[X_1, \dots, X_n] = \dim_v A[X_1, \dots, X_n] + \text{tr.deg.}_A B + 1 =$ $\dim A[X_1, \dots, X_n] + \text{tr.deg.}_A B + 1$, (where $\dim_v A[X_1, \dots, X_n] = \dim A[X_1, \dots, X_n]$ since $n \geq \dim_v A - 1$, [2, Definition-Theorem 0.1]) and thus $\dim R[X_1, \dots, X_n] =$ $\dim_v R[X_1, \dots, X_n]$.

(b) Let $n < \text{tr.deg.}_A B$, by the same reason as in **(a)** we get $ht_{R[X_1, \dots, X_n]} XB[X_1, \dots, X_n][X] \geq n + 1$. We deduce that $1 + n + \dim A[X_1, \dots, X_n]$ $\leq \dim R[X_1, \dots, X_n] \leq \dim A[X_1, \dots, X_n] + \dim B[X_1, \dots, X_n][X]$. Therefore, if B is a field, then $\dim B[X_1, \dots, X_n][X] = n + 1$ and $\dim R[X_1, \dots, X_n] =$ $\dim A[X_1, \dots, X_n] + n + 1 < \dim_v A[X_1, \dots, X_n] + \text{tr.deg.}_A B + 1 =$ $\dim_v R[X_1, \dots, X_n]$. As a consequence, we obtain that $R[X_1, \dots, X_n]$ is not a Jaffard domain. ∎

REMARK 2.12. It is known by [13, Theorem 3.5 (a)] that D is a Jaffard domain if and only if $D + XD_S[X]$ is a Jaffard domain and $\dim(D + XD_S[X]) = 1 + \dim D$. This result can not be directely generalized to the general constructions $A + XB[X]$ with $\text{tr.deg.}_A B > 0$ (cf. Examples 3.4 and 3.5).

3. EXAMPLES AND COUNTER-EXAMPLES

In this section we construct several examples showing the limits of the results proved in Sections 1 and 2. We give also a few counter-examples showing that some results concerning the domains of the type $D + XD_S[X]$ can not be extended to the general constructions $A + XB[X]$.

EXAMPLE 3.1. Let K be a field and let X, X_1, X_2, X_3, X_4 be indeterminants over K. Set,

$A := K[X_1]_{(X_1)} + X_4K(X_1, X_2, X_3)[X_4]_{(X_4)}$,

$B := K(X_1)[X_2]_{(X_2)} + X_3K(X_1, X_2)[X_3]_{(X_3)} + X_4K(X_1, X_2, X_3)[X_4]_{(X_4)}$,

$R := A + XB[X]$.

Then :

(a) $\text{Max}\{\dim A + \text{ht}_R\, XB[X], \dim B[X]\} < \dim R < \dim A + \dim B[X]$.

(b) $\dim A[X] < \dim R$, (notice that for the construction $D + XD_S[X]$, it happens that $\dim D + XD_S[X] \leq \dim D[X]$, [13, Proposition 3.1]).

(c) R shows that [13, Theorem 3.5 (b) (j) $\Longrightarrow$ (jj)], concerning the domains $D + XD_S[X]$, can not be extended to the constructions $A + XB[X]$.

(d) R is a locally Jaffard domain, even if A is not a locally Jaffard domain (cf. Theorem 2.8 and [2, Proposition 2.16]).

As a matter of fact, set $L := K(X_1, X_2, X_3)$, $k := K(X_1, X_2)$, $\mathfrak{M} := X_4L[X_4]_{(X_4)}$, $\mathfrak{N} := X_3k[X_3]_{(X_3)}$, $V := L + \mathfrak{M}$, $V_1 := k + \mathfrak{N}$, $D := K(X_1,)[X_2]_{(X_2)} + \mathfrak{N}$, $B := D + \mathfrak{M}$ and $A := K[X_1]_{(X_1)} + \mathfrak{M}$. By some well known result concerning the $D + \mathfrak{M}$ domains, by [2, Corollary 2.8] and [12, Proposition 2.1 (5) and Theorem 2.4 (1)] we obtain that V, V_1, D and B are valuation domains of dimensions 1, 1, 2 and 3 respectively. Moreover :

- $\dim B[X] = \dim B + 1 = 4$,
- $\dim A = \dim K[X_1]_{(X_1)} + \dim V = 2$ [2, Corollary 2.8],
- $\dim_v A = \dim_v K[X_1]_{(X_1)} + \dim V + \text{tr.deg.}_{K(X_1)} L = 4$ [2, Proposition 2.14 (a)],
- $\dim A[X] = \dim V + \dim K[X_1]_{(X_1)}[X] + \text{Min}\{1, \text{tr.deg.}_{K(X_1)} L\} = 4$ [2, Corollary 2.8],
- $\text{Spec}(B) = \{(0)\, ;\ \mathfrak{M}\, ;\ \mathfrak{P}_1 := \mathfrak{N} + \mathfrak{M}\, ;\ \mathfrak{P}_2 := X_2K(X_1,)[X_2]_{(X_2)} + \mathfrak{P}_1\}$,
- $\text{Spec}(A) = \{(0)\, ;\ \mathfrak{M}\, ;\ \mathfrak{Q} = X_1K[X_1]_{(X_1)} + \mathfrak{M}\}$,
- $\mathfrak{M} \cap A = \mathfrak{P}_1 \cap A = \mathfrak{P}_2 \cap A = \mathfrak{M}$ [12, Theorem 1.4].

(a) and **(b)**. We notice that $\mathrm{qf}(A) = \mathrm{qf}(B) = \mathrm{qf}(V)$, since they A, B and V have the ideal $\mathfrak{M}$ in common. Inside $\mathrm{Spec}(R)$ we have the following chain of prime ideals :

© $(0) \subset \mathfrak{M}[X] \cap R \subset \mathfrak{P}_1[X] \cap R \subset \mathfrak{P}_2[X] \cap R \subset \mathfrak{M} + XB[X] \subset \mathfrak{Q} + XB[X]$.

Therefore $\dim R \geq 5$. By Theorem 2.3 (a), we deduce that $\dim_v R = \dim_v A + \mathrm{tr.deg.}_A B + 1 = 5$, thus $\dim R = 5 = \dim_v R$. As a consequence, we have :

$$(\dim A[X] = 4) < (\dim R = 5) < (\dim A + \dim B[X] = 6).$$

On the other hand (Theorem 1.2 (b)), we have

$$(\dim A + \mathrm{ht}_R\, XB[X] = 2 + 1 = 3) < (\dim B[X] = 4) < (\dim R = 5).$$

(c) Since $\dim R = \dim_v R = 5$, R is then a Jaffard domain. Furthermore, $\dim A[X] = 4$ and $\dim_v A[X] = \dim_v A + 1 = 5$, then $A[X]$ is not a Jaffard domain. This example shows that [13, Theorem 3.5 (b) (j) $\Longrightarrow$ (jj)] can not be extended to the constructions $A + XB[X]$.

(d) The domain A is not a locally Jaffard domain [2, Proposition 1.5 (b)], since it is not a Jaffard domain. In order to show that R is a locally Jaffard domain, it is sufficient to see what happens for the prime ideals of the type $\mathfrak{p} + XB[X]$ with $\mathfrak{p} \in \mathrm{Spec}(A)$. As a matter of fact, if $\mathfrak{P} \in \mathrm{Spec}(R)$ and if $X \notin \mathfrak{P}$, by setting $S := \{ X^n \mid n \geq 0 \}$, we get that $R_{\mathfrak{P}} = (S^{-1}R)_{S^{-1}\mathfrak{P}} = B[X, X^{-1}]_{S^{-1}\mathfrak{P}}$ which is a Jaffard domain (because B is a valuation domain). By the proof of Theorem 2.8 (Case1), we deduce that $\dim_v R_{(\mathfrak{p}+XB[X])} = \dim_v A_{\mathfrak{p}} + \mathrm{tr.deg.}_A B + 1$. We claim that $R_{(\mathfrak{p}+XB[X])}$ is a Jaffard domain, for each $\mathfrak{p} \in \mathrm{Spec}(A)$.

- $\mathfrak{P} = XB[X]$, i.e. $\mathfrak{p} = (0)$. In this case, $\dim R_{\mathfrak{P}} = \dim R_{XB[X]} = \mathrm{ht}_R\, XB[X] = 1 = \dim_v A_{(0)} + \mathrm{tr.deg.}_A B + 1 = \dim_v R_{\mathfrak{P}}$.
- $\mathfrak{P} = \mathfrak{M} + XB[X]$. In the present situation, the chain © shows that $\mathrm{ht}\, \mathfrak{P} = 4$ (since $\dim R = 5$), hence $\dim R_{\mathfrak{P}} = 4 = \dim_v K(X_1) + \dim V + \mathrm{tr.deg.}_{K(X_1)} L + 1 = \dim_v A_{\mathfrak{M}} + 1 = \dim_v R_{\mathfrak{P}}$.
- $\mathfrak{P} = \mathfrak{Q} + XB[X]$. In this case, $\dim R_{\mathfrak{P}} = \mathrm{ht}\, \mathfrak{P} = 5$ (see the chain ©) and $\dim_v R_{\mathfrak{P}} = \dim_v A_{\mathfrak{Q}} + 1 = \dim_v K[X_1]_{(X_1)} + \dim V + \mathrm{tr.deg.}_{K(X_1)} L + 1 = 5$, since $A_{\mathfrak{Q}} = A$ (cf. also [2, Theorem 2.6 (a)])

In all the cases, $R_{\mathfrak{P}}$ is a Jaffard domain. ∎

Theorem 2.3 shows the way to construct new classes of Jaffard domains.

EXAMPLE 3.2. Let,

- $A_1 := \mathbb{Z}$, $B_1 := \mathbb{Z}^\nabla$ and $R_1 := \mathbb{Z} + X\mathbb{Z}^\nabla[X]$,

where $\mathbb{Z}^\nabla$ is the integral closure of $\mathbb{Z}$ inside an algebraic extension of $\mathbb{Q}$.
Since A_1 is a Jaffard domain, then by Theorem 2.3 (b) we deduce that R_1 is a Jaffard domain and $\dim R_1 = 1 + \dim A_1 = 2 < \dim A_1 + \dim B_1[X] = 1 + 2 = 3$ (cf. also Theorem 2.1 (a)). It is possible to show that R_1 is a Noetherian domain if and only if $\mathbb{Z}^\nabla$ is the integral closure of $\mathbb{Z}$ inside a finite extension of $\mathbb{Q}$.

We consider

- $A_2 := \mathbb{R}$, $B_2 := \mathbb{C}[Y]$ and $R_2 := \mathbb{R} + X\mathbb{C}[Y][X]$,

Since $\mathrm{qf}(A_2) \subset B_2$ then, by Theorem 2.1 (b), we deduce that $\dim R_2 = \dim A_2 + \dim B_2[X] = 2$ and, by Theorem 2.3 (a), that $\dim_v R_2 = \dim_v A_2 + \mathrm{tr.deg.}_{A_2} B_2 + 1 = 2$. Therefore R_2 is a Jaffard domain. Moreover $A_2[X] = \mathbb{R}[X]$ is obviously a Jaffard domain, but $\dim R_2 \neq \dim A_2[X]$. (Notice that, for the domain R_2 , the inequalities of Theorem 2.1 (a) are both equalities.) ∎

The following two examples show the limits of some of the results established in Section 2.

EXAMPLE 3.3. Let :

- $A := \mathbb{Z}$, $B := \mathbb{Q}(Y)$ and $R := \mathbb{Z} + X\mathbb{Q}(Y)[X]$.

Then :

(a) The bounds of the inequalities established in Theorem 2.1 (a) can be effectively reached.

(b) R shows that [13, Theorem 3.5 (a), (i) $\Longrightarrow$ (ii)] and [13, Theorem 3.5 (b), (jj) $\Longrightarrow$ (j)] can not be extended to the case $A + B[X]$.
As a matter of fact , let $N = A \setminus \{0\}$:

(a) $\dim R = \dim A + \dim B[X] = 2$ (Theorem 2.1 (b)) and $\dim N^{-1}B[X] + \dim A = 1 + 1 = 2$. Therefore, $\dim N^{-1}B[X] = 1$, since $\dim N^{-1}B[X] + \dim A \leq \dim R$.

Henceforth, $\text{Max}\{\dim N^{-1}B[X] + \dim A\ ,\ \dim B[X]\} = \dim N^{-1}B[X] + \dim A = \dim R = \dim A + \dim B[X] = 2$, thus the bounds of the inequalities established in Theorem 2.1 (a) can be effectively reached.

(b) By (a), $\dim R = 2$ and Theorem 2.3 (a) shows that $\dim_v R = \dim_v \mathbb{Z} + \text{tr.deg.}_{\mathbb{Q}}\ \mathbb{Q}(Y) + 1 = 3$, thus R is not a Jaffard domain. However, A is a Jaffard domain [13, Theorem 3.5 (a), (i) $\Longrightarrow$ (ii)]. Moreover, $A[X]$ is a Jaffard domain and $\dim R = \dim A[X]$, in contrast with [13, Theorem 3.5 (b), (jj) $\Longrightarrow$ (j)] for the domains of the type $D + XD_S[X]$. ∎

EXAMPLE 3.4. Let :

- $A := \mathbb{Z}$, $B := \mathbb{Q}[Y]$ and $R := \mathbb{Z} + X\mathbb{Q}[Y][X]$.

Alors,

(a) R shows that [13, Theorem 3.5 (a), (i) $\Longrightarrow$ (ii)], [13, Theorem 3.5 (b), (j) $\Longrightarrow$ (jj)] and [2, Proposition 2.15 (a)] can not be extended to the case $A + B[X]$.

(b) $\dim R > \dim A[X]$ with $\text{tr.deg.}_A\ B > 0$.

As a matter of fact,

(a) $\text{qf}(A) \subset B$, then by Theorem 2.1 (b) we deduce that $\dim R = \dim A + \dim B[X] = 1 + 2 = 3$, and hence Theorem 2.3 (a) shows that $\dim_v R = \dim_v A + \text{tr.deg.}_A\ B + 1 = 3$. We deduce that R is a Jaffard domain. Since $A = \mathbb{Z}$ is a Jaffard domain, but $\dim R \neq \dim A + 1$, the domain R shows that [13, Theorem 3.5 (a), (i) $\Longrightarrow$ (ii)] can not be extended from the case $D + XD_S[X]$ to the general situation $A + B[X]$.

The rings R and $A[X]$ are Jaffard domains, but $\dim R \neq \dim A[X]$, hence [13, Theorem 3.5 (b), (j) $\Longrightarrow$ (jj)] can not be extended to the general construction $A + B[X]$.

If A and R are the Jaffard domains introduced above, since $\text{tr.deg.}_A\ B \neq 0$, then it is clear that [2,Proposition 2.15] does not hold for the general construction $A + B[X]$.

(b) We know that $3 = \dim R > \dim A[X] = 2$, like in Example 3.1 (b), but in this case $\text{tr.deg.}_A\ B \neq 0$. ∎

EXAMPLE 3.5. Let K be a field and let X, Y, Z, W be indeterminants over K . Set :

$A := K[Y]_{(Y)}$, $B := K[Y]_{(Y)} + ZK(Y)[Z]_{(Z)} + WK(Y, Z)[W]_{(W)}$ and $R := A + XB[X]$.

Then,

(a) $\text{ht}_R\, XB[X] = 1 + \text{tr.deg.}_A\, B > 1$.

(b) $\dim R = \dim A + \text{ht}_R\, XB[X]$.

(c) R is a locally Jaffard domain, different from all the examples already known.

As a matter of fact, set :

$V := K(X, Y)[W]_{(W)}$, $\mathfrak{M} := WK(X, Y)[W]_{(W)}$, $V' := K[Y]_{(Y)} + ZK(Y)[Z]_{(Z)}$,

then,

(a) $A \subset B = V' + \mathfrak{M}$ are both valuation domains, with $\dim A = 1$ and $\dim B = 3$ (cf. [12, Proposition 2.1 (5), Theorem 2.4 (1)]), and $d := \text{tr.deg.}_A\, B = 2$. Moreover :

$$\text{Spec}(B) = \{(0)\,;\ \mathfrak{M}\,;\ \mathfrak{P} := ZK(Y)[Z]_{(Z)} + \mathfrak{M}\,;\ \mathfrak{Q} := YK[Y]_{(Y)} + \mathfrak{P}\},$$

$$\text{Spec}(A) = \{(0)\,;\ YK[Y]_{(Y)}\},$$

$$\mathfrak{M} \cap A = \mathfrak{P} \cap A = (0), \quad \text{et} \quad \mathfrak{Q} \cap A = YK[Y]_{(Y)},$$

thus $\text{ht}_R\, XB[X] = 3 = 1 + \text{tr.deg.}_A\, B$ (Theorem 1.2 (b) and Lemma 1.3) .

(b) We notice that $\dim R \geq \dim A + \text{ht}_R\, XB[X] = 1 + 3 = 4$ (Theorem 1.2 (a)), and that $\dim_v R = \dim_v A + \text{deg.tr.}_A\, B + 1 = 4$ (Theorem 2.3 (a)). Therefore R is a Jaffard domain with Krull dimension $4 = \dim A + \text{ht}_R\, XB[X] < \dim A + \dim B[X] = 5$ (thus the hypothesis $\text{qf}(A) \subset B$ in Corollary 1.4 (a1) is essential).

(c) We notice that A and B are locally Jaffard domains, since they are both valuation domains. It is clear that $B[X]$ is a locally Jaffard domain [9, Proposition 1 (i)] and by Theorem 2.8 it follows that R is a locally Jaffard domain, since $\text{ht}_R\, XB[X] = 1 + \text{tr.deg.}_A\, B$ (cf. (a)) . ∎

REFERENCES

[1] **Anderson, D. D., Anderson, D. F., and Zafrullah, M.** (1991). Rings between $D[X]$ and $K[X]$, Houston J. Math., 17: 109.

[2] **Anderson, D. F., Bouvier, A., Dobbs, D., Fontana, M., and Kabbaj, S.** (1988). On Jaffard domains, Expositiones Mathematicæ, 6: 145.

[3] **Ayache, A.** (1991). Inégalité ou formule de la dimension et produits fibrés. Thèse de doctorat en sciences, Université d'Aix-Marseille .

[4] **Ayache, A., et Cahen, P.-J.** (to appear). Anneaux vérifiant absolument l'inégalité et la formule de la dimension, Boll. Un. Mat. Ital.

[5] **Bastida, E., and Gilmer, R.** (1973). Overrings and divisorial ideals of rings of the form $D+M$, Michigan Math. J., 20: 79.

[6] **Bouvier, A., Dobbs, D., and Fontana, M.** (1988). Universally catenarian integral domains, Adv. Math., 72: 211.

[7] **Bouvier, A., and Kabbaj, S.** (1988). Examples of Jaffard domains, J. Pure Appl. Algebra, 54 : 155.

[8] **Brewer, J. W., and Rutter, E. A.** (1976). $D+M$ construction with general overrings, Michigan Math. J., 23: 33.

[9] **Cahen, P.-J.** (1990). Construction B, I, D et anneaux localement ou résiduellement de Jaffard, Arch. Math., 54 : 125.

[10] **Costa, D., Mott, J. L., and Zafrullah, M.** (1978). The construction $D+XD_S[X]$, J. Algebra, 53: 423.

[11] **Costa, D., Mott, J. L., and Zafrullah, M.** (1986). Overrings and dimension of general $D+M$ construction, J. Natur. Sci. & Math., 26: 7.

[12] **Fontana, M.** (1980). Topologically defined classes of commutative rings, Ann. Mat. Pura Appl., 123: 331.

[13] **Fontana, M., and Kabbaj, S.** (1990). On the Krull and valuative dimension of $D+XD_S[X]$ domains, J. Pure Appl. Algebra, 63: 231.

[14] **Gilmer, R.** (1972). *Multiplicative Ideal Theory*, Marcel Dekker, New York.

[15] **Jaffard, P.** (1960). Théorie de la dimension dans les anneaux de polynômes, Mém. Sc. Math., 146, Gauthier-Villars, Paris.

[16] **Kabbaj, S.** (1986). La formule de la dimension pour les S-domaines forts universels, Boll. Un. Mat. Ital., Algebra e Geometria, 5: 145.

[17] **Kabbaj, S.** (1991).Sur les S-domaines forts de Kaplansky, J. Algebra, 137: 400.

13
Divisorial Ideals and Class Groups of Mori Domains

STEFANIA GABELLI Dipartimento di Matematica, Università di Roma "La Sapienza", Piazzale A. Moro 5, 00185 Roma , Italy; e-mail MARTA@ITCASPUR.bitnet.

The properties of divisorial ideals and class groups of noetherian integrally closed domains, more generally of Krull domains, have been extensively studied in the past and are now well known [F] . More recently, the notion of class group has been introduced for any domain in [Bv] and [BvZ] and its general properties have been studied by several authors (see for example [A], [AA], [AAZ], [ARy], [G], [GRt], [NA], [Ry]). However the structure of the class group is known for only a few special families of domains.

A first class of domains for which this investigation has been carried out is that of Mori domains, namely those domains with the ascending chain condition on divisorial ideals [Rl], [BG1], [BG2], [BGR2]. A motivation to consider this kind of domain is that noetherian and Krull domains are Mori. Thus, the results obtained hold in particular for noetherian domains and moreover, since a Krull domain is a Mori completely integrally closed domain, this study puts in evidence which properties of Krull domains depend uniquely on the ascending chain condition on divisorial ideals and which ones depend also on the condition of being completely integrally closed.

In this paper we will survey recent and older results on this subject.

Let R be an integral domain and K its quotient field.

An ideal I of R is *divisorial* if $I = I_v := R:(R:I) = \cap\{xR : x \in K , xR \supset I\}$ and a divisorial ideal I is v-*finite* if there is a finitely generated ideal J such that $I = J_v$. We denote by D(R) the set of all divisorial ideals of R and by $D_f(R)$ the set of all v-finite ideals of R . The sets D(R) and $D_f(R)$ are semigroups, with unit R , with respect to the operation $I*J = (IJ)_v$. The group P(R) of principal ideals of R is a subgroup of $D_f(R)$ and

$D_f(A) = P(A)$ if and only if any two elements of R have a greatest common divisor, that is, R is a GCD-domain.

If R is a Krull domain, for example an integrally closed noetherian domain, then $D(R) = D_f(R)$ is a group, in fact a free group on the set of its height one primes, and the class group of R is defined as $C(R) = D(R)/P(R)$. The study of $C(R)$ gives a good deal of information about R [F]. For example, $C(R) = (0)$ if and only if R is a unique factorization domain .

If R is any domain, the semigroup D(R) may not be a group. A domain R such that D(R) is a group is called *completely integrally closed* (for short c.i.c.)[1]. Also, if D(R) is a group, then $D_f(R)$ need not be a group[2]. In case $D_f(R)$ is a group R is called a *Prüfer* v-*multiplication domain* (for short a PvMD). A PvMD is not necessarily c.i.c.[3] and a Krull domain is a c.i.c. PvMD.

In general the class group of a domain R is defined as $C(R) = T(R)/P(R)$, where T(R)[4] is the largest subgroup of $D_f(R)$ [BvZ] .

It is clear that a PvMD R is a GCD domain if and only if $C(R) = (0)$ [Ry]. In particular, if R is a Prüfer domain, that is, if finitely generated ideal is invertible, then T(R) coincides with the group of invertible ideals of R and hence in this case $C(R) = \mathrm{Pic}(R) = (0)$ if and only if R is a Bezout domain.

A Mori domain is a domain in which the ascending chain condition holds for integral divisorial ideals. Noetherian and Krull domains are Mori domains and an intersection with finite character of Mori domains is Mori[5].

When R is a Mori domain, then $D(R) = D_f(R)$[6] and so T(R) is just the group of units of D(R) . A unit of D(R) is called a v-*invertible* ideal.

The larger T(R) is, the closer R is to being c.i.c. , that is, Krull[7]. How far is a Mori domain from being Krull can be measured by means of its maximal divisorial ideals, that is,

---

[1] This is equivalent to the condition that $I:I = R$ for any ideal I of R . A domain R is integrally closed (i.c.) if and only if $I:I = R$ for any finitely generated ideal I of R . Thus a c.i.c. domain is i.c. and the converse holds if R is noetherian.

[2] If, for example, R is a one-dimensional quasilocal c.i.c. domain that is not a valuation domain (cf. [N1] and [R]), then $D_f(R)$ is not a group.

[3] Take for example a valuation domain of dimension greater than one.

[4] We use this notation because T(R) equals the group of t-invertible t-ideals of R [BvZ].

[5]A finite intersection of noetherian domains need not be noetherian. For example, let F be a field and X, Y indeterminates over F . Consider $R_1 := F(X^2) + YF(X)[[Y]]$ and $R_2 := F(X+X^2) + YF(X)[[Y]]$, R_1 and R_2 are noetherian (not integrally closed), but $R_1 \cap R_2 = F + YF(X)[[Y]]$ is not noetherian (and it is integrally closed).

[6]The converse is not true. For example, any divisorial ideal of the domain $R := \mathbb{Z}+X\mathbb{R}[[X]]$ is v-finite, but R does not even have the ascending chain condition on principal ideals.

[7]This observation may lead to the question of whether the complete integral closure of a Mori domain is Krull. The answer is positive in the noetherian case by the Mori-Nagata theorem (1955) and in several other cases (see for example [B2], [BGR1]). However it is negative in general. Indeed the complete integral closure of a Mori domain need be neither c.i.c. nor Mori [Rt1].

those ideals which are maximal among the integral divisorial ideals. Indeed, denoting by $D_m(R)$ the set of the maximal divisorial ideals[8] of R, we have:

PROPOSITION 1 [BG1] Let R be a Mori domain. Then:

(a) An ideal in $D_m(R)$ is a prime ideal;

(b) If P is a prime ideal, then either P is divisorial of height one or $P(R:P) = P$.

PROPOSITION 2 [BG2] Let R be a Mori domain and P a prime ideal. Then P is v-invertible if and only if $P \in D_m(R)$ and R_P is a DVR (in this case height $P = 1$).

An ideal I of a domain R such that $I(R:I) = I$ is a *strong* ideal. A strong divisorial ideal is also called *strongly divisorial*. A strongly divisorial ideal of a Mori domain can have any height. For example if $R := \mathbb{Q}+\mathbf{X}\mathbb{R}[\mathbf{X}]$, where $\mathbf{X} := \{X_1, \dots, X_n\}$ is a set of indeterminates, then the ideal $P := \mathbf{X}\mathbb{R}[\mathbf{X}]$ is a strongly divisorial ideal of R of height n. An example of a Mori domain with an infinite decreasing chain of (strongly) divisorial prime ideals is given in [HLV].

Denote by $\mathcal{I}(R)$ the set of the v-invertible divisorial prime ideals of a Mori domain R and by $\mathcal{S}(R)$ the set of the strong maximal divisorial ideals of R. Then $\mathcal{I}(R) \subset D_m(A)$, $D_m(A) = \mathcal{I}(R) \cup \mathcal{S}(R)$ and $\mathcal{I}(R) \cap \mathcal{S}(R) = \emptyset$. It is clear that R is a Krull domain if and only if $\mathcal{S}(R) = \emptyset$, that is $D_m(A) = \mathcal{I}(R)$. We say that R is *strongly Mori* if $\mathcal{I}(R) = \emptyset$, that is, $D_m(R) = \mathcal{S}(R)$ or equivalently any divisorial prime is strong. A strongly Mori domain is the farthest possible from being a Krull domain (in the noetherian case from being integrally closed).

THEOREM 3 [BG1] If R is Mori, then:

(a) $R = \cap\{R_P : P \in D_m(R)\}$ with finite character;

(b) $R_1 := \cap\{R_P : P \in \mathcal{I}(A)\}$ is a Krull domain, $R_2 := \cap\{A_Q : Q \in \mathcal{S}(A)\}$ is a strongly Mori domain and $R = R_1 \cap R_2$;

(c) $D_m(R_1) = \{PR_P \cap R_1 : P \in \mathcal{I}(R)\}$ and $D_m(R_2) = \{QA_Q \cap R_2 : Q \in \mathcal{S}(R)\}$;

(d) $T(R) = T(R_1) \oplus T(R_2)$.

Observe that, since R_1 is Krull, then $T(R_1) = D(R_1)$ is free on the set of its height-one primes, equivalently on the set $\mathcal{I}(R)$. Moreover, if a divisorial ideal I of R is contained just in v-invertible maximal divisorial ideals, then I behaves like a divisorial ideal in a Krull domain. For example $I = P_1^{(e_1)} \cap \dots \cap P_n^{(e_n)}$, where the P_i's are the maximal divisorial

[8] In general $D_m(A)$ can be empty. For example, when A is a one-dimensional valuation domain with a non-principal maximal ideal, then $D_m(A) = \emptyset$.

ideals containing I . On the other hand, when R is strongly Mori, then in T(R) there are no primes, because a strongly Mori domain has no v-invertible divisorial primes.

A decomposition similar to the one expressed in Theorem 3 (d) does not hold for C(R) . Indeed we can have $C(R_1) = C(R_2) = (0)$ while $C(R) \neq (0)$ [G] .

A relation among the class groups of R , R_1 , and R_2 is given by the two exact sequences

$$0 \longrightarrow T(R_1)/H(R_1) \longrightarrow C(R) \longrightarrow C(R_2) \longrightarrow 0$$
$$0 \longrightarrow T(R_2)/H(R_2) \longrightarrow C(R) \longrightarrow C(R_1) \longrightarrow 0$$

where $H(R_1)$ and $H(R_2)$ consist of principal ideals [BG2].

It follows that, if $C(R) = C(R_1)$ then $C(R_2) = (0)$. Similarly, when $C(R) = C(R_2)$ then $C(R_1) = (0)$ [BGR2].

Two related questions arise:

QUESTION 1 When is $C(R) = (0)$?

QUESTION 2 When is $C(R) = C(R_1)$ or $C(R) = C(R_2)$?

It is interesting to observe that, in case every divisorial maximal ideal of R has height one, the condition $C(R) = (0)$ means that every non-invertible element of R can be expressed as a product of primary elements [AZ]. The meaning of this condition in the general case is not known.

A first answer to Question 1 is the following.

THEOREM 4 [BG2] If a Mori domain R has just finitely many maximal divisorial ideals, then R is semilocal and $C(R) = (0)$.

This theorem generalizes the well known result that a semilocal Dedekind domain is factorial.

As for Question 2, an answer to the second case is the following.

THEOREM 5 [BGR 2] If R is Mori, the following are equivalent:

(i) P is principal for every $P \in \mathcal{I}(R)$;

(ii) $T := R \setminus \cup\{Q : Q \in \mathcal{S}(R)\}$ is generated by primes and $T^{-1}R = R_2$;

(iii) $C(R) = C(R_2)$.

Under these conditions, we have $C(R_1) = (0)$. However the example in [G] shows that we can have $C(R_1) = (0)$ while $C(R) \neq C(R_2)$.

COROLLARY 6 If R is Mori, then $C(R) = (0)$ if and only if $C(R_2) = (0)$ and P is principal for every $P \in \mathcal{I}(R)$.

COROLLARY 7 If R is Mori and has a finite number of strong maximal divisorial ideals, then C(R) = (0) if and only if P is principal for every $P \in \mathcal{I}(R)$.

The case $Cl(R) = C(R_1)$ is in some way more interesting to study, because R_1 is a Krull domain and the structure of the class group of a Krull domain is well known.

A necessary condition in order to have $C(R) = C(R_1)$ is that $C(R_2) = (0)$. This condition is certainly satisfied when $\mathcal{S}(R)$ is finite by Theorems 3 (c) and 4 .

What happens when R has an infinite number of strong maximal divisorial ideals?

This investigation has led to the following result

THEOREM 8 [BGR2] Given any abelian group G , there is an integrally closed strongly Mori domain S such that C(S) = G .

The theorem is proved constructively. Given any integrally closed Mori domain R with class group G, we construct an integrally closed strongly Mori domain S with the same class group of R . We recall that a Dedekind domain with class group G exists by Claborn's Theorem [C].

Observe that:

- There is no way to compute directly the class group of a (strongly) Mori domain, even though in some cases it is possible to compare the class groups of two given domains ([A], [ARy], [G], [GRt]);

- Mori domains with infinitely many strong maximal divisorial ideals have zero conductor in their complete integral closure. Very few explicit examples are known of this kind of domains. For the noetherian case an example is given in [N2], while the domains we construct are not noetherian, being integrally closed and not Krull.

To construct a strongly Mori domain starting with a given Mori domain R , we have to convert the v-invertible divisorial primes of R into strongly divisorial primes.

In [BG1] a method is given to do that using the so called *Mori pull-backs*.

Take a Mori domain R and a v-invertible divisorial prime P . Let $V := R_P$ and $M := PR_P$ its maximal ideal. Then V is a DVR . Let $\pi : V \longrightarrow V/M$ be the canonical projection and let k be a subfield of V/M .

Then, $T := \pi^{-1}(k)$ is a one-dimensional quasilocal Mori domain and its maximal ideal M is strongly divisorial [B1].

We have:

$$\begin{array}{ccc} T := \pi^{-1}(k) & \longrightarrow & k \\ \downarrow & & \downarrow \\ \pi : V & \longrightarrow & V/M \end{array}$$

Consider $R' := R \cap T$. Then T is a Mori domain as an intersection of two Mori domains. Moreover

- the v-invertible divisorial primes of R' are the ideals $Q \cap R' = Q \cap T$, with Q a v-invertible divisorial prime of R and $Q \neq P$;
- the strongly divisorial primes of R' are the ideals $Q \cap R' = Q \cap T$, with Q a strongly divisorial prime of R and also $P' = M \cap R'$.

Note that R' is a pull-back of R. Indeed, if $\mu : R\backslash P \longrightarrow R_P/PR_P = V/M$ is the canonical injection, we have:

$$\begin{array}{ccccc} R' := \pi^{-1}\mu^{-1}(k) & \longrightarrow & \mu^{-1}(k) & \longrightarrow & k \\ \downarrow & & \downarrow & & \downarrow \\ R & \longrightarrow & R/P & \longrightarrow & R_P/PR_P = V/M \end{array}$$

Under more restrictive conditions, this costruction can be carried out simultaneously for any set of v-invertible primes [BG1]. However, if we try to build with this method a strongly Mori domain starting from a Mori domain with infinitely many v-invertible primes, the needed conditions may not be satisfied.

For example, take $R = \mathbb{C}[X]$ and, for any P of height one of R, consider the pull-back $\mathbb{R}+PR_P$ of R_P. The intersection of all these pull-backs gives $\mathbb{R}$ and so the construction fails.

To construct a strongly Mori domain S starting from any integrally closed Mori domain R and to control at the same time its class group, we modified the previous construction in order to get $R \subset S \subset R[\mathbf{X}]$, where $\mathbf{X}$ is a suitable set of infinitely many indeterminates.

The hypothesis that R is integrally closed is needed because in this case $R[\mathbf{X}]$ is still a Mori domain and moreover $C(R) = C(R[\mathbf{X}])$[9] [Q].

The first step of the construction is the following [BGR2]:

Let R be an integrally closed Mori domain with quotient field K, and let P be a maximal divisorial ideal of R. Set $T := R + P[X]$, where X is an indeterminate over R. Observe that T is the pull-back of the injection $R/P \longrightarrow (R/P)[X]$ with respect to the natural surjection $R[X] \longrightarrow R[X]/P[X]$ (via the natural isomorphism $(R/P)[X] \approx R[X]/P[X]$) :

[9]If R is a Mori domain, then $R[X]$ is not always Mori [Rt2]. On the other hand, given any domain R, we have $C(R) = C(R[\mathbf{X}])$ if and only if R is integrally closed [G].

$$\begin{array}{ccc} T & \longrightarrow & R/P \\ \downarrow & & \downarrow \\ R[X] & \longrightarrow & (R/P)[X] \approx R[X]/P[X] \end{array}$$

In this situation:

(a) T is an integrally closed Mori domain;

(b) The maximal divisorial ideals of T intersecting R in a proper ideal are all the ideals of type $Q^* := (QT)_v$, where Q is a maximal divisorial ideal of R different from P , and also P[X] .

Moreover the strong maximal divisorial ideals of T are the ideals Q^* , where Q is strongly divisorial, and P[X] ;

(c) Cl(R) = Cl(T) .

Hence, in this way, if P is a v-invertible prime of R , then P[X] is a strongly divisorial prime of T . However T has more v-invertible maximal divisorial ideals than R , namely the maximal divisorial ideals which intersect R in the zero ideal. These ideals are of type $fK[X] \cap R$, with f an irreducible polynomial.

Nevertheless, if Q is a v-invertible divisorial prime ideal of R , then Q^* is v-invertible in T and we can repeat this construction for T and Q^* .

Set S(R; P) := T , and S(T; Q*) = S((S(R;P) ; Q*) := S(R; P, Q) .

By induction, we can define for any set $P_1,\ldots, P_n$ of v-invertible divisorial primes of R , the domain $S(R; P_1,\ldots, P_n)$. Clearly the family of all these domains is directed and we set $S(R) := \cup\{(S(R; P_1,\ldots, P_n)\}$. We have $R \subset S(R) \subset R[\mathbf{X}]$ where $\mathbf{X} = \{X_P ; P\in \mathcal{I}(R)\}$ is a set of indeterminates.

Then:

(a) S(R) is an integrally closed Mori domain;

(b) All the maximal divisorial ideals of S(R) intersecting R in a proper ideal are strongly divisorial;

(c) Cl(R) = Cl(S(R)) .

To prove the last statement, we have first established some technical results concerning the class groups of a directed family of domains and that of their union.

S(R) is not strongly Mori, because the maximal divisorial ideals intersecting R in the zero ideal are v-invertible.

To get a strongly Mori domain, we have to repeat again the construction for S(R).

Set $S^1(R) := S(R)$, $S^n(R) := S(S^{n-1}(R))$ for $n \geq 1$ and consider $S^\infty(R) = \cup\{(S^n(R)$; $n \geq 1\}$. Then:

(a) $S^\infty(R)$ is an integrally closed Mori domain;

(b) All the maximal divisorial ideals of $S^\infty(R)$ are strongly divisorial;

(c) $Cl(R) = Cl(S^\infty(R))$.

To prove that $S^\infty(R)$ is a strongly Mori domain, we observe that every maximal divisorial ideal of $S^\infty(R)$ intersects $S^n(R)$ for some $n \geq 1$ and so is strongly divisorial in $S^m(R)$, for any $m > n$.

It is still an open problem to contruct a noetherian strongly Mori domain with a given class group.

REFERENCES

[A] D. F. Anderson: A general theory of class groups, *Comm. Algebra* **16** (1988), 805-847.

[AA] D. D. Anderson and D.F. Anderson, Some remarks on star operations and the class group, *J. Pure Appl. Algebra* **16** (1988), 805-847.

[AAZ] D. D. Anderson, D. F. Anderson and M.Zafrullah, Splitting the t-class group, *J. Pure Appl. Algebra* **74** (1991), 17-37.

[AZ] D.D. Anderson and M. Zafrullah, Weakly factorial domains and groups of divisibility, *Proc. Amer. Math. Soc.* **109** (1990), 907-913.

[ARy] D. F. Anderson and A. Ryckaert: The class group of D+M , *J. Pure Appl. Algebra* **52** (1988), 199-212.

[B1] V. Barucci : On a class of Mori domains, *Comm. Algebra* **11** (1983), 1989-2001.

[B1] V. Barucci : Seminormal Mori domains, preprint.

[BG1] V. Barucci and S. Gabelli: How far is a Mori domain from being a Krull domain? *J. Pure Appl. Algebra* **45** (1987), 101-112.

[BG2] V. Barucci and S. Gabelli: On the class group of a Mori domain, *J. Algebra* **108** (1987), 161- 173.

[BGR1] V. Barucci, S. Gabelli and M. Roitman: On semi-Krull domains, *J. Algebra* **145** (1992), 306-328.

[BGR2] V. Barucci, S. Gabelli and M. Roitman: The class group of a strongly Mori domain, *Comm. Algebra*, to appear.

[Bv] A. Bouvier: Le groupe des classes d'un anneau intègre, 107 Congrès national des sociétés savantes, Brest (1982), Sciences, fasc. IV, 85-92.

[BvZ] A. Bouvier and M. Zafrullah: On some class groups of an integral domain, *Bull. Soc. Math. Grèce* (N.S.) **29** (1988), 45-59.

[C] L. Claborn: Every abelian group is a class group, *Pac. J. Math.* **18** (1966), 219-222.

[F] R. Fossum: *The divisor class group of a Krull domain*, Springer-Verlag, 1973.

[G] S. Gabelli: On divisorial ideals in polynomal rings over Mori domains, *Comm. Algebra* **15** (1987), 2349-2370.

[GRt] S. Gabelli and M. Roitman: On Nagata's theorem, *J. Pure Appl. Algebra* **66** (1990), 31-42.

[HLV] E. G. Houston, T. G. Lucas and T. M. Viswanathan: Primary decomposition of divisorial ideals in Mori domains, *J. Algebra* **117** (1988), 327-342.

[N1] M. Nagata: On Krull's conjecture concerning valuation rings, *Nagoya Math. J.* **4** (1952), 29-33.

[N2] M. Nagata: *Local rings*, Interscience, New York, 1962.

[NA] D. Nour el Abidine: *Groupe des classes de certains anneaux intègres et idéaux transformes*, Thèse, Université Claude Bernard, Lyon 1, 1992.

[Q] J. Querré: Idéaux divisoriels d'un anneau de polynomes, *J. Algebra* **64** (1980), 270-284.

[R] P. Ribenboim: Sur une note de Nagata relative à un problème de Krull, *Math Z.* **64** (1956), 159-168.

[Rl] N. Raillard (Dessagne): *Sur les anneaux de Mori*, Thèse, Paris VI, 1976.

[Rt1] M. Roitman: On the complete integral closure of a Mori domain, *J. Pure applied Algebra* **66** (1990), 55-79.

[Rt2] M. Roitman: On polynomial extensions of Mori domains over countable fields, *J. Pure Appl. Algebra* **64** (1990), 315-328.

[Ry] A. Ryckaert, *Sur le groupe des classes d'un anneaux intègre*, Thèse, Université Claude Bernard, Lyon 1, 1986.

14
t-Invertibility and Comparability

ROBERT GILMER Department of Mathematics, Florida State University, Tallahassee, FL 32306-3027 USA. email: gilmer@math.fsu.edu.

JOE MOTT Department of Mathematics, Florida State University, Tallahassee, FL 32306-3027 USA. email: mott@math.fsu.edu.

MUHAMMAD ZAFRULLAH* Department of Mathematics, Winthrop College, Rock Hill, SC 29733 USA.

Let D be an integral domain with quotient field K and let $F(D)$ denote the set of nonzero fractional ideals of D. For $A \in F(D)$ we define $A^{-1} = \{x \in K | xA \subseteq D\}$. An ideal $A \in F(D)$ is t-invertible if (i) $(AA^{-1})^{-1} = D$, (ii) there exists $x_1, x_2, \ldots, x_n \in A$ such that $(x_1, \ldots, x_n)^{-1} = A^{-1}$, and (iii) $A^{-1} = ((y_1, \ldots, y_m)^{-1})^{-1}$ for some $y_1, \ldots, y_m \in K$. The notion of t-invertibility has been useful in characterizing Krull domains in various ways [HZ], [J], [K] and [MZ].

Houston and Zafrullah in [HZ] discussed situations in which t-invertibility arises naturally. They showed that if a nonzero integral ideal A of a polynomial ring $D[X]$ contains a polynomial f with $A_f^{-1} = D$, then A is t-invertible. Here A_f denotes the content of $f(X)$. This result led us to ask, in a general setting, what nonzero elements x of a domain D are *t-invertibility elements* of D in the sense that each ideal of D containing x is t-invertible. We obtain a characterization of such elements (and more generally, of the so-called t-invertibility ideals of D; see Section 1 for the definition) in Theorem 1.3. The proof of (1.3) uses some of the following terminology and results; our main references are [G, Sections 32, 34] and [AA].

I. A star operation is a function $F \to F^\star$ from $F(D)$ to $F(D)$ with the following properties: if $A, B \in F(D)$ and $a \in K \setminus \{0\}$, then
   (i) $(a)^\star = (a)$ and $(aA)^\star = aA^\star$.
   (ii) $A \subseteq A^\star$ and if $A \subseteq B$, then $A^\star \subseteq B^\star$.
   (iii) $(A^\star)^\star = A^\star$.

The mapping $A \to (A^{-1})^{-1} = A_v$ on $F(D)$ is a star operation called the *v-operation* on D. Similarly, $A \to A_t = \cup F_v$, where F ranges over the set of finitely generated

* The third author spoke at the Fez Conference on a preliminary version of this paper, entitled *t-Invertibility*. This paper extends results of that preliminary version.

nonzero subideals of A, is a star operation called the *t-operation* on D. The identity mapping $F \to F$ on $F(D)$ is a star operation called the *d-operation* on D. A star operation $F \to F^\star$ is said to be *of finite character* if $A^\star = \cup_i A_i^\star$ for each $A \in F(D)$, where $\{A_i\}$ is the family of nonzero finitely generated fractional ideals of D contained in A. Given any star operation $F \to F^\star$ on D, the function $A \to A^{\star_s} = \cup_i A_i^\star$, where $\{A_i\}$ is the family of nonzero finitely generated fractional ideals of D contained in A, is a star operation of finite character on D, and $B^\star = B^{\star_s}$ for each finitely generated $B \in F(D)$. The d-operation on D is of finite character, and the t-operation is the v_s-operation.

II. For all $A, B \in F(D)$ and for each star operation $\star$, $(AB)^\star = (AB^\star)^\star = (A^\star B^\star)^\star$, $(A^{-1})^\star = A^{-1}$, and $(A^\star)_v = A_v = (A_v)^\star$.

III. An ideal A of D is said to be a $\star$-ideal, for a given star operation $\star$, if $A^\star = A$. Thus from II above $A^\star$, A^{-1} and A_v are $\star$-ideals. Likewise we define v-ideals and t-ideals and note that $A^\star \subseteq A_v$ for any star operation $\star$.

IV. An integral ideal A of D is said to be a maximal $\star$-ideal, for a given star operation $\star$, if $A \neq D$ and A is maximal with respect to being a $\star$-ideal. If $\star$ is of finite character, then every proper integral $\star$-ideal is contained in a maximal $\star$-ideal. A maximal $\star$-ideal is necessarily a prime ideal [J, p. 30].

V. Let $\{D_i\}_{i \in I}$ be a family of overrings of D such that $D = \bigcap_{i \in I} D_i$. Then the operation on $F(D)$ defined by $A \mapsto \cap AD_i$ is a star operation. This star operation is said to be *induced by* $\{D_i\}$ [G], [A].

VI. An ideal $A \in F(D)$ is *$\star$-invertible* if there exists $B \in F(D)$ such that $(AB)^\star = D$. In this case, $B^\star = A^{-1}$. This yields the definitions of v-invertibility and t-invertibility of A. The definition of the t-operation implies that if A is t-invertible, then there exists a finitely generated ideal B contained in AA^{-1} such that $B_t = B_v = D$; hence the definition of t-invertibility in the first paragraph.

In Section 2 we consider the notion of a *comparable element* of a domain D, as defined by Anderson and Zafrullah in [AZ2] (the definition is that $d \in D\backslash(0)$ is comparable if (d) compares with each ideal of D under inclusion). Theorem 2.3 characterizes domains that contain a nonunit comparable element. In Section 3 we relate the concepts of t-invertibility and t-local comparability.

1. A CHARACTERIZATION OF t-INVERTIBILITY ELEMENTS AND IDEALS

Our main goal in this section is to characterize elements or ideals of a domain D such that each ideal of D in which they are contained is t-invertible. We achieve this goal in Theorem 1.3. Our first result deals, however, with ordinary invertibility of ideals of a commutative ring.

THEOREM 1.1. *Suppose A is an ideal of the commutative unitary ring R. Each ideal of R containing A is invertible if and only if A is a finite product of invertible maximal ideals of R.*

Proof. Suppose each ideal of R containing A is invertible. Since invertible ideals are finitely generated, R/A is Noetherian. Thus if some ideal of R containing A is not a finite product of maximal ideals, there is an ideal B maximal with respect to this property. Then B is not maximal, so let M be a maximal ideal of R containing B. Since M is invertible, $B = MC$ for some ideal $C \supseteq B$. Because B is invertible and

$M \neq R$, we have $B < C$. Hence C is a finite product of maximal ideals, and so is B. This contradiction shows that each ideal of R containing A is a finite product of invertible maximal ideals, and in particular, A has this property.

Conversely, if $A = M_1 M_2 \cdots M_k$ is a finite product of invertible maximal ideals M_i, then the ideals M_i are the only prime ideals of R that contain A. Hence each prime ideal of R containing A is invertible, and this implies, in the usual fashion (cf. [Kp, Exer. 10, p. 11]), that each ideal of R containing A is invertible. ■

REMARK 1.2. If the conditions of Theorem 1.1 are satisfied, then A is invertible and we can see that R/A is a zero-dimensional PIR by the following argument: That R/A is zero-dimensional with only finitely many maximal ideals is clear; to see that R/A is a PIR, it suffices to show that M/A is principal for each maximal ideal M of R containing A. Let $\{M_i\}_{i=1}^n$ be the (finite) set of maximal ideals of R distinct from M. Choose $m \in M - [M^2 \cup (\cup_{i=1}^n M_i)]$. Then (A, m) is invertible, so $(A, m) = MB$ for some ideal B of R containing A. Moreover, B is contained in neither M nor any M_i by choice of m. Hence $B = R$, $M = (A, m)$, and M/A is principal as we wished to show.

The converse of the observation just made fails. For example, if t is an indeterminate over the field K and if $R = K[t^2, t^3]$, then $A = t^2 R$ is invertible and $R/A \simeq K[X]/(X^2)$ is a zero-dimensional PIR. However, the ideal (t^2, t^3) of R contains A, but is not invertible.

The proof of Theorem 1.1 can be adapted to give a characterization of t-invertibility elements, as indicated in Theorem 1.3.

THEOREM 1.3. *Suppose A is a nonzero proper ideal of the domain D. Each ideal of D containing A is t-invertible if and only if A_t is a finite t-product of maximal t-ideals of D, each of which is t-invertible.*

Proof. Suppose each ideal of D containing A is t-invertible. We show that each t-ideal B of D containing A is a finite t-product of maximal t-ideals, each of which is t-invertible. Because the t-operation is of finite character, a t-invertible t-ideal is of finite type [J, p. 30], [Kr], so each ideal of D containing A is of finite t-type. Moreover, the union of a chain of t-ideals is again a t-ideal. It then follows by the usual argument that the set of t-ideals of D containing A is inductive under $\subseteq$. Hence, if there exists a t-ideal B of D containing A such that B is not a finite t-product of maximal t-ideals, there is a maximal such B. Then B is not a maximal t-ideal. Let M be a maximal t-ideal containing B. If $C = BM^{-1}$, then C is an integral ideal of D containing B, and because M is t-invertible, $B = (MC)_t$. We cannot have $B = C$, for t-invertibility of B would then imply that $M = M_t = D$. Hence C_t is a t-product of maximal t-ideals, as is B. This contradiction establishes the desired assertion, and in particular, A is a finite product of t-invertible maximal t-ideals.

Conversely, if $A_t = (M_1 M_2 \cdots M_k)_t$ is a finite t-product of t-invertible maximal t-ideals M_i, then since $M_1 \cdots M_k \subseteq A_t$, the ideals M_i are the only proper prime t-ideals of D that contain A_t. Hence each prime t-ideal of D containing A is t-invertible, and this implies [MZ, Prop. 2.1] that each ideal of D containing A is t-invertible. ■

An examination of the proof of Theorem 1.3 reveals that the only special property of the t-operation, among star operations, used in the proof is that the t-operation is of finite character. Hence Theorem 1.3 generalizes to the following result.

THEOREM 1.3′. *Suppose $F \to F^\star$ is a star operation of finite character on the domain D. Let A be a nonzero proper ideal of D. Each ideal of D containing A is $\star$-invertible if and only if A is a finite $\star$-product of $\star$-invertible maximal $\star$-ideals.*

Since the t-operation and the d-operation on a domain D are of finite character, Theorem 1.3 and the case of Theorem 1.1 where R is an integral domain could be obtained from Theorem 1.3′. Theorem 1.1 itself seems, however, not to be a direct corollary of Theorem 1.3′.

Suppose $F \to F^\star$ is a star operation of finite character on D. We call a nonzero ideal A of D a *$\star$-invertibility ideal* if either $A^\star = D$ or else A satisfies the equivalent conditions of Theorem 1.3′; an element d of D is a *$\star$-invertibility element* if (d) is a $\star$-invertibility ideal. Theorem 1.3′ gives rise to the following corollary.

COROLLARY 1.4. *Let the notation and hypothesis be as in the preceding paragraph and let S be the set of $\star$-invertibility ideals of D. Then S is closed under multiplication of ideals and under taking overideals within D. In particular, the set S of $\star$-invertibility elements of D is a saturated multiplicative system in D.*

Proof. Theorem 1.3′ implies that S is closed under multiplication, and it follows from the proof of Theorem 1.3 that if $A \in$ S and if B is an overideal of A in D, then $B \in$ S.∎

REMARK 1.5. If $F \to F^\star$ is a $\star$-operation on D of finite character, the proof of Theorem 1.3 can easily be adapted to show that a nonzero ideal A of D is a $\star$-invertibility ideal if and only if A satisfies the following three conditions: (i) A belongs to only finitely many maximal t-ideals $M_1, \ldots, M_s$; (ii) each M_i is t-invertible; (iii) each M_i is minimal in the set of prime t-ideals of D containing A. From this characterization and from Corollary 1.4 it follows that if S is the set of $\star$-invertibility ideals of D and if A and B are ideals of D, then (a) $\mathrm{rad}(A) \in$ S if and only if $A \in$ S, and (b) $A \cap B \in$ S if and only if $A, B \in$ S.

Houston and Zafrullah [HZ] proved a special case of the "only if" part of Theorem 1.3. To wit, let X be an indeterminate over D and let $f(X) \in D[X]$ be such that $A_f^{-1} = D$. Then according to [HZ]:

1. If A is an (integral) ideal of $D[X]$ with $f(X) \in A$, then A is t-invertible.
2. Every prime upper to (0) in $D[X]$ that contains $f(X)$ is a t-invertible t-ideal, and hence a maximal t-ideal.
3. Using (1), (2), and some localization, it follows that $(f(X)) = P_1^{(e_1)} \cap \ldots \cap P_r^{(e_r)} = (P_1^{e_1} \cdots P_r^{e_r})_t$, where $\{P_1, \ldots, P_r\}$ is the set of prime uppers to (0) that contain $f(X)$.

From these observations we conclude:

(A) If $(A_f)_v = D$, then $(f(X))$ is a t-product of primes.

(B) If $(A_f)_v = D$, then $f(X)$ belongs to a finite number of maximal t-ideals $P_1, P_2, \ldots, P_r$, and for each of these, $D[X]_{P_i}$ is a discrete rank-one valuation domain. (Because the ideals P_i are prime uppers to (0).)

In our terminology, (1) merely says that f is a t-invertibility element of $D[X]$ if $A_f^{-1} = D$. We proceed to show, in an abstract setting, that an element of a domain satisfying conditions analogous to those of (B) is a t-invertibility element. For the sake of brevity, we say that a nonzero element d of a domain D is a *t-element* of D if either d is a unit of D, or else d belongs to only finitely many maximal t-ideals $P_1, \ldots, P_r$ of

D and each D_{P_i} is a rank-one discrete valuation domain. We show in Theorem 1.6 that t-elements are t-invertibility elements. Recall that an ideal A of D is *strictly v-finite* if there exists a finitely generated ideal $B \subseteq A$ such that $A_v = B_v$.

THEOREM 1.6. *If d is a t-element of the domain D, then d is a t-invertibility element of D.*

Proof. The statement is clear if d is a unit of D. If d is a nonunit, we show that each ideal A of D containing d is t-invertible. Thus, let $P_1, \ldots, P_r$ be the maximal t-ideals of D containing d. If $\{P_\alpha\}$ is the set of all maximal t-ideals of D, then $D = \bigcap_\alpha D_{P_\alpha}$ [Gr, Prop. 4], and AD_{P_i} is principal for $1 \leq i \leq r$. Hence Lemma 1.10 of [MMZ] shows that A is strictly v-finite, and [MMZ, Cor. 1.16] then implies that A is t-invertible. ∎

The converse of Theorem 1.6 fails. For example, in the classical $D + M$ construction $\mathbb{Z} + X\mathbb{Q}[[X]]$, 2 is a t-invertibility element, but not a t-element. The problem in this case is that the unique maximal t-ideal containing 2 has height greater than 1. We show in Theorem 1.9 that a t-invertibility element without this defect is a t-element.

LEMMA 1.7. *Suppose A is a t-invertible ideal of the domain D. If S is a multiplicative system in D, then AD_S is t-invertible in D_S.*

Proof. Both A and A^{-1} are of finite type — say $A_t = B_t$ and $(A^{-1})_t = C_t$, where B and C are finitely generated subideals of A and A^{-1}, respectively. Then by Lemma 4 of [Z], $(BCD_S)_t = (BCD_S)_v = ((BC)_v D_S)_v = (DD_S)_v = D_S$. Hence BD_S is t-invertible. Again using [Z, Lemma 4], we have $(AD_S)_t \subseteq (A_t D_S)_t = (B_t D_S)_t = (BD_S)_t \subseteq (AD_S)_t$. Therefore $(AD_S)_t = (BD_S)_t$ and AD_S is t-invertible, as asserted. ∎

LEMMA 1.8. *Let (D, M) be a one-dimensional quasilocal domain. If M is t-invertible, then D is a* DVR.

Proof. Because M is t-invertible, it is of finite type. Also, M is the unique maximal t-ideal of D. Corollary 1.6 of [MMZ] then implies that $MD_M = M$ is principal, so D is a DVR, as we wished to show. ∎

THEOREM 1.9. *Suppose d is a t-invertibility element of the domain D such that $(d) = (P_1P_2 \cdots P_r)_t$, where each P_i is a maximal t-ideal of D of height one. Then d is a t-element of D.*

Proof. It is clear that $P_1, \ldots, P_r$ are the only maximal t-ideals of D containing d. To see that each D_{P_i} is a DVR, we observe that P_i is t-invertible since $(P_1 \cdots P_r)_t = (d)$ is t-invertible. Lemma 1.7 shows that $P_i D_{P_i}$ is t-invertible in D_{P_i}, and Lemma 1.8 then implies that D_{P_i} is a DVR. ∎

To return briefly to results of [HZ], we remark that while each $f \in D[X]$ such that $A_f^{-1} = D$ is a t-invertibility element of $D[X]$, there may exist others. For example, if $D[X]$ is a Krull domain (that is, if D is a Krull domain), then each nonzero element of $D[X]$ is a t-invertibility element. (Conversely, if E is an integral domain with set $E - \{0\}$ of t-invertibility elements, then E is a Krull domain.)

2. COMPARABLE ELEMENTS

Suppose D is an integral domain. Recall that Anderson and Zafrullah in [AZ2] called a nonzero element d of D *comparable* if (d) compares with each ideal of D under inclusion. In Theorem 3.1 we establish a connection between the concepts of comparability and t-invertibility, but this section is devoted primarily to a determination of those domains D that admit a nonunit comparable element (Theorem 2.3). We record the following useful result of Anderson and Zafrullah that is established in the proof of Theorem 2 of [AZ2].

PROPOSITION 2.1. (Anderson-Zafrullah) *The set of comparable elements of an integral domain D is a saturated multiplicative system in D.*

Before proving Theorem 2.3, we state and prove Proposition 2.2. While this result should be known, we have been unable to locate a satisfactory reference in the literature (cf. [ABDFK, Lemma 2.1]).

PROPOSITION 2.2. *Suppose (D, M) is a quasilocal domain. Let $\varphi : D \to D/M$ be the canonical homomorphism, and let (J, P) be a quasilocal subring of D/M. Then $\varphi^{-1}(J) = R$ is quasilocal with maximal ideal $\varphi^{-1}(P)$.*

Proof. It is clear that $\varphi^{-1}(P)$ is a maximal ideal of R, that $M \subseteq \varphi^{-1}(P)$, and that $R \subseteq D$. We note that $1 + M \subseteq \mathrm{U}(R)$, the set of units of R. To see this, let $x = 1 + m$, where $m \in M$; x is a unit of D and its inverse y is necessarily of the form $1 + n$ for some $n \in M$ since $\varphi(x) = \varphi(1)$. Hence $y \in R$, and x is a unit of R, as asserted. To show $\varphi^{-1}(P)$ is the unique maximal ideal of R, take $r \in R - \varphi^{-1}(P)$. Then $\varphi(r) \in J - P$, so $\varphi(r)$ is a unit of J. Thus there exists $s \in R$ such that $\varphi(rs) = \varphi(1)$, so $rs \in 1 + M \subseteq \mathrm{U}(R)$. Therefore $r \in \mathrm{U}(R)$, as we wished to show. ∎

In the statement of Theorem 2.3, we call an ideal A of a ring R a *comparable ideal* of R if A compares with each ideal of R under inclusion.

THEOREM 2.3. *Suppose the integral domain D contains a nonzero nonunit comparable element; let Y be the set of nonzero comparable elements of D. Then:*
(1) $P = \cap\{(c) \mid c \in Y\}$ is a prime ideal of D, and $D \backslash P = Y$.
(2) D/P is a valuation domain.
(3) $P = PD_P$.
(4) D is quasilocal, P is a comparable ideal of D, and $\dim D = \dim(D/P) + \dim(D_P)$.

Moreover, if J is any integral domain such that there exists a nonmaximal prime ideal Q of J such that (a) J/Q is a valuation domain, and (b) $Q = QJ_Q$, then each element of $J \backslash Q$ is comparable. If, in addition, Q is minimal with respect to properties (a) and (b), then $J \backslash Q$ is the set of nonzero comparable elements of J.

Proof. To prove (1), choose $x, y \in D \backslash P$. There exist elements c and d of Y such that $x \notin (c)$ and $y \notin (d)$. Hence $(x) > (c)$, $(y) > (d)$, and consequently, $(xy) > (cy) > (cd)$. Because $cd \in Y$ by Proposition 2.1, it follows that $xy \notin P$, so P is prime in D. We note that if $x \in D \backslash P$, then x divides an element of Y, and because the multiplicative system Y is saturated, $x \in Y$. Therefore $D \backslash P \subseteq Y$. If $y \in Y$, then $(y) > (y^2) \supseteq P$, so $y \notin P$ and the equality $D \backslash P = Y$ holds.

It follows from (1) that the set of principal ideals of D/P is linearly ordered under inclusion. Consequently, D/P is a valuation domain.

To prove (3), take an element a/x of PD_P, where $a \in P$ and $x \in D\backslash P$. Then $(a) < (x)$, so $a = xy$ for some $y \in D$ that is necessarily in P. Thus $a/x = y \in P$ and $PD_P \subseteq P$.

(4): Since D/P and D_P are quasilocal, Proposition 2.2 shows that D is also quasilocal. To conclude that P is comparable, we need only show that $P \subseteq (x)$ for each $x \in D\backslash P$, and this is immediate from (1). Finally, the equality $\dim D = \dim(D/P) + \dim(D_P)$ follows from the fact that each prime ideal of D compares with P under inclusion.

Assume now that J is an integral domain satisfying (a) and (b). If $s \in Q$ and $t \in J\backslash Q$, then $s/t \in QJ_Q = Q$, and hence $s \in tQ \subseteq (t)$. Thus, to show t is comparable, we need only show that (t) compares with (u) for each $u \in J\backslash Q$. This follows because $Q \subseteq (t) \cap (u)$ and because the ideals $(t)/Q$ and $(u)/Q$ of the valuation domain J/Q are comparable. Therefore each element of $J\backslash Q$ is comparable. Let Q_0 be the intersection of the family $\{(b) \mid b \text{ is a nonzero comparable element of } J\}$. It follows from (1) that $J\backslash Q \subseteq J\backslash Q_0$, so $Q_0 \subseteq Q$. Moreover, (2) and (3) show that Q_0 satisfies (a) and (b). Thus, if Q is minimal with respect to satisfying (a) and (b), then $Q = Q_0$ and $J\backslash Q$ is the set of nonzero comparable elements of J. This completes the proof of Theorem 2.3. ∎

Let D, Y, and P be as in the statement of Theorem 2.3. In the literature, the domain D is called the *pullback* in D_P of the domain $D/P = D_P/PD_P$ [D], or the *composite* of D/P and D_P over PD_P [O], [MS]. Theorem 2.3 implies that the domains that admit nonzero nonunit comparable elements are precisely the pullbacks in R of nontrivial valuation domains on R/M, where (R, M) is any quasilocal domain (see also [GO, Prop. 5.1(f)]). Under the notation and hypothesis of Theorem 2.3, we note that if $P \neq (0)$, then neither D nor D_P is a valuation domain. For D this is clear, and for D_P the statement follows from the fact that the composite of two valuation domains is again a valuation domain [ZS, p.43], [N, p.35].

THEOREM 2.4. *In a quasilocal domain (D, M), the following conditions are equivalent for a nonzero element $x \in D$.*

(1) x is comparable.

(2) Each finitely generated ideal of D containing x is principal.

(3) Each strictly v-finite ideal of D containing x is principal.

Proof. (1)⇒(2): It suffices to show that $(x, d_1, \ldots, d_n)$ is principal for $d_1, \ldots, d_n \in D$. For $n = 1$, this follows since x is comparable. If $(x, d_1, \ldots, d_{n-1}) = (y)$ is principal, then y is comparable since y divides x, and by the case $n = 1$, $(x, d_1, \ldots, d_n) = (y, d_n)$ is also principal.

(2)⇒(3): Let A be a strictly v-finite ideal of D containing x. Thus there exists a finitely generated ideal $B \subseteq A$ such that $A_v = B_v$. Without loss of generality we can assume that $x \in B$, whence $B = (b)$ is principal by (2). Then $A_v = (b)_v = (b) \subseteq A \subseteq A_v$ and $A_v = (b)$ is principal.

(3)⇒(1): Pick an element $y \in D$. The ideal (x, y) is strictly v-finite, hence principal. Because D is quasilocal, it follows that $(x, y) = (x)$ or $(x, y) = (y)$ [G, Prop. 7.4]; consequently, $(y) \subseteq (x)$ or $(x) \subseteq (y)$ — that is, x is comparable. ∎

COROLLARY 2.5. *If the integral domain D contains a nonzero nonunit comparable element, then the maximal ideal of D is a t-ideal.*

Proof. Theorem 2.3 shows that D has a unique maximal ideal M. Let c be a nonzero nonunit comparable element of M. If $x_1, x_2, \ldots, x_n \in M$, then $(x_1, x_2, \ldots, x_n, c) = (m)$ is principal by Theorem 2.4, and $(x_1, \ldots, x_n)_v \subseteq (m) \subseteq M$. Therefore M is a t-ideal. ∎

We remark that to the equivalent conditions of Theorem 2.4, we could add: (4) M is a t-ideal, and each integral v-ideal of finite type containing x is principal.

COROLLARY 2.6. *A GCD-domain D contains a nonzero nonunit comparable element if and only if D is a valuation domain.*

Proof. We need only show that if D contains a nonzero nonunit comparable element c, then D is a valuation domain. Let M be the maximal ideal of D. Take nonzero elements x, y of M and let $(x, y)_v = (d)$. Write $x = x_1 d$, $y = y_1 d$. Then $(x_1, y_1)_v = D$. If $(x_1, y_1) \subseteq M$, then by Theorem 2.4, (x_1, y_1, c) is a proper principal ideal of D containing (x_1, y_1), and hence $(x_1, y_1)_v \subseteq (x_1, y_1, c) < D$. This contradiction shows that $(x_1, y_1) = D$, so x_1 or y_1 is a unit of D. We conclude that $(x, y) = (x)$ or $(x, y) = (y)$. Consequently, D is a valuation domain. ∎

3. t-INVERTIBILITY AND t-LOCAL COMPARABILITY

Suppose D is an integral domain and $d \in D - (0)$. We say that d is *t-locally comparable* if d is a comparable element of D_M for each maximal t-ideal M of D. Theorem 3.1 provides a connection between the concepts of t-invertibility and t-local comparability.

THEOREM 3.1. *Suppose D is an integral domain.*
(1) If A is a strictly v-finite ideal of D such that A contains a t-locally comparable element d, then A is t-invertible.
(2) Conversely, if $r \in D - (0)$ is such that each strictly v-finite ideal A of D containing r is t-invertible, then r is t-locally comparable.

Proof. (1): Suppose M is a maximal t-ideal of D containing A, and let B be a finitely generated ideal contained in A such that $A_v = B_v$. Without loss of generality we assume that $d \in B$. Then $((A_v)D_M)_v = (B_v D_M)_v = (BD_M)_v$ by [Z, Lemma 4]. Since d is a comparable element in D_M, Theorem 2.4 implies that $BD_M = aD_M$, where $a \in A$. Hence $aD_M \subseteq AD_M \subseteq (A_v D_M)_v = (BD_M)_v = (aD_M)_v = aD_M$. Therefore AD_M is principal for each maximal t-ideal containing A, and Corollary 1.6 of [MMZ] implies that A is t-invertible.

(2): Suppose M is a maximal t-ideal of D containing r. To show that r is a comparable element of D_M, it suffices, by Theorem 2.4, to show that each finitely generated ideal B of D_M containing r is principal. Now $B = AD_M$, where A is a finitely generated ideal of D containing r. By hypothesis, A is t-invertible, so $(AA^{-1})_t = D$. Since M is a maximal t-ideal, $AA^{-1} \nsubseteq M$. Therefore $AA^{-1}D_M = D_M$, so AD_M is invertible, and hence principal [G, Prop. 7.4]. ∎

Recall that an integral domain D is a *Prüfer v-multiplication domain* (PVMD) if the set of v-ideals of D of finite type form a group under v-multiplication. It is well known [Gr, Thm. 5] that D is a PVMD if and only if D_M is a valuation domain for each maximal t-ideal M of D. Theorem 3.2 provides a characterization of PVMD's in terms of t-local comparability.

THEOREM 3.2. *An integral domain D is a* PVMD *if and only if each nonzero prime ideal of D contains a t-locally comparable element.*

Proof. If D is a PVMD, then each nonzero element of D is t-locally comparable, so the condition holds. Conversely, suppose that each nonzero prime ideal of D meets S, the set of t-locally comparable elements of D. Proposition 2.1 implies that S is a saturated multiplicative system in D; hence each nonzero ideal of D meets S. Suppose A is a v-ideal of D of finite type. Then $A = B_v = B_t$ for some finitely generated ideal B. Because B is strictly v-finite, it is t-invertible by (1) of Theorem 3.1. Therefore $D = (BB^{-1})_t \subseteq (BB^{-1})_v = (AB^{-1})_v$, and A is v-invertible. Consequently, D is a PVMD. ■

Recall that a nonzero element d of D is a *t-valuation element* if D_M is a valuation domain for each maximal t-ideal M containing d. Clearly a t-valuation element is t-locally comparable, and since each nonzero element of a PVMD is a t-valuation element, Theorem 3.2 has the following corollary.

COROLLARY 3.3. *An integral domain D is a* PVMD *if and only if each nonzero prime ideal of D contains a t-valuation element.*

Added in Proof. The referee has kindly pointed out to us that Proposition 2.2 is a direct consequence of results on glueing in the paper *Topologically defined classes of commutative rings*, Annali di Mat. Pura Applic. 123(1980), 331-355, by M. Fontana. ■

REFERENCES

[A] D.D. Anderson, Star operations induced by overrings, Comm. Algebra 16(1988), 2535-2553.

[AA] D.D. Anderson and D.F. Anderson, Some remarks on star operations and the class group, J. Pure Appl. Algebra 51(1988), 27-33.

[AZ1] D.D. Anderson and M. Zafrullah, On t-invertibility III, Comm. Algebra. (to appear).

[AZ2] D.D. Anderson and M. Zafrullah, On a theorem of Kaplansky. (preprint)

[ABDFK] D.F. Anderson, A. Bouvier, D.E. Dobbs, M. Fontana, and S. Kabbaj, On Jaffard domains, Exposit. Math. 6(1988), 145-175.

[D] D.E. Dobbs, On locally divided domains and CPI-overrings, Internat. J. Math. and Math. Sci. 1(1981), 119-135.

[G] R. Gilmer, Multiplicative Ideal Theory, Queen's Papers Pure Appl. Math. Vol. 90, 1992.

[GO] R. Gilmer and J. Ohm, Primary ideals and valuation ideals, Trans. Amer. Math. Soc. 117(1965), 237-250.

[Gr] M. Griffin, Some results on v-multiplication rings, Canad. J. Math. 19(1967), 710-722.
[HZ] E. Houston and M. Zafrullah, On t-invertibility II, Comm. Algebra 17(1989), 1955-1969.
[J] P. Jaffard, Les Systemes d'Ideaux, Dunod, Paris, 1960.
[K] B.G. Kang, On the converse of a well known fact about Krull domains, J. Algebra 124(1989), 284-299.
[Kp] I. Kaplansky, Commutative Rings, Allyn and Bacon, Boston, 1970.
[Kr] W. Krull, Ein Hauptsatz über umkehrbare Ideale, Math. Zeit. 31(1930), 558.
[MMZ] S. Malik, J. Mott and M. Zafrullah, On t-invertibility, Comm. Algebra 16(1988), 149-170.
[MS] J. Mott and M. Schexnayder, Exact sequences of semi-value groups, J. Reine Angew. Math. 283/284(1976), 388-401.
[MZ] J. Mott and M. Zafrullah, On Krull domains, Arch. Math. 56(1991), 559-568.
[N] M. Nagata, Local Rings, Interscience, New York, 1962.
[O] J. Ohm, Semi-valuations and groups of divisibility, Canad. J. Math. 21(1969), 576-591.
[Z] M. Zafrullah, On finite conductor domains, manuscripta mathematica 24(1978), 191-204.
[ZS] O. Zariski and P. Samuel, Commutative Algebra, Vol. II, Van Nostrand, Princeton, N.J., 1960.

15
AF-Rings and Locally Jaffard Rings

FLORIDA GIROLAMI Dipartimento di Matematica, Università di Roma "La Sapienza", Piazzale A. Moro 5, 00185 Roma, Italy ; e-mail GIROLAMI@ITCASPUR.bitnet.

0 INTRODUCTION

AF-rings have been introduced by A.R. Wadsworth in [W] for computing the (Krull) dimension of the tensor product of A_1 and A_2, $\dim(A_1 \otimes_k A_2)$, where A_1 and A_2 are commutative algebras over a field k. Previously R.Y.Sharp proved in [S] that if K_1 and K_2 are extension fields of k, then

$$\dim(K_1 \otimes_k K_2) = \min\{\text{t.d.}(K_1:k)),\ \text{t.d.}(K_2:k)\}.$$

Since, for each prime ideal P of a k-algebra A, $\text{ht}(P) + \text{t.d.}(A/P:k) \leq \text{t.d.}(A_P:k)$ (cf.[ZS, p.10] and [W, p.392]), for giving a generalization of Sharp's result, A.R. Wadsworth has studied the k-algebras A which satisfy the altitude formula over k, that is

$$\text{ht}(P) + \text{t.d.}(A/P:k) = \text{t.d.}(A_P:k)$$

for each prime ideal P of A and called these algebras AF-rings. In [W, Theorem 3.8.] it is shown that if D_1 and D_2 are AF-domains, then

$$\dim(D_1 \otimes_k D_2) = \min\{\text{t.d.}(D_1:k) + \dim(D_2),\ \dim(D_1) + \text{t.d.}(D_2:k)\}.$$

On the other hand in [ABDFK] Jaffard domains have been introduced and studied. We recall from [ABDFK, Definition 0.2.] that a finite-dimensional domain D is a Jaffard domain if $\dim(D[X_1, \ldots, X_n]) = n + \dim(D)$ for each nonnegative integer n. The class of Jaffard domains is not stable under localizations and also in [ABDFK, Definition 1.4.] a domain D is defined to be a locally Jaffard domain if D_P is a Jaffard domain for each prime ideal P of D.

Analogous definitions are given in [C] for a finite-dimensional ring. An important class of locally Jaffard domains is the class of the stably strongS-domains (cf.[K, Theorem 1.6.]).

By [W, Corollary 3.2.] an AF-ring is a locally Jaffard ring. In section 1, after recalling relevant definitions and facts on the class of AF-rings, we compare the class of AF-domains with the class of k-algebras which are stably strong S-domains and give an example of an AF- domain which is not a stably strong S-domain. In section 2 we study the behaviour of the class of AF-domains with respect to certain pull-back type constructions and apply the results to pseudo-valuation domains and to the D + M construction. In section 3 we give an upper bound for the valuative dimension of the tensor product of two k-algebras, more exactly : if A_1 and A_2 are k-algebras with t.d.$(A_1:k) < \infty$ and t.d.$(A_2 :k) < \infty$, then

$$\dim_v (A_1 \otimes_k A_2) \leq \min\{\dim_v (A_1) + \text{t.d.}(A_2:k) , \dim_v (A_2) + \text{t.d.}(A_1:k)\}.$$

By this upper bound and by a lower bound established in [W, Theorem4.1.] for $\dim(D \otimes_k D)$, where D is a domain , we prove that if D is a Jaffard domain with t.d.$(D:k) < \infty$, then $D \otimes_k D$ is a Jaffard ring. At last we prove that if A is a locally Jaffard ring with t.d.$(A:k) < \infty$ and C an integral extension of a localization of $k[X_1, ..., X_n]$, then $A \otimes_k C$ is a Jaffard ring.

All rings considered in this work are commutative, with identity and all ring-homomorphisms are unital.

1 LOCALLY JAFFARD RINGS, AF-RINGS AND STABLY STRONG S-DOMAINS

We recall the definition and some properties of valuative dimension of a ring A introduced and studied by P. Jaffard [J].

DEFINITION 1.1. *If D is a domain which is not a field, D is said to have valuative dimension n (in short,* $\dim_v (D) = n$*) if each valuation overring of D has dimension at most n and if there exists a valuation overring of D of dimension* n. *If no such integer* n *exists, D is said to have infinite valuative dimension (in short,* $\dim_v (D) = \infty$*). Each field is assigned valuative dimension 0.*

REMARK 1.2. The valuative dimension of a domain D satisfies the following properties:

a) $\dim(D) \leq \dim_v (D)$.

b) $\dim_v (D/P) \leq \dim_v (D)$ for each $P \in \text{Spec}(D)$ and $\dim_v (D/P) \leq \dim_v (D)-1$ if $P \neq (0)$.

c) For each $P \in \text{Spec}(D)$, $\dim_v (D_P) + \dim_v (D/P) \leq \dim_v (D)$.

d) $\dim_v (D) = \sup\{\dim_v (D_P) : P \in \text{Spec}(D)\} = \sup\{\dim_v (D_M) : M \in \text{Max}(D)\}$.

e) If D contains a field k, then $\dim_v (D) \leq \text{t.d.}(D: k)$.

DEFINITION 1.3. *Let* A *be a ring, then* $\dim_v(A) = \sup\{\dim_v(A/P) : P \in \mathrm{Spec}(A)\}$.

REMARK 1.4. The valuative dimension of a ring A satisfies the following properties:

a) $\dim(A) \le \dim_v(A)$.

b) $\dim_v(A) = \sup\{\dim_v(A/P) : P \in \mathrm{Min}(A)\}$.

c) $\dim_v(A) = \sup\{\dim_v(A_P) : P \in \mathrm{Spec}(A)\} = \sup\{\dim_v(A_M) : M \in \mathrm{Max}(A)\}$.

d) If A contains a field k, then $\dim_v(A) \le \mathrm{t.d.}(A{:}k)$.

THEOREM 1.5. (cf.[J, Corollary 1 p.67]) *For a ring A the following conditions are equivalent:*

i) $\dim_v(A) = \dim(A)$.

ii) $\dim(A[X_1, ..., X_n]) = n + \dim(A)$ *for each nonnegative integer* n.

In [C] the definition of Jaffard domains given in [ABDFK] is extended to rings.

DEFINITION 1.6. *A finite-dimensional ring A is said to be a Jaffard ring if* $\dim(A) = \dim_v(A)$.

PROPOSITION 1.7. (cf.[J, Proposition 4 p.58]) *Let* $A \subseteq B$ *be an integral extension of rings. Then* $\dim_v(A) = \dim_v(B)$.

THEOREM 1.8. *For a finite dimensional ring A, the following conditions are equivalent:*

i) For each $P \in \mathrm{Spec}(A)$, A_P *is a Jaffard ring.*

ii) For each multiplicative subset S *of* A, $S^{-1}A$ *is a Jaffard ring.*

iii) For each $P \in \mathrm{Spec}(A)$ *and for each nonnegative integer n,* $\mathrm{ht}(P) = \mathrm{ht}(P[X_1, ..., X_n])$.

PROOF. i) $\Longrightarrow$ ii) For each multiplicative subset S of A $\dim(S^{-1}A) \le \dim(A) < \infty$ and

$$\begin{aligned}\dim_v(S^{-1}A) &= \sup\{\dim_v((S^{-1}A)_Q) : Q \in \mathrm{Spec}(S^{-1}A)\}\\ &= \sup\{\dim_v(A_P) : P \in \mathrm{Spec}(A) \text{ and } P \cap S = \emptyset\}\\ &= \sup\{\dim(A_P) : P \in \mathrm{Spec}(A) \text{ and } P \cap S = \emptyset\} = \dim(S^{-1}A).\end{aligned}$$

Hence $S^{-1}A$ is a Jaffard ring.

ii) $\Longrightarrow$ i) Trivial.

i) $\Longleftrightarrow$ iii) Let P be any prime ideal of A ; then by [BMRH, Theorem1], for each nonnegative integer n, $\dim(A_P[X_1, ..., X_n]) = \mathrm{ht}(P[X_1, ..., X_n]) + n$. By Theorem 1.5. A_P is a Jaffard ring if and only if for each nonnegative integer n $\dim(A_P[X_1, ..., X_n]) = n + \dim(A_P)$; so A_P is a Jaffard ring if and only if, for each nonnegative integer n, $\mathrm{ht}(P[X_1, ..., X_n]) = \mathrm{ht}(P)$.

DEFINITION 1.9. *A finite dimensional ring A is said to be locally Jaffard if it satisfies the equivalent conditions of Theorem 1.8.*

REMARK 1.10. If A is a locally Jaffard ring, then A is a Jaffard ring. Moreover there are Jaffard rings not locally Jaffard rings (cf.[ABDFK, Example 3.2.]).

For the class of locally Jaffard domains there are another two characterizations.

PROPOSITION 1.11. *For a finite-dimensional domain D the following conditions are equivalent:*

i) D is locally Jaffard.

ii) For each valuation overring (V, N) of D

$$\text{ht}(N) + \text{t.d.}(\mathbf{k}(N):\mathbf{k}(N \cap D)) \leq \text{ht}(N \cap D).$$

iii) D satisfies the altitude inequality formula, that is, for each finite-type D-algebra S (domain) containing D and each prime ideal Q of S

$$\text{ht}(Q) + \text{t.d.}(\mathbf{k}(Q):\mathbf{k}(Q \cap D)) \leq \text{ht}(Q \cap D) + \text{t.d.}(S:D).$$

PROOF. i)$\Longleftrightarrow$ ii) is the Proposition 9.3. of [BDF] ; i)$\Longleftrightarrow$ iii) follows from Lemma 1.4. of [K] and Theorem 1.8.

For a given field k, let $\mathfrak{C}$ be the class of commutative algebras A over k with identity and with t.d.$(A:k) < \infty$. We recall from [W] the following:

DEFINITION 1.12. *A ring $A \in \mathfrak{C}$ is said to be an AF-ring if for each prime ideal P of A*

$$\text{ht}(P) + \text{t.d.}(A/P:k) = \text{t.d.}(A_P:k).$$

Let $\mathfrak{F}$ be the class of algebras of $\mathfrak{C}$ which are AF-rings. Any finitely generated algebra over k or any integral extension of such an algebra is an AF-ring (cf.[W, pag.395]). The class $\mathfrak{F}$ satisfies the following properties, all proved in [W]:

a) If $A \in \mathfrak{F}$ and S is a multiplicative subset A, then $S^{-1}A \in \mathfrak{F}$.

b) If $A_1,.., A_n \in \mathfrak{F}$, then $A_1 \otimes_k ... \otimes_k A_n \in \mathfrak{F}$.

c) If $A_1, A_2 \in \mathfrak{F}$, then $A_1 \times A_2 \in \mathfrak{F}$.

d) If $A \in \mathfrak{F}$, then $A[X] \in \mathfrak{F}$ and for each prime ideal P of A $\text{ht}(P) = \text{ht}(P[X])$.

e) If D_1 and D_2 are AF-domains, then

$$\dim(D_1 \otimes_k D_2) = \min\{\text{t.d.}(D_1:k) + \dim(D_2), \dim(D_1) + \text{t.d.}(D_2:k)\}.$$

f) The class $\mathfrak{F}$ is not stable under factor domains.

PROPOSITION 1.13. *An AF-ring A is a locally Jaffard ring.*

PROOF. Let P be any prime ideal of A. By [W, Corollary 3.2.] $ht(P) = ht(P[X])$ and $A[X]$ is an AF-ring and so, for each positive integer n, $ht(P) = ht(P[X_1, ..., X_n])$. Hence by Theorem 1.8. A is a locally Jaffard ring.

REMARK 1.14. In $\mathbb{C}$ there are locally Jaffard rings which are not AF-rings.
Let s be an element of $k[[t]]$ which is transcendental over $k(t)$ and which is of the form $s = t + a_2t^2 + a_3t^6 + ... + a_it^{i!} + ...$ where infinitely many of the coefficients are not 0 [G,p.507]; if f is the application from $k[X,Y]$ in $k[[t]]$ which sends X to t, Y to s and k identically onto itself, then f can be extended to the quotient fields of $k[X,Y]$ and $k[[t]]$. Let V be the inverse image of $k[[t]]$; V is a DVR with quotient field $k(X,Y)$; if M is the maximal ideal of V, then $V/M = (f^{-1}(k[[t]]))/(f^{-1}((t))) \cong k$; so $ht(M) + t.d.(V/M : k) < t.d.(V : k)$.

We recall from [Kap, p.26] the following two definitions.

DEFINITION 1.15. *A domain D is said to be an S-domain if for each prime ideal P of D of height one, the extended ideal P[X] in the polynomial ring D[X] is also of height one.*

DEFINITION 1.16. *A ring A is said to be a strong S-ring if the residue class ring A/P is an S-domain for each prime ideal P of A.*

Being the class of strong S-rings not stable under polynomial ring extensions , in [MM,p.250] it is given the following definition:

DEFINITION 1.17. *A ring A is said to be a stably strong S-ring if* $A[X_1, ..., X_n]$ *is a strong S-ring for each nonnegative integer n.*

REMARK 1.18. The most important examples of stably strong S-rings are arbitrary noetherian domains, arbitrary Prüfer domains [MM,Theorem 3.5.] and arbitrary universally catenarian domains [BDF, Theorem 2.4.]. Moreover, S. Kabbaj showed in [K, Theorem 1.6.] that any stably strong S- domain satisfies the altitude inequality formula and so any finite dimensional stably strong S-domain is a locally Jaffard ring. The DVR V of the Remark 1.14. is an example of a stably strong S-domain which is not an AF-domain. Now we give an example of an AF-domain which is not a stably strong S-domain.

REMARK 1.19. We consider from [BMRH] and [K] the following example of a locally Jaffard domain which is not a stably strong S-domain. Let $V = k(X)[Y]_{(Y)} = k(X) + Yk(X)[Y]_{(Y)} =$ $= k(X) + M$. Let $D = k + M$; D is a one-dimensional integrally closed pseudo-valuation domain . Being $k(X)$ a transcendental extension of k, $\dim_v(D) = 2$ (cf.[ABDFK, Proposition 2.9.]); by [MM, Theorem 5.1.] D is not a strong S-domain and so $D[Z]$ is not a stably strong

S-domain ; moreover $D[Z]$ is not catenarian by [BDF,Lemma 2.3.]. To show that $D[Z]$ is an AF-domain we examine the few cases which arise. If $P = M[Z]$, then $ht(P) = 2$ and $D[Z]/P$ is isomorphic to $k[Z]$ and so $ht(P)+t.d.(D[Z]/P:k) = 2+1 = t.d.(D[Z]:k)$. If $P \cap D \neq (0)$ and if $P \neq M[Z]$, then $ht(P) = 3 = t.d.(D[Z]:k)$. If $P \cap D = (0)$ and $P \neq (0)$, then $ht(P) = 1$; moreover $D \longrightarrow D[Z]/P$ is an algebraic extension and so $t.d.(D[Z]/P:k) = t.d.(D:k) = 2$; in conclusion, $ht(P) + t.d.((D[Z]/P:k)) = 1+2 = t.d.(D[Z]:k)$.

We observe that if $P = ZD[Z]$, then $D[Z]/P \cong D$ is not an AF-domain , since D is not a Jaffard domain. Hence the class $\mathfrak{F}$ is not stable under factor domains. We remark that $D[Z]$ is not catenarian.

PROPOSITION 1.20. *If D is a catenarian AF-domain, then for each prime ideal P of D, D/P is an AF-domain.*

PROOF. Each prime ideal of D/P is of type P'/P with P and P' prime ideals of D and $P \subseteq P'$. Being D a catenarian finite dimensional AF-domain, it is :

$$ht(P'/P) = ht(P') - ht(P) = t.d.(D_{P'}:k) - t.d.(D/P':k) - t.d.(D_P:k) + t.d.(D/P:k) =$$
$$= t.d.((D/P)_{P'/P}:k) - t.d.((D/P)/(P'/P):k).$$

2 PULLBACKS AND AF-DOMAINS

Now we give necessary and sufficient conditions for certain pull-back type constructions to be AF-domains. We consider pullbacks of commutative rings

$$\begin{array}{ccc} R & \longrightarrow & D \\ \downarrow & & \downarrow \\ T & \longrightarrow & K \end{array}$$

where T is a domain , φ is a homomorphism from T onto a field K with $\ker(\varphi) = M$, D a proper subring of K and $R = \varphi^{-1}(D)$. R is a k -algebra if and only if T and D are k -algebras.

PROPOSITION 2.1. *Let T be a domain of $\mathfrak{C}$ with maximal ideal M , $K = T/M$ and $\varphi : T \longrightarrow K$ the canonical surjection. Let D be a proper subring of K belonging to $\mathfrak{C}$ with quotient field F . Put $R = \varphi^{-1}(D)$. Then R is an AF-domain if and only if T and D are AF-domains and K is algebraic over F.*

PROOF. We recall from [ABDFK, Lemma 2.1 (a)] that $M = (R:T)$ and the k -algebras R/M and D are k-isomorphic. Moreover for [ABDFK, Lemma 2.1 (g)] M has the same height in T and in R and for [F, Corollary 1.5.(6)] R and T have the same quotient field.

If R is an AF-domain, then $\mathrm{ht}(M) + \mathrm{t.d.}(R/M:k) = \mathrm{t.d.}(R:k)$; and so $\mathrm{t.d.}(D:k) = \mathrm{t.d.}(T:k) + - \mathrm{ht}(M)$. Then, since it is always true that $\mathrm{ht}(M) + \mathrm{t.d.}(T/M:k) \leq \mathrm{t.d.}(T:k)$, we have $\mathrm{t.d.}(T/M:k) \leq \mathrm{t.d.}(D:k)$; therefore the extension $F \subseteq K$ is algebraic and $\mathrm{ht}(M) + \mathrm{t.d.}(T/M:k) = \mathrm{t.d.}(T:k)$. Now we prove that T is an AF-domain. Let Q be a prime ideal of T and $Q \neq M$; put $P = R \cap Q$; then $P \neq M$ and for [ABDFK, Lemma 2.1 (e)] $T_Q = R_P$; moreover, since R is an AF-domain, $\mathrm{ht}(P) + \mathrm{t.d.}(R/P:k) = \mathrm{t.d.}(R:k)$ and so $\mathrm{ht}(Q) + \mathrm{t.d.}(T/Q:k) = \mathrm{t.d.}(T:k)$. We show next that D is an AF-domain; let Q be a prime ideal of D and $P = \varphi^{-1}(Q)$. Then for [ABDFK, Lemma 2.1 (g) and (d)] $\mathrm{ht}(P) = \mathrm{ht}(M) + \mathrm{ht}(Q)$ and R/P and D/Q are isomorphic k -algebras. Therefore, being R an AF-domain,

$\mathrm{ht}(Q) + \mathrm{t.d.}(D/Q:k) = \mathrm{ht}(P) - \mathrm{ht}(M)+\mathrm{t.d.}(R/P:k) = \mathrm{t.d.}(R:k) - \mathrm{ht}(M) = \mathrm{t.d.}(R/M:k) = \mathrm{t.d.}(D:k)$.

Suppose conversely that T and D are AF-domains and $F \subseteq K$ is an algebraic extension. Let P be a prime ideal of R. If $M \not\subseteq P$, by [ABDFK, Lemma 2.1 (e)] there is a unique prime ideal Q of T such that $P = R \cap Q$, and this Q satisfies $T_Q = R_P$; then

$$\mathrm{ht}(P) + \mathrm{t.d.}(R/P:k) = \mathrm{ht}(Q) + \mathrm{t.d.}(T/Q:k) = \mathrm{t.d.}(T:k) = \mathrm{t.d.}(R:k).$$

If $M \subseteq P$, by [ABDFK, Lemma 2.1 (e)] there is a unique prime ideal Q of D such that $P = \varphi^{-1}(Q)$ and $\mathrm{ht}(P) = \mathrm{ht}(Q) + \mathrm{ht}(M)$. Then

$$\mathrm{ht}(P) + \mathrm{t.d.}(R/P:k) = \mathrm{ht}(Q) + \mathrm{ht}(M) + \mathrm{t.d.}(D/Q:k) = \mathrm{t.d.}(D:k) + \mathrm{t.d.}(T:k) - \mathrm{t.d.}(T/M:k) =$$
$$= \mathrm{t.d.}(R:k).$$

We apply previous result to pseudo-valuation domains (PVD's). As in [HH] we say that a prime ideal P of a domain D is *strongly prime* if $xy \in P$ with x and y in the quotient field of D implies that either $x \in P$ or $y \in P$; and that D is a *pseudo-valuation domain* if each of its prime ideals is strongly prime. A quasi-local domain D with maximal ideal M is a PVD if and only if M is also the maximal ideal of some valuation overring V of D. In this case V is uniquely determined as the conductor $(M:M)$ and is called the valuation domain associated to D; and, by [AD], D may be recovered by V as a pullback : $D = V \times_{\mathbf{k}(V)} \mathbf{k}(D)$. As immediate consequence of Proposition 2.1. we obtain the following:

COROLLARY 2.2. *Let D be a PVD belonging to $\mathbf{C}$ with maximal ideal M and residue field $F = D/M$. Let $V = (M:M)$ the associated valuation domain of D, with residue field $K = V/M$. Then D is an AF- domain if and only if V is an AF-domain and $F \subseteq K$ is an algebraic extension.*

We now consider the $D + M$ construction and apply the Proposition 2.1.

COROLLARY 2.3. *Let V be a (nontrivial) valuation domain of the form $V = K + M$, where K is a field containing k and M the maximal ideal of V . Let $R = D + M$, where D is a proper subring of K containing k . Then R is an AF-domain if and only if V and D are AF-domains and K is algebraic over the quotient field of D.*

REMARK 2.4. For each positive integer n, there exists a n-dimensional domain V of the form $V = k + M$ which is an AF-domain. For n = 1 it suffices to consider $k\,[X]_{(X)}$; for n > 1, if $V_{n-1} = k + M_{n-1}$ is the required valuation domain and K_{n-1} is its quotient field, then by Corollary 2.3. $V_n = V_{n-1} + XK_{n-1}\,[X]_{(X)}$ is an AF-domain of the form $V_n = k + M_n$.

3 TENSOR PRODUCT OF LOCALLY JAFFARD DOMAINS

All rings A considered in this section belong to $\mathbb{C}$, that is, for a given field k , are k - algebras with t.d.$(A{:}k) < \infty$. Now we establish an upper bound for the valuative dimension of the tensor product of two rings .

PROPOSITION 3.1. *If A_1 and A_2 are rings, then*

$$\dim_v (A_1 \otimes_k A_2) \leq \min\{\dim_v (A_1) + \text{t.d.}(A_2{:}k)\ ,\ \dim_v (A_2) + \text{t.d.}(A_1{:}k)\}.$$

PROOF. Let Q be any minimal prime of $A_1 \otimes_k A_2$; identifying A_1 with $A_1 \otimes_k k$ and A_2 with $k \otimes_k A_2$, since the extensions $A_1 \subseteq A_1 \otimes_k A_2$ and $A_2 \subseteq A_1 \otimes_k A_2$ are flat and so satisfy the going-down property, then $P_1 = Q \cap A_1$ is a minimal prime of A_1 and $P_2 = Q \cap A_2$ is a minimal prime of A_2. Therefore we may consider $(A_1 \otimes_k A_2)/Q$ as an extension of A_1/P_1 and so we have

$$\dim_v ((A_1 \otimes_k A_2)/Q) \leq \dim_v (A_1/P_1) + \text{t.d.}((A_1 \otimes_k A_2)/Q : A_1/P_1).$$

Since $\text{t.d.}((A_1 \otimes_k A_2)/Q{:}k) = \text{t.d.}((A_1/P_1){:}k) + \text{t.d.}((A_2/P_2){:}k)$ (cf.[W, Proposition 2.3.]), then $\text{t.d.}((A_1 \otimes_k A_2)/Q{:}\,(A_1/P_1)) = \text{t.d.}((A_2/P_2){:}k)$. Hence

$$\dim_v ((A_1 \otimes_k A_2)/Q) \leq \dim_v (A_1/P_1) + \text{t.d.}((A_2/P_2){:}k)$$

and by Remark 1.4. it follows that $\dim_v ((A_1 \otimes_k A_2)/Q) \leq \dim_v (A_1) + \text{t.d.}(A_2{:}k)$. Considering $(A_1 \otimes_k A_2)/Q$ as an extension of A_2/P_2 , it is $\dim_v ((A_1 \otimes_k A_2)/Q) \leq \dim_v (A_2) + \text{t.d.}(A_1{:}k)$ and so the result follows.

COROLLARY 3.2. *If A_1 and A_2 are Jaffard rings, then*

$$\dim_v (A_1 \otimes_k A_2) \leq \min\{\dim(A_1) + \text{t.d.}(A_2{:}k)\ ,\ \dim(A_2) + \text{t.d.}(A_1{:}k)\}.$$

PROOF. Being $\dim(A_1) = \dim_v(A_1)$ and $\dim(A_2) = \dim_v(A_2)$, the result follows immediately from Propositon 3.1.

COROLLARY 3.3. *Let A be a ring, then*

$$\dim_v(A \otimes_k A) \leq \dim_v(A) + \text{t.d.}(A:k).$$

PROOF. An immediate consequence of Proposition 3.1.

COROLLARY 3.4. *Let D be a domain, then*

$$\dim(D) + \text{t.d.}(D:k) \leq \dim(D \otimes_k D) \leq \dim_v(D \otimes_k D) \leq \dim_v(D) + \text{t.d.}(D:k).$$

PROOF. The first inequality is proved in [W, Theorem 4.1.], the second is always true and the third is established in Corollary 3.3.

COROLLARY 3.5. *If D is a Jaffard domain, then* $D\otimes_k D$ *is a Jaffard ring.*

PROOF. If D is a Jaffard domain, then $\dim(D) = \dim_v(D)$ and so the result follows from Corollary 3.4.

REMARK 3.6. In [W] it is remarked that the lower bound for $\dim(D\otimes_k D)$ used in Corollary 3.4. need not hold for a ring, neither for an AF-ring ; but, if A is a ring such that $\text{t.d.}((A/P:k) = \text{t.d.}(A:k)$ for each minimal prime P of A, then

$$\dim(A) + \text{t.d.}(A:k) \leq \dim(A\otimes_k A).$$

This inequality also holds if A is a ring such that $\dim(A) = \dim(A/P)$ for each minimal prime P of A.

LEMMA 3.7. *Let A a ring such that* $\dim(A) = \dim(A/P)$ *for each of its minimal primes P ; then* $\dim(A) + \text{t.d.}(A:k) \leq \dim(A\otimes_k A)$.

PROOF. For each minimal prime P of A, since $(A/P)\otimes_k(A/P)$ is a homomorphic image of $A\otimes_k A$, the inequality $\dim(A\otimes_k A) \geq \dim((A/P)\otimes_k(A/P))$ holds; being for [W, Theorem 4.1.] $\dim((A/P)\otimes_k(A/P)) \geq \dim(A/P) + \text{t.d.}((A/P):k)$, we obtain, by assumption on A, that $\dim(A \otimes_k A) \geq \dim(A) + \text{t.d.}((A/P):k)$ for each minimal prime P of A and so the result follows.

COROLLARY 3.8. *If A is a Jaffard ring such that* $\dim(A) = \dim(A/P)$ *for each of its minimal primes P , then* $A\otimes_k A$ *is a Jaffard ring.*

THEOREM 3.9. *Let A be a locally Jaffard ring and C an integral extension of*

$S^{-1}(k\,[X_1, ..., X_n])$, *where* n *is a positive integer and* S *a multiplicative subset of* $k\,[X_1, ..., X_n]$, *then* $A \otimes_k C$ *is a Jaffard ring.*

PROOF. There is a natural isomorphism of k-algebras between $A[X_1, ..., X_n]$ and $A \otimes_k k\,[X_1, ..., X_n]$; moreover $A \otimes_k S^{-1}(k\,[X_1, ..., X_n])$ is a localization of $A \otimes_k k\,[X_1, ..., X_n]$. Then, being A and also $A[X_1, ..., X_n]$ locally Jaffard rings, by Theorem 1.8., $A \otimes_k S^{-1}(k\,[X_1, ..., X_n])$ is a Jaffard ring. Moreover, since C is an integral extension of $S^{-1}(k\,[X_1, ..., X_n])$, the extension ring $A \otimes_k S^{-1}k\,[X_1, ..., X_n] \subseteq A \otimes_k C$ is integral and so by Proposition 1.7. $A \otimes_k C$ is a Jaffard ring.

PROPOSITION 3 10. *If A is a locally Jaffard ring and K is an extension field of k with* $t.d.(K:k) < \infty$, *then* $A \otimes K$ *is a Jaffard ring.*

PROOF. Let $v_1, v_2, ..., v_n$ be n elements of K which are algebraically independent over k and such that K is algebraic over $k(v_1, v_2, ..., v_n)$. Applying Theorem 3.9. to $S^{-1}(k\,[v_1, v_2, ..., v_n])$ with $S = k\,[v_1, v_2, ..., v_n] - \{0\}$,.the result follows.

REMARK 3.11. Let A be a locally Jaffard ring and K an extension field of k with t.d.$(K:k)$ = n; then by Proposition 3.10. $\dim(A \otimes_k K) = \dim_v (A \otimes_k K)$; moreover, using [W, Corollary 3.6.] it is $\dim(A \otimes_k K) = \max\{ht(P) + \min(t.d.(K:k), t.d.(A/P:k)) : P \in Spec(A)\}$.

REFERENCES

[ABDFK] Anderson, D.F., Bouvier, A., Dobbs, D. E., Fontana, M., and Kabbaj, S., (1988). On Jaffard domains, Expo. Math., 6 : 145-175 .

[AD] Anderson, D.F., and Dobbs, D. E., (1980). Pairs of rings with the same prime ideals, Canad. J. Math., 32 : 362-384.

[BDF] Bouvier, A., Dobbs, D.E., and Fontana, M., (1988). Universally catenarian integral domains, Advances in Math., 72 : 211-238.

[BK] Bouvier, A., and Kabbaj, S., (1988). Examples of Jaffard domains, J. Pure Appl. Algebra , 54 : 155-165 .

[BMRH] Brewer, J.W., Montgomery, P.R., Rutter, E.A., and Heinzer,W.J., (1972). Krull dimension of polynomial rings, Lecture Notes in Math. No. 311, Springer Berlin-New York , 26-45.

[C] Cahen, J.-P., (1990). Construction B, I, D et anneaux localement ou résiduellement de Jaffard, Arch Math., 54 : 125-141..

[F] Fontana, M., (1980). Topologically defined classes of commutative rings, Ann. Mat.Pura Appl., 123 : 331-355.

[G] Gilmer, R., (1972). Multiplicative ideal theory, Dekker, New York .

[HH] Hedstrom, J.R.,and Houston, E.G., (1978). Pseudo-valuation domains, Pacific J. Math., 75 : 137-147.

[J] Jaffard, P., (1960). Théorie de la dimension dans les anneaux de polynômes, Mém.Sc. Math. , 146, Gauthier-Villars, Paris .

[K] Kabbaj, S., (1980). La formule de la dimension pour les S-domaines forts universels, Boll.Un.Mat.Ital.Algebra e Geometria , 5 : 145-161.

[Kap] Kaplansky, I., (1974). Commutative Rings, Allyn and Bacon, Boston .

[MM] Malik, S., and Mott, J.L., (1983). Strong S-domains, J. Pure Appl. Algebra , 28 : 249-264

[S] Sharp, R.Y., (1977). The dimension of the tensor product of two field extensions, Bull.London Math. Soc , 9 : 42-48.

[SV] Sharp, R.Y., and Vamos, P., (1977). The dimension of the tensor product of a finite number of field extensions, J. Pure Appl. Algebra, 10 : 249-252.

[W] Wadsworth, A.R., (1979). The Krull dimension of tensor products of commutative algebras over a field, J.London Math.Soc., 19 : 391-401.

[ZS] Zariski, O., and Samuel, P., (1960). Commutative Algebra, Vol. II, Van Nostrand, New York .

16
Prime t-Ideals in R[X]

Evan G. Houston
Department of Mathematics
University of North Carolina at Charlotte
Charlotte, NC 28223 USA

Let R be an integral domain with quotient field K, and let X be an indeterminate. It is well known that (prime) t-ideals of R extend to (prime) t-ideals of R[X] and that every prime upper to zero $fK[X] \cap R[X]$ is a t-ideal of R[X]. We give a method for constructing prime t-ideals which assume neither of these forms; that is, we produce examples of prime t-ideals P in R[X] for which $0 \neq (P \cap R)[X] \neq P$. It is hoped that this construction will prove useful in investigating the t-dimension of polynomial rings.

1 INTRODUCTION AND PRELIMINARY REMARKS

Let R be an integral domain with quotient field K. Recall that for a nonzero fractional ideal I of R, $I^{-1} = \{u \in K \mid uI \subseteq R\}$ and $I_v = (I^{-1})^{-1}$. The ideal I is called a v-ideal or divisorial if $I = I_v$. Finally, $I_t = \cup J_v$, where the union is taken over all nonzero finitely generated subideals J of I; the ideal I is called a t-ideal if $I = I_t$. The v- and t-operations are examples of star-operations, and we refer the reader to [G] for information on their properties (which we shall use freely). We shall be especially interested in prime t-ideals (t-primes). It is well known that height-one primes are t-primes. Thus, in particular, if $Q = fK[X] \cap R[X]$ (f an irreducible polynomial of K[X]) is an upper to zero in R[X], then Q is a t-prime. Moveover, if P is a t-prime in R, then P[X] is a t-prime in R[X] [HH, Proposition 4.3]. It is natural to ask if there are any other prime t-ideals in R[X]. This is especially interesting if one is interested in relating the t-dimension of R[X] to that of R. We define the t-dimension of a domain R to be the supremum of the lengths of chains of t-primes in R; for the purposes of this definition, we include (0) as a t-prime (although, technically, it is not). We shall denote the t-dimension of R by t-dim(R).

It turns out that there is a connection between the existence of "non-obvious" t-primes in R[X] and uppers to zero which (are t-primes but) are not almost principal; as in [HHJ] we call an upper to zero $Q = fK[X] \cap R[X]$ almost principal if there is a nonzero element a in R with $aQ \subseteq fR[X]$. Since every element of Q may be written in the form fk for some $k \in K[X]$, this amounts to requiring that there be a single element a in R that simultaneously clears the denominators of all polynomials $k \in K[X]$ for which $fk \in R[X]$. Clearly, this set of polynomials is Qf^{-1}. We give some equivalent conditions for an upper to zero to be almost principal.

PROPOSITION 1.1 Let Q be a prime upper to zero in R[X]. The following statements are equivalent.

(1) Q is almost principal.

(2) $Q^{-1} \not\subseteq K[X]$.

(3) $Q^{-1} \neq [Q:Q]$ ($= \{u \in K(X) \mid uQ \subseteq Q\}$).

(4) If $Q = fK[X] \cap R[X]$ with $f \in Q$, then $Q^{-1}f \not\subseteq Q$.

Proof: Statements (1), (2), and (3) are equivalent by [HHJ, Proposition 1.15]. It is easy to see that (4) $\Rightarrow$ (3). Assume (3). By [HHJ, Proposition 1.15 (5) $\Rightarrow$ (6)], $\exists$ $g \in R[X] \setminus Q$ with $gQ \subseteq fR[X]$; clearly, $g \in Q^{-1}f \setminus Q$, so that (3) $\Rightarrow$ (4). □

Examples of almost principal ideals abound; indeed, some effort is required to produce an example of an upper to zero which is not almost principal. For example, every upper to zero in R[X] is almost principal in case R is Noetherian or integrally closed. And an individual upper Q is almost principal if (i) Q contains a generating set of bounded degree, (ii) Q is the kernel of a map sending X to an element which is almost integral, or (iii) Q contains a polynomial g for which $c(g)^{-1} = R$. (For proofs of these facts, see [HHJ]). It is easy to see that an almost principal upper $Q = fK[X] \cap R[X]$ is divisorial. We may assume that $f \in Q$. If $aQ \subseteq fR[X]$, $a \neq 0$, then $aQ_v \subseteq fR[X] \subseteq Q$; since $a \notin Q$, we have $Q_v \subseteq Q$. However, in [HHJ] the authors were unable to produce an example of a divisorial upper which is not almost principal, and, as we will see later, this bears on the question of the existence of divisorial primes in R[X] which are neither uppers to zero nor extended from R[X].

But there do exist examples of uppers which are not almost principal. The first such example was produced by Jimmy Arnold for a different--but related--reason. Perhaps the easiest way to understand this example is by means of the notion of a spotty element. An element $t \in K$ is said to be spotty over R if t is not almost integral over R but $\exists\, a \in R$ with $at^i \in R$ for infinitely many positive integers i. We shall show that if t is spotty over R, then $Q = \ker(X \mapsto t)$ is an upper to zero in R[X] which is not almost principal. (This result is proved in a roundabout way in [HHJ]; we include a direct proof here for completeness.)

PROPOSITION 1.2 Let R be an integral domain with quotient field K, and let $t \in K$ be spotty over R. Then $Q = \ker(X \mapsto t)$ is an upper to zero in R[X] which is not almost principal.

Proof: Since t is spotty, there is an element $a \in R$ with $at^i \in R$ for infinitely many positive integers i. Suppose that Q is almost principal, and choose $c \in R$ with $cQ \subseteq (X - t)R[X]$. Pick i with $at^i \in R$, and let $k(X) = a(X^{i-1} + tX^{i-2} + \cdots + t^{i-1})$. Then $(X - t)k(X) = aX^i - at^i \in Q$. Hence $ck(X) \in R[X]$; that is, $cat^j \in R$ for $j = 1, \ldots, i$. Since i can be made arbitrarily large, this contradicts the fact that t is not almost integral. □

Thus to produce an upper to zero which is not almost principal, it suffices to produce a spotty element. We record the following example from [HHJ].

EXAMPLE 1.3 Let s, t be indeterminates over the field F, and let $R = F[s, \{st^{2^n} \mid n = 0,1,2,\ldots\}]$. Then the element t is spotty over R, and $Q = \ker(X \mapsto t)$ is an upper to zero in R[X] which is not almost principal. □

REMARK 1.4 It is easy to see that the integral closure of R in Example 1.3 is $R' = F[\{st^n \mid n = 0,1,2,\ldots\}]$, a two-dimensional integrally closed (Mori) domain. In particular, R is two dimensional. I do not know an example of a one-dimensional domain with an upper to zero which is not almost principal. As we shall see, the lack of a one-dimensional example will be a roadblock to producing examples of domains R for which the t-dimension of R[X] is substantially greater than that of R. □

In the example above, it is easy to see that the complete integral closure R^* of R is not completely integrally closed, that is, that $(R^*)^* \neq R^*$. This is typical of spotty elements, as we now show.

PROPOSITION 1.5 Let R be an integral domain with quotient field K, and let $t \in K$ be spotty over R. Then the root closure of R contains at^i for each positive integer i; in particular, $(R^*)^* \neq R^*$.

Proof: Pick $a \in R$ with $at^j \in R$ for infinitely many positive integers j. For any i we may choose $j > i$ with $at^j \in R$. Then $(at^i)^j = a^{j-i}(at^j)^i \in R$. Thus at^i lies in the root closure $\overline{R}$ of R for each i. The "in particular" statement follows since $at^i \in \overline{R} \subseteq R^*$ for each i shows that $t \in (R^*)^*$ while the spottiness of t implies that $t \notin R^*$. □

2 THE CONSTRUCTION

This construction is inspired by [O]. Let $S = K + P$ be an integral domain, where K is a subring of S, P is a nonzero prime ideal of S, and $K \cap P = (0)$. Let A be a subring of K, and put $R = A + P$. Also, denote the quotient field of S by L. We need several rather technical lemmas about inverses of ideals in R. Our first result is a slight generalization of [AR, Proposition 2.4 (1)].

LEMMA 2.1 (cf. [AR, Proposition 2.4]) Let I be a nonzero fractional ideal of A for which $I \subseteq K$ and $I^{-1} \subseteq K$. Then $(I+P)^{-1} \supseteq I^{-1}+P$. If $(I+P)^{-1} \subseteq S$, then $(I+P)^{-1} = I^{-1}+P$. (Here, $(I+P)^{-1}$ is computed with respect to R, and I^{-1} is computed with respect to A.)

Proof: The inclusion $\supseteq$ is straightforward (but does require $I, I^{-1} \subseteq K$). Let $v \in (I+P)^{-1} \subseteq S$. Write $v = u+p$, where $u \in K$ and $p \in P$. For $i \in I$, we have $ui = vi - pi \in R \cap K = A$, whence $u \in I^{-1}$ and $v \in I^{-1}+P$. □

LEMMA 2.2 Assume that K is a field. Let I be a nonzero fractional ideal of A[X] with $I, I^{-1} \subseteq K[X]$. Then $(I+P[X])^{-1} = I^{-1}+P[X]$.

Proof: We have $R[X] = A[X]+P[X]$ and $S[X] = K[X]+P[X]$. The inclusion $(I+P[X])^{-1} \supseteq I^{-1}+P[X]$ follows immediately from Lemma 2.1. The other inclusion will also follow from Lemma 2.1 once we show that $(I+P[X])^{-1} \subseteq S[X]$. Accordingly, let $h \in (I+P[X])^{-1}$. For any nonzero element $b \in P$, we have $bh \in R[X]$, whence $h \in L[X]$. Pick $0 \neq f \in I \cap A[X]$. By the content formula [G, Proposition 28.1] there is a positive integer m with $c(f)^{m+1}c(h) = c(f)^m c(fh) \subseteq R$. (Here c(f) denotes the content ideal of f, i.e., the ideal of R generated by the coefficients of f.) Let a be a nonzero element of $c(f)^{m+1} \cap A$. Then $ah \in R[X] \subseteq S[X]$. Since K is a field, a is a unit of S, and so $h \in S[X]$, as desired. □

LEMMA 2.3 Assume that K is a field. Let Q be an upper to zero in A[X], and set $Q_1 = Q+P[X]$, so that Q_1 is an upper to P in R[X] (that is, $Q_1 \cap R = P$ but $Q_1 \neq P[X]$). Let F denote the quotient field of A. Write $Q = fF[X] \cap A[X]$ with $f \in A[X]$, and let b be a nonzero element of P. Then $(b,f)^{-1} = (fA[X]+P[X])^{-1} = Qf^{-1}+P[X]$, and $(b,f)_v = (fA[X]+P[X])_v = Q^{-1}f+P[X]$.

Proof: Since $Qf^{-1} \subseteq F[X] \subseteq S[X]$, we have $Qf^{-1}P[X] \subseteq P[X] \subseteq R[X]$. Also, $Qf^{-1} \cdot fA[X] = Q \subseteq R[X]$. It follows that $(b,f)^{-1} \supseteq (fA[X] + P[X])^{-1} \supseteq Qf^{-1} + P[X]$. Let $h \in (b,f)^{-1}$. Since $bh \in R[X]$, we have $h \in L[X]$. Since $fh \in R[X]$, we may use the content formula as above to get $h \in S[X]$. Write $h = k + g$ with $k \in K[X]$ and $g \in P[X]$. Then $kf = hf - gf \in R[X] \cap K[X] = A[X]$. Thus $kf \in fK[X] \cap A[X] = fF[X] \cap A[X] = Q$, and $k \in Qf^{-1}$. Hence $h \in Qf^{-1} + P[X]$. Therefore, $(b,f)^{-1} = (fA[X] + P[X])^{-1} = Qf^{-1} + P[X]$. To complete the proof, we need only show that $(Qf^{-1} + P[X])^{-1} = Q^{-1}f + P[X]$. However, since $Qf^{-1}, Q^{-1}f \subseteq F[X] \subseteq K[X]$, this follows from Lemma 2.2. □

THEOREM 2.4 With the hypotheses of the preceding lemma, Q_1 is a t-prime of $R[X]$ if and only if Q is not almost principal in $A[X]$; and Q_1 is divisorial if and only if Q is divisorial but not almost principal.

Proof: Suppose that Q is not almost principal. Let $h_1, \ldots, h_n \in Q_1$. We wish to show that $(h_1, \ldots, h_n)_v \subseteq Q_1$. Write $h_i = f_i + g_i$ with $f_i \in Q$ and $g_i \in P[X]$. Choose $a \neq 0$ in A with $af_i \in fA[X]$ for each i. Then $a(h_1, \ldots, h_n)_v \subseteq (fA[X] + P[X])_v$; since $a \notin Q_1$, it suffices to show that $(fA[X] + P[X])_v \subseteq Q_1$. However, by Lemma 2.3 and Proposition 1.1, $(fA[X] + P[X])_v = Q^{-1}f + P[X] \subseteq Q + P[X] = Q_1$.

Conversely, suppose that Q is almost principal. By Proposition 1.1, $Q^{-1}f \not\subseteq Q$. Let b be a nonzero element of P. By Lemma 2.3 $(b,f)_v = Q^{-1}f + P[X]$, from which it follows easily that $(b,f)_v \not\subseteq Q_1$, and Q_1 is not a t-ideal.

For the divisoriality conclusions, we may assume that Q is not almost principal. Thus by Proposition 1.1 $Q^{-1} \subseteq F[X] \subseteq K[X]$. Two applications of Lemma 2.2 then yield $(Q_1)_v = Q_v + P[X]$, from which the desired conclusions follow easily. □

EXAMPLE 2.5 Let A be a domain with quotient field K, and assume that $A[X]$ contains an upper to zero Q which is not almost principal; e.g., let A be the example in 1.3. Let S be any domain of the form $K + P$, where K is a field containing A as a subring, and put $R = A + P$. Then $Q_1 = Q + P[X]$ is a t-prime in R[X] which is not an upper to zero and is not extended from any ideal of R. □

We close with some brief remarks on the t-dimension of R[X]. Since t-ideals of R extend to t-ideals of R[X], we have $\text{t-dim}(R[X]) \geq \text{t-dim}(R)$, and it follows easily from [HZ, Proposition 1.1] that $\text{t-dim}(R[X]) \leq 2 \cdot \text{t-dim}(R)$; this corresponds to the classical inequalities for Krull dimension of A. Seidenberg [S]. Since uppers to zero are t-primes, it is easy to obtain examples for which $\text{t-dim}(R[X]) = \text{t-dim}(R) + 1$. To obtain examples with t-dim(R[X]) assuming values between $\text{t-dim}(R) + 1$ and $2 \cdot \text{t-dim}(R)$, however, one must find an R with adjacent t-primes $0 \neq P_1 \subseteq P_2$ for which there is an upper Q_1 to P_1 in R[X] such that Q_1 is a t-prime and $Q_1 \subseteq P_2[X]$. In the example above, Q_1 is an upper to $P_1 = P$. Since $R/P \simeq A$, the prime ideal structure of R above P is that of A (and the prime ideal structure of R[X] above P[X] is that of A[X]). Thus the desired prime P_2 would correspond to a height-one prime of A whose extension to A[X] contains the upper to zero Q. Unfortunately, A is two dimensional, and it is easy to show that Q is not contained in the extension to A[X] of any height-one prime ideal. (Q does lie in the extension of a (t-)height-two t-prime of A.) Moreover, as mentioned above (Remark 1.4), I do not know of a one-dimensional domain A for which A[X] contains an upper to zero which is not almost principal. Thus, although our construction produces non-extended, non-upper t-primes, it does not produce an R with $\text{t-dim}(R[X]) > \text{t-dim}(R) + 1$.

Our construction also does not produce an example of a non-extended non-upper divisorial prime ideal, for, according to Theorem 2.4, such an example would require that A[X] contain an upper to zero which is divisorial but not almost principal; but the known examples of non-almost principal uppers to zero are also not divisorial (including the upper to zero of Example 2.3).

REFERENCES

[AR] D. F. Anderson and A. Ryckaert, The class group of D+M, J. Pure Appl. Algebra 52 (1988), 199-212.

[G] R. Gilmer, Multiplicative Ideal Theory, Marcel Dekker, New York, 1972.

[HH] J. Hedstrom and E. Houston, Some remarks on star-operations, J. Pure Appl. Algebra 18 (1980), 37-44.

[HHJ] E. Hamann, E. Houston, and J. Johnson, Properties of uppers to zero in R[x], Pacific J. Math. 135 (1988), 65-79.

[HZ] E. Houston and M. Zafrullah, On t-invertibility II, Comm. Algebra 17 (1989), 1955-1969.

[O] A. Ouertani, Exemples de dimension de Krull d'anneaux de polynômes, Canad. Math. Bull. 33 (1990), 135-138.

[S] A. Seidenberg, A note on the dimension theory of rings, Pacific J. Math. 3 (1953), 505-512.

17
A Characterization of Semi-Artinian Rings

ABDELOUAHAB IDELHADJ Département de Mathématiques, Faculté des Sciences de Tétouan, Tétouan, Morocco

AMINE KAIDI Département de Mathématiques, Faculté des Sciences de Rabat, Rabat, Morocco

INTRODUCTION

Let A be a commutative ring and M an A-module. The socle of M is the sum of simple submodules of M, the socle of A is the socle of the A-module A. The socle of an A-module is an invariant of its isomorphy class but does not determine it. That is to say, if M and N are two isomorphic A-modules then soc M and soc N are isomorphic, but the inverse is not true in general. So in the category of groups considered as Z-modules we have :

$$\mathrm{Soc}\left(Q_p / Z\right) \approx \mathrm{Soc}\left(Z / p^n Z\right) \approx (Z / pZ) \quad \text{and} \quad \mathrm{Soc}\, Q \approx \mathrm{Soc}\, Z$$

where Q is the set of rational numbers, Z is the set of integers, p is a prime number, n is a positive integer, and $Q_p = \left\{ \frac{a}{p^n}, a \in Z, n \in N \right\}$.

However, we observe that we can characterize by socle the isomorphy class of each group involved in these examples, where the modules are restricted to a particular class of Z-modules. Therefore :

- Q_p/Z is the only indecomposable Z-module of infinite length with socle Z/pZ .

- Z/p^nZ is the only indecomposable Z-module of finite length with socle Z/pZ .
- Q is the only divisible indecomposable non-zero Z-module with zero socle .
- Z is the only non-zero cyclic Z-module with zero socle .
As a consequence of the foregoing observation, we are giving a definition which will render the formulation more precise .

DEFINITION. We say that a class of A-modules is socle fine if for all M and N in this class, M and N are isomorphic if and only if socle of M and socle of N are isomorphic.

1 DIVISIBLE MODULES OVER PRINCIPAL DOMAINS RINGS

PROPOSITION 1.1 Let A be a principal domain ring and K its field of fractions . For two divisible A-modules M and N the following conditions are equivalent :

1. Soc M ≈ Soc N .
2. There is a set J such that :

$$M \approx N \oplus K^{(J)} \text{ or } N \approx M \oplus K^{(J)}$$

Proof : Let A be a principal domain ring and let p be an irreducible element of A. We write :
A_p = {m/n , m ∈ A , n ∈ A* , p and n are relatively prime} and
$A_{p^\infty} = K/A_p$
Let M and N two divisible modules over the principal domain ring A, we have ;

$$M \approx \bigoplus_{p \in w} A_{p^\infty} \oplus K^{(I)} \quad \text{and} \quad N \approx \bigoplus_{p' \in w'} A_{p'^\infty} \oplus K^{(I')}$$

where w and w' are subsets of irreductible elements of A, I and I' are two sets .
For every irreductible p of A, A_{p^∞} is the injective hull E(A/pA) of A/pA hence soc $A_{p^\infty} \approx A/pA$ and since soc $K^{(I)} = \text{soc } K^{(I')} = 0$, then ;

$$\text{soc } M_1 \approx \text{soc } (\bigoplus_{p \in w} A_{p^\infty}) \approx \bigoplus_{p \in w} A^* / pA^*$$

likewise $\operatorname{soc} M_2 \approx \bigoplus_{p' \in w'} A/p'A$. If $\operatorname{soc} M_1 \approx \operatorname{soc} M_2$ we will have

$\bigoplus_{p \in w} A/pA \approx \bigoplus_{p' \in w'} A/p'A$. According to Krull-Remak-Schmidt theorem for every p in w there is a p' in w' such that $A/pA \approx A/p'A$, therefore $E(A/pA) \approx E(A/p'A)$ and $Ap\infty \approx Ap'\infty$ hence

$\bigoplus_{p \in w} Ap\infty \approx \bigoplus_{p' \in w'} Ap'\infty$. If we take

$H = \bigoplus_{p \in w} A/pA \approx \bigoplus_{p' \in w'} A/p'A$ we will get :

$M_1 \approx H \oplus K^{(I)}$ or $M_2 \approx H \oplus K^{(I')}$ hence the existence of the set J .

COROLLARY 1.2 If A is a principal domain ring then the class of the divisible indecomposable A-modules is socle-fine .

PROPOSITION 1.3 If A is a principal domain ring then the class of indecomposable A-modules with the same finite length n is socle-fine.
Proof: let M and N be two indecomposable A-modules with the same finite length n over a principal domain ring A. There exists two irreductible elements p and q of A such that $M \approx A/p^nA$ and $N \approx A/q^nA$. Then $\operatorname{soc}M \approx \operatorname{soc}N$ implies $A/pA \approx A/qA$, hence $pA \approx qA$ therefore $M \approx N$.

2 THE SOCLE-FINE CLASS OF MODULES OVER SPECIFIED RINGS

In all the what follows the rings A are not necessarily commutative, and the A-modules will be left A-modules. We notice that the socle-fine property of certain classes of modules can give some properties of the ring A, the two following propositions give a result of this kind .

PROPOSITION 2.1 For a non-zero local ring A, the following properties are equivalent .

1. The class of finitely generated projective A-modules is socle fine.
2. Soc A is a non zero finitely generated A-module.

Proof:1) $\Rightarrow$ 2) socA $\neq 0$, if not we will have socA = socO and since A is projective then A = 0, which contradicts the fact that A is a non-zero ring. Let S = A/J(A) the only simple A-module, with J(A) the Jacobson radical of A. Soc A= $S^{(I)}$, if I is infinite we will have $S^{(I)} \oplus S^{(I)} \approx S^{(I)}$ and soc (A $\oplus$ A) $\approx$ soc(A) and then A $\approx$ A $\oplus$ A that is absurde since A is local hence indecomposable .

2/ $\Rightarrow$ 1/ Let P and P' two projective finitely generated A-modules, since A is local P and P' are free and there exist, two integers n and m such that $P \approx A^n$ and $P' \approx A^m$, if soc P $\approx$ soc P' we will have $(\text{soc } A)^n \approx (\text{soc } A)^m$ or there exists an integer r such that soc A $\approx S^r$. $(\text{soc } A)^n \approx (\text{soc } A)^m$ implies $S^{rn} \approx S^{rm}$ hence n = m and P $\approx$ P' .

We know that any module has an injective hull, but any module has a projective hull is not true; it is true if the ring is perfect, it characterises even the perfect rings. We say that a ring A is semiperfect if any finitely generated A-module has a projective hull. The principal results on semiperfect ring structures are :

if A is a semiperfect ring then there exists integers $k_1, k_2, \ldots, k_m$ such that, $A \approx (A e_1)^{k_1} \oplus \ldots \oplus (A e_m)^{k_m}$ where the Ae_i form a complete set of representatives of indecomposable projective A-modules and $S_i \approx Ae_i/Je_i$ a complete set of simple A-modules. Assume $\text{Soc } A e_i = S^{\alpha_{i,1}} \oplus \ldots \oplus S_m^{\alpha_{i,m}}$ and let $\text{Det}\left(\alpha_{i,j}\right)_{\substack{1\leq i\leq m\\ 1\leq j\leq m}}$ denote the determinant of the matrix $\left(\alpha_{i,j}\right)_{\substack{1\leq i\leq m\\ 1\leq j\leq m}}$. We have the following proposition :

PROPOSITION 2.2 Let A be a semiperfect ring . The following properties are equivalent :

1. $\text{Det}\left(\alpha_{i,j}\right)_{\substack{1\leq i\leq m\\ 1\leq j\leq m}} \neq 0$
2. The class of finitely generated projective A-modules is socle-fine .

Proof : Let P a finitely generated projective A-module , we have :

$$P \approx Ae_1^{l_1} \oplus \ldots \oplus Ae_m^{l_m} \quad \text{where } l_1, \ldots, l_m \text{ in } N$$

and $\operatorname{Soc}P \approx (\operatorname{Soc} Ae_1)^{l_1} \oplus \ldots \oplus (\operatorname{Soc} Ae_m)^{l_m}$

with $\operatorname{Soc} Ae_i = S_1^{\alpha_{i,1}} \oplus \ldots \oplus S_m^{\alpha_{i,m}}$ for $1 \le i \le m$, then :

$$\operatorname{Soc} P \approx (S_1^{\alpha_{1.1}} \oplus \ldots \oplus S_m^{\alpha_{1.m}})^{l_1} \oplus (S_1^{\alpha_{2.1}} \oplus \ldots \oplus S_m^{\alpha_{2.m}})^{l_2}$$

$$\oplus \ldots \oplus (S_1^{\alpha_{m.1}} \oplus \ldots \oplus S_m^{\alpha_{m.m}})^{l_m}$$

hence: $\operatorname{Soc} P \approx (S_1^{\alpha_{1.1} l_1 + \ldots + \alpha_{m.1} l_m}) \oplus (S_1^{\alpha_{1.2} l_1 + \ldots + \alpha_{m.2} l_m})$

$$\oplus \ldots \oplus (S_1^{\alpha_{1.m} l_1 + \ldots + \alpha_{m.m} l_m}).$$

Let P' another finitely generated projective A-module with

$$P' \approx Ae_1^{t_1} \oplus \ldots \oplus Ae_m^{t_m} \quad \text{where } t_1, \ldots, t_m \text{ in } N$$

we will have likewise :

$$\operatorname{Soc}P \approx (S_1^{\alpha_{1.1} t_1 + \alpha_{2.1} t_2 + \ldots + \alpha_{m.1} t_m}) \oplus \ldots \oplus (S_m^{\alpha_{1.m} t_1 + \ldots + \alpha_{m.m} t_m})$$

if $\operatorname{Soc} P \approx \operatorname{soc} P'$ we will have:

$$\alpha_{1.1} l_1 + \alpha_{21} l_2 + \ldots + \alpha_{m.1} l_m = \alpha_{1\,1} t_1 + \alpha_{21} t_2 + \ldots + \alpha_{m.1} t_m$$

$$\alpha_{1\,2} l_1 + \alpha_{22} l_2 + \ldots + \alpha_{m.2} l_m = \alpha_{1\,2} t_1 + \alpha_{22} t_2 + \ldots + \alpha_{m.2} t_m$$

$$\vdots \qquad\qquad \vdots$$

$$\alpha_{1.m} l_1 + \alpha_{2.m} l_2 + \ldots + \alpha_{m.m} l_m = \alpha_{1.m} t_1 + \alpha_{2.m} t_2 + \ldots + \alpha_{m.m} t_m$$

this system is equivalent to the following

$$\alpha_{1.1}(l_1 - t_1) + \alpha_{2.1}(l_2 - t_2) + \ldots\ldots + \alpha_{m.1}(l_m - t_m) = 0$$

$$\alpha_{1.2}(l_1 - t_1) + \alpha_{2.2}(l_2 - t_2) + \ldots\ldots + \alpha_{m.2}(l_m - t_m) = 0$$

$$\vdots \qquad\qquad \vdots$$

$$\alpha_{1.m}(l_1 - t_1) + \alpha_{2.m}(l_2 - t_2) + \ldots\ldots + \alpha_{m.m}(l_m - t_m) = 0$$

then if $\mathrm{Det}\left(\alpha_{i,j}\right)_{\substack{1\le i\le m\\ 1\le j\le m}} \neq 0$ we will have $l_1 = t_1$, $l_2 = t_2, \ldots$, et $l_m = t_m$ and therefore $P \approx P'$.

2) $\Rightarrow$ 1) Let u the endomorphism of the Z-module Z^m defined by the matrix $\left(\alpha_{i,j}\right)_{\substack{1\le i\le m\\ 1\le j\le m}}$. Let $(X_1, \ldots, X_m)$ and $(X'_1, \ldots, X'_m)$ in N^m such that $u(X_1, \ldots, X_m) = u(X'_1, \ldots, X'_m)$ we will have

$$\alpha_{1.1}X_1 + \ldots + \alpha_{m.1}X_m = \alpha_{1\,1}X_1 + \ldots + \alpha_{m.1}X_m$$

$$\vdots$$

$$\alpha_{1.m}X_1 + \ldots + \alpha_{m.m}X_m = \alpha_{1\,m}X_1 + \ldots + \alpha_{m.m}X_m$$

then $\mathrm{soc}\left(A e_1^{X_1} \oplus \ldots \oplus A e_m^{X_m}\right) = \mathrm{soc}\left(A e_1^{X'_1} \oplus \ldots \oplus A e_m^{X'_m}\right)$

and since A verifies 2) then :

$$A e_1^{X_1} \oplus \ldots \oplus A e_m^{X_m} = A e_1^{X'_1} \oplus \ldots \oplus A e_m^{X'_m}$$

we have $\mathrm{End}_A(Ae_i)$ is local for every i in $\{1, \ldots, n\}$ hence $X_1 = X'_1, \ldots, X_m = X'_m$ and $(X_1, \ldots, X_m) = (X'_1, \ldots, X'_m)$ hence u is monomorphism and we have according the following lemma :

LEMMA 2.3 : ([2] ch.3). Let A be a commutative ring and $u : A^{(m)} \longrightarrow A^{(m)}$ an endomorphism (m in N). Let M_u the matrix of u, we have; u is monomorphism if and only if $\mathrm{Det}\, M_u \neq 0$.

3 INJECTIVE MODULES OVER SEMI-ARTINIAN RINGS

Recall that a module M is semi-artinian (Stenström [6]. Prop 6.2.5) if and only if every non-zero quotient module of M has non zero socle. A ring A is called semi-artinian if A is semi-artinian as a left A-module. There are many characterizations (see Stenström [6] ; in Faith [3] semi-ar tinian = socular) .

PROPOSITION 3.1 If A is an artinian ring, then the class of injective A-modules is socle-fine.

Proof: Since A an artinian ring, every injective A-module is a direct sum of injective hulls of simple A-modules. Let M and N be two injective A-modules, we have $M \approx \bigoplus_{i \in I} E(S_i)$ and $N \approx \bigoplus_{j \in J} E(S'_j)$ where S_i and S'_i are the simple submodules respectively of M and N. If soc M $\approx$ soc N we will have $\bigoplus_{i \in I} S_i \approx \bigoplus_{j \in J} S'_J$ According to the Krull-Remak Shmidt there exists a bijection $\sigma : I \longrightarrow J$ such that $S_i \approx S_{\sigma(i)}$ and $E(Si) \approx E(S_{\sigma(i)})$ then $\bigoplus_{i \in I} S_i \approx \bigoplus_{j \in J} S'_J$ hence $M \approx N$.

THEOREM.3.2 The following are equivalent for a ring A :

1. A is semi-artinian .
2. The class of injective A-modules is socle-fine .

Proof : 1) $\Rightarrow$ 2) If A is semi-artinian, then for every cyclic A-module (x) , soc(x)$\neq$0, because (x) =A/Ann(x). If M is an A-module and $x \neq 0$ in M, we have soc(x) = socM $\cap$(x), so socM$\neq$0. Let K a submodule of M we have soc(x) = socM$\cap$(x) $\subset$ socM$\cap$K for all x in K, so socM is essential in M for all A-module M, and then, the injective hull, E(M) = E(socM). Let M and M' be two injective A-modules, then E(socM)=M and E(socM')=M', so if socM$\approx$socM' you'll have M $\approx$ M' .

2) $\Rightarrow$1) Let M be an A-module such that soc(M)=0, then;

soc(M) = soc(0) = soc(E(M)), E(M) and (0) are injective, assume 2) then E(M)=0 , so M=0. Then for every A-module M$\neq$0 we have soc(M) is non-zero, then A is semi-artinian .

COROLLARY. 3. 3 For a noetherian ring A the following conditions are equivalent ;

1. A is artinian .
2. the class of injective A-modules is socle-fine .

Proof: Assume 1) Then A is semi-artinian, and according to the Theorem 3.1. we have 2).
Assume 2) and let M a non-zero injective A-module. Then soc(M)$\neq$0, and

$socM = \bigoplus_{i \in I} S_i$ (S_i the simple submodules of M). We have

$soc(\bigoplus_{i \in I} E(S_i)) = \bigoplus_{i \in I} socE(S_i) = \bigoplus_{i \in I} S_i$ =soc(M).Since A is noetherian then

$\bigoplus_{i \in I} E(Si)$ is injective and M is injective, then $M = \bigoplus_{i \in I} E(Si)$, so every

injective A-module is a direct sum of injective hulls of simple A-modules. Then A is artinian .

REMARK 3.4 Since a ring A is left noetherian if, and only if, every injective A-module is a direct sum of indecomposable injective A-modules; and an indecomposable injective A-module with a non-zero socle is an injective hull of a simple A-module, then for a noetherian ring we have the following proposition :

PROPOSITION 3.5 For any left notherian ring A the following properties are equivalent :

1. A is artinian .
2. the socle of any indecomposable injective A-module is non-zero .

REMARK 3.6 Let A be a ring with Jacobson radical J. If A is a left artinian ring, then J is right T-nilpotent and A/J is semi-simple. The inverse is not true. But if A/J is semi-simple and J is right nilpotent then A is semi-artinian, hence the class of injective A-modules is socle-fine. According to remark 3.4.we will have the Hopkins-theorem ([1] p : 72) .

COROLLARY.3.7 For a ring A with Jacobson radical J, the following properties are equivalent ;

1. A is left artinian .
2. A is left noetherian, J is T-nilpotent and A/J is semi-simple .

REFERENCES

1. Anderson,F. and Fuller,K. (1974). Rings and categories of modules, Springer-Verlag. Berlin. Heidelberg. New York.

2. Bourbaki, N. (1981). Elements de Mathématique ch.3, Masson. Paris.

3. Faith,C.(1976). Algebra II, Rings Theory Springer-Verlag. Berlin.New York.

4. Kent, R. and Fuller, K. (1969). On indecomposable injectives over artinian, Pac.J.Math.vol.29 N°1.

5. Muller, B. (1970). On semi-perfect rings, J.Math. 14. 464-467

6. Stenström,B.(1975). Rings of Quotients, Springer-Verlag Berlin Heidelberg New York .

18
The Ring of Finite Fractions

THOMAS G. LUCAS Department of Mathematics, University of North Carolina at Charlotte, Charlotte, North Carolina

1 INTRODUCTION

Let R be a commutative ring with unit having total quotient ring $T(R)$. The total quotient ring of R is made up of the "simple" fractions of the form r/s where both r and s are in R and s is regular; i.e., s is not a zero divisor. (As in forming the quotient field of an integral domain, two fractions r/s and t/u are equivalent if $ru = ts$.) The fraction r/s can also be viewed as the R-module homomorphism from the principal ideal sR into R which maps s to r. Of course for a nonzero ideal I of an integral domain D, each D-module homomorphism mapping I into D can be viewed as an element of the quotient field of D. For example, if $a \in I \backslash \{0\}$ and $f \in Hom_D(I, D)$ with $f(a) = b$, then for each $c \in I$ $af(c) = cf(a) = cb$. Hence $f(c) = cb/a$. For rings with nonzero zero divisors a similar thing happens for the regular ideals; i.e., those which contain regular elements. But for ideals which contain only zero divisors there may be homomorphisms which cannot be defined simply by multiplication by elements of the total quotient ring. The homomorphisms we will be concerned with are those which are defined on finitely generated faithful ideals (an ideal I is "faithful" if $rI = (0)$ implies $r = 0$). For reasons to appear later we call such homomorphisms *finite fractions*.

Let D be an integral domain and let $\mathcal{P}$ be a nonempty subset of $Spec(D)$. Let $\mathcal{A}$ be an index set for $\mathcal{P}$ and let $\mathcal{I} = \mathcal{A} \times \mathbf{N}$ where $\mathbf{N}$ is the set of natural numbers. For each $i = (\alpha, n)$ in $\mathcal{I}$, let K_i be the quotient field of D/P_α. Construct the D-module $B = \sum K_i$ and form the ring $R = D + B$ from the direct sum of D and B by defining multiplication as $(r, b)(s, c) = (rs, rc + sb + bc)$. Here we are viewing B

both as a D-module (for the products rc and sb) and as a ring without unit (for the product bc).

The total quotient ring of R can be identified with $D_S + B$ where $S = \{r \in D : r \notin \bigcup P_\alpha\}$. Also an ideal I of R is faithful if and only if $I = J + B$ where J is an ideal of D which is not contained in any P_α. For more on this construction see [13, §§ 26 and 27] and [19, §1].

EXAMPLE 1. *Let K be a field and let $D_1 = K[Y^2, Y^3, YZ, Z]$. Construct the ring $R_1 = D_1 + B$ using $\mathcal{P} = Spec(D_1)\backslash\{(Y^2, Y^3, YZ, Z)\}$. In this case $S = K$ so $T(R_1) = R_1$. Thus R_1 is integrally closed. However the polynomial ring $R_1[X]$ is not integrally closed.*

(Note that when we say that a ring is integrally closed we mean that it is integrally closed in its total quotient.)

For ease of notation we shall use lower case letters to denote Y and Z as elements of R and continue to use upper case when we consider them in D. Since the ideal (Y^2, Z) is not contained in any of the prime ideals in $\mathcal{P}$, the corresponding ideal (y^2, z) is a finitely generated faithful ideal of R_1. Thus the polynomial $y^2X + z$ is a regular element of $R_1[X]$. Hence y "$=$"$(y^3X + yz)/(y^2X + z) \in T(R_1[X])\backslash R_1[X]$ since $Y \notin D_1$. But $((y^3X + yz)/(y^2X + z))^2 = y^2 \in R_1$. Thus $R_1[X]$ is not integrally closed.

With a slight modification we get a ring R_2 where $R_2[X]$ is integrally closed.

EXAMPLE 2. *Let K be a field and let $D_2 = K[Y^2, Y^3, Z]$. Construct the ring $R_2 = D_2 + B$ using $\mathcal{P} = Spec(D)\backslash\{(Y^2, Y^3, Z)\}$. As in Example 1, $S = K$ so that $T(R_2) = R_2$. However, unlike the ring R_1, $R_2[X]$ is integrally closed.*

Since $YZ \notin D_2$, $(D_2 : Y) = (Y^2, Y^3) \in \mathcal{P}$. Thus if $YJ \subset D_2$, then the ideal $J + B$ is not faithful. It follows that y cannot be represented as a quotient of polynomials in $T(R_2[X])$ even though (y^2, z) is a faithful ideal of R_2. (There are a number reasons why $R_2[X]$ is integrally closed. For example for a ring of the form $R = D + B$, if $D = \bigcap\{D_{P_\alpha} : P_\alpha \in \mathcal{P}\}$, then $R[X]$ is integrlly closed [19, Theorem 1.3]. Thus since $J^{-1} \neq D_2$ for an ideal J of D_2 only if $J \subset P_\alpha$ for some $P_\alpha \in \mathcal{P}$, $R_2[X]$ is integrally closed.)

2 INTEGRALITY AND R[X]

There are many examples in the literature of integrally closed reduced rings for which the corresponding polynomial rings are not integrally closed. (See for example [1] ,[5] and [17].) It turns out that in each case there is an element of $T(R[X])\backslash R[X]$ which is integral over R. In R. Gilmer's book [8] there is an exercise which asks the reader to "determine necessary and sufficient conditions in order that $R[\{X_\lambda\}]$ be integrally closed" [8, Exercise 11, page 100]. Conditions which are obviously necessary (but not sufficient) are that R be reduced and integrally closed (in $T(R)$). However, requiring R to be reduced and integrally closed in $T(R[X])$ is both necessary and sufficient for $R[X]$ to be integrally closed.

Before proving the above result we need to introduce the ring of finite fractions, which we denote by $Q_0(R)$. An ideal I is said to be ***semi-regular*** if it contains a finitely generated faithful ideal. The set $\mathcal{F}$ of all semi-regular ideals is closed to finite products and intersections. Let $I_1, I_2 \in \mathcal{F}$ and let $f_1 \in Hom_R(I_1, R), f_2 \in$

$Hom_R(I_2, R)$. By restricting the domains of both f_1 and f_2 to I_1I_2, we can easily define both the product f_1f_2 and the sum $f_1 + f_2$ as homomorphisms on I_1I_2. Moreover, a relatively straightforward argument shows that if a pair of such homomorphisms agree on some faithful ideal, then they agree on every faithful ideal on which they are both defined (see for example [15, Lemma 1, page 38]). This leads to the construction of an equivalence relation on the set of homomorphisms defined on semi-regular ideals. Since the product of two semi-regular ideal is again semi-regular, we can form the ring $Q_0(R)$ from the equivalence classes of homomorphisms defined on semi-regular ideals of R.

EXAMPLE 3. *Let D_1 and $\mathcal{P}$ be as in Example 1 and let R be the "idealized" ring $D_1(+)N$ where $N = \sum K_\alpha$. Then $R = T(R)$ (as in Example 1) and $Q_0(R)$ can be identified with the ring $K[Y, Z](+)N$.*

PROOF: The elements of R are of the form (r, a) where $r \in D_1$ and $a \in N$. Addition and multiplication are defined as follows:

For elements (r, a) and (s, b) in $D_1 \oplus N$, $(r, a)+(s, b) = (r+s, a+b)$ and $(r, a)(s, b) = (rs, ra + sb)$.

For an ideal I of R, the set $J = \{r \in D_1 : (r, b) \in I$ for some $b \in N\}$ is an ideal of D_1. Moreover, $I \subset J(+)N$. It follows that the maximal ideals of R are all of the form $M'(+)N$ where M' is a maximal ideal of D_1. Let $M = (Y^2, Y^3, YZ, Z)$. If $M' \neq M$, then $M' \in \mathcal{P}$ and, hence $Ann(M'(+)N) \neq (0)$ (since one of the K_α's equals D_1/M'). It follows that if I is a faithful ideal of R, then the corresponding ideal J of D_1 must be M-primary. But if J is M-primary, then $JN = N$ since J survives in each K_α. Thus the faithful ideals of R are all of the form $J(+)N$ where J is an M-primary ideal of D_1. Since the integral closure of D_1 is $K[Y, Z]$ (and M is height two), if J is M-primary, then $J^{-1} = K[Y, Z]$. It follows that $Hom_R(J(+)N, R)$ can be identified with the ring $K[Y, Z](+)N$ (see [21, Theorem 11]). Hence $Q_0(R) = K[Y, Z](+)N$.

For the ring R_1 of the first example, multiplication by y defines an R_1-module homomorphism from the ideal (y^2, y^3, yz, z) into R_1. The proof of Theorem 27.1 in [13] can be easily adapted to show that a finitely generated ideal J of R_1 is faithful if and only if $J = A + B = AR_1$ for some finitely generated ideal A of D_1 such that A is not contained in any $P_\alpha \in \mathcal{P}$. Thus as in Example 3, if $J = A + B = AR_1$ is semi-regular, then $Hom_{R_1}(AR_1, R_1)$ can be identified with "$K[Y, Z] + B$". Hence we can identify $Q_0(R_1)$ with $K[Y, Z] + B$ (or if one prefers $K[y, z] + B$).

LEMMA 4. *Let $s \in Q_0(R)$. Then there exist polynomials f and g in $R[X]$ with $c(g)$ faithful in R such that $s \in Hom_R(c(g), R)$ and for each coefficient g_j of g, $s(g_j) = f_j$. In particular, $Q_0(R)$ can be considered as a subring of $T(R[X])$.*

PROOF: As $s \in Q_0(R)$ there is a finitely generated faithful ideal $J = (g_0, g_1, \ldots, g_n)$ such that $s \in Hom_R(J, R)$. For each $j = 0, 1, \ldots, n$ set $s(g_j) = f_j, f = \sum f_j X^j$ and $g = \sum g_j X^j$. Then for each i and $j, f_i g_j = f_j g_i$ since $g_i s(g_j) = g_j s(g_i)$. Hence,$(f/g)g_j = f_j$for each j and we can consider s as an element of $T(R[X])$.

For a reduced ring R, the complete ring of quotients, $Q(R)$, is always a von Neumann regular ring. (Recall that the complete ring of quotients of R consists of the equivalence classes of those homomorphisms which are defined on faithful ideals of R. A good source for information on $Q(R)$ is [15, pages 36–46].) Since von Neumann regular rings are locally fields, $Q(R)[X]$ is integrally closed when R is reduced. We have the following containment relations: $R \subset T(R) \subset Q_0(R) \subset Q(R)$ and $Q_0(R) \subset T(R[X])$. Hence $R[X] \subset Q_0(R)[X] \subset T(R[X]) = T(Q_0(R)[X]) \subset T(Q(R)[X])$.

LEMMA 5. *Let $s \in Q(R)$. If $sJ \subset Q_0(R)$ for some finitely generated faithful ideal J of R, then $s \in Q_0(R)$.*

PROOF: Let $J = (a_0, a_1, \ldots, a_n)$ be a faithful ideal of R such that $sa_j \in Q_0(R)$ for each j. By Lemma 4 there exist polynomials F_i and G_i in $R[X]$ with G_i regular such that $sa_i = F_i/G_i$. Let t be the maximum degree of the G_i's and define polynomials F and G by

$$F(X) = a_0F_0 + a_1F_1X^{t+1} + a_2F_2X^{2t+2} + \cdots + a_nF_nX^{nt+n}$$

and

$$G(X) = a_0G_0 + a_1G_1X^{t+1} + a_2G_2X^{2t+2} + \cdots + a_nG_nX^{nt+n}.$$

Since J is a faithful ideal of R and each of the G_i's is regular, G is a regular element of $R[X]$. Moreover, since $sa_i = F_i/G_i$, $F_i = sa_iG_i$ and thus $sG = F$ and $s = F/G \in T(R[X])$.

LEMMA 6. *If $s(X) \in Q(R)[X] \cap T(R[X])$, then $s(X) \in Q_0(R)[X]$.*

PROOF: Let $s(X) \in Q(R)[X] \cap T(R[X])$. Then $s(X) = f(X)/g(X)$ for some $f(X)$ and $g(X)$ is $R[X]$ with $g(X)$ regular. If $s(X)$ is not in $Q_0(R)[X]$, we may assume it is of minimal degree. Write $s(X) = s_kX^k + \cdots + s_0$ and $g(X) = g_mX^m + \cdots + g_0$. By Lemma 5 we may assume $k > 0$. Since $Q_0(R) \subset T(R[X])$, if $s_k \in Q_0(R)$, then $s(X) - s_kX^k \in Q(R)[X] \cap T(R[X])$ contradicting the minimality of degree of $s(X)$. Thus $s_k \in Q(R)\backslash Q_0(R)$. However $s_kg_m \in Q_0(R)$ so that $g_ms(X)$ is a polynomial in $Q_0(R)[X]$. In particular, $s_{k-1}g_m$ is in $Q_0(R)$ so that s_kg_{m-1} must also be in $Q_0(R)$ since the sum of the two is in R. Thus by minimality, $g_{m-1}s(X)$ is a polynomial in $Q_0(R)[X]$. Inductively we get $g_js(X) \in Q_0(R)[X]$ for $j = m, m-1, \ldots, 0$. In particular, $s_kg_j \in Q_0(R)$ for all j which by Lemma 5 implies $s_k \in Q_0(R)$ contradicting the minimality of degree of $s(X)$. Therefore, $s(X) \in Q_0(R)[X]$.

Note that since $Q_0(R) \subset Q(R)$, we actually have that $Q(R)[X] \cap T(R[X]) = Q_0(R)[X]$.

Recall that for an integral domain D with quotient field K, the integral closure of D is the ring $D' = \bigcup\{(I, I) : I$ is a finitely generated ideal of $D\}$. From the lemmas above we get a similar characterization of the integral closure of R in $Q_0(R)$.

THEOREM 7. *Let R be a reduced ring and let $R^\flat = \varinjlim\{Hom_R(I, I) : I$ is a finitely generated faithful ideal of $R\}$. Then $R^\flat$ is the integral closure of R in $Q_0(R)$ and $R^\flat[X]$ is the integral closure of $R[X]$ in $T(R[X])$.*

PROOF: For any pair of rings $D \subset E$ and indeterminate X, if D' is the integral closure of D in E, then $D'[X]$ is the integral closure of $D[X]$ in $E[X]$ [8,Theorem 10.7].

Let $s \in Q_0(R)$ be integral over R. Then there is a finitely generated faithful ideal $I = (a_0, a_1, \ldots, a_m)$ such that $s \in Hom_R(I, R)$. Moreover we can write $s = (b_mX^m + \cdots + b_0)/(a_mX^m + \cdots + a_0)$ where $sa_k = b_k$ for each k. Let $f(Z) = Z^{n+1} + f_nZ^n + \cdots + f_0$ be an equation of integrality of s over R and let $C = c(a^n) + c(a^{n-1}b) + \cdots + c(b^n)$ where $a = a_mX^m + \cdots + a_0$ and $b = b_mX^m + \cdots + b_0$. We will show that $s \in Hom_R(C,C)$. For each $j > 0$, we have $sa^j b^{n-j} = a^{j-1}b^{n-j+1} \in R[X]$ and $sc(a^j b^{n-j}) = c(a^{j-1}b^{n-j+1})$ since $s = b/a$ and $sa_k = b_k$. To see that $sc(b^n) \subset C$ we need to use the integrality equation. Write $s^{n+1} = -(f_n s^n + \cdots + f_0)$. Replacing each s^j by b^j/a^j we get $s(b^n/a^n) = -(f_n b^n + f_{n-1}b^{n-1}a + \cdots + f_0 a^n)/a^n$. Hence $sb^n = -(f_n b^n + \cdots + f_0 a^n)$. Whence, $sc(b^n) \subset C$ and $sC \subset C$ as desired. Thus R^b contains the integral closure of R in $Q_0(R)$.

Conversely, since $Q_0(R) \subset T(R[X])$, no nonzero element of $Q_0(R)$ can annihilate a finitely generated faithful ideal of$R$. Hence as in the integral domain case, if $sI \subset I$ for some finitely generated faithful ideal of R, Cramer's Rule implies s is integral over R. Whence R^b is the integral closure of R in $Q_0(R)$.

Since $Q(R)[X]$ is integrally closed and $T(R[X]) \subset T(Q(R)[X])$, the integral closure of $R[X]$ is a subset of $Q(R)[X] \cap T(R[X])$. Thus by Lemma 6, the integral closure of $R[X]$ is contained in $Q_0(R)[X]$ (and $Q_0(R)[X]$ is integrally closed). Thus $R^b[X]$ is the integral closure of $R[X]$.

The following corollary is immediate.

COROLLARY 8. *For a reduced ring R, $R[X]$ is integrally closed if and only if R is integrally closed in $Q_0(R)$.*

A ring R is said to have ***property A*** if every finitely generated faithful ideal is regular. Since $Hom_R(I,R) \subset T(R)$ for each regular ideal I, if R has property A, $Q_0(R) = T(R)$. The following result was first proved by T. Akiba in [1].

COROLLARY 9. *Let R be a ring with property A. Then $R[X]$ is integrally closed if and only if R is integrally closed (in $T(R)$).*

Every polynomial ring has property A. Thus $R[\{X_\lambda\}]$ is integrally closed if and only if $R[X]$ is. Thus the answer to Gilmer's question is that $R[\{X_\lambda\}]$ is integrally closed if and only if R is integrally closed in $Q_0(R)$.

Using Theorems 1 and 2 of [5], it is possible to characterize when $R[X]$ is n-root closed, root closed and seminormal. Basically $R[X]$ has a particular type of closure property if and only if R has the same closure property in $Q_0(R)$ (and R is reduced) (see [18]).

For a pair of rings $R \subset T$, Gilmer defines an element $t \in T$ to be ***almost integral over*** R (as an element of T) if there exists a finitely generated R-submodule M of T which contains all positive powers of t. The set of all elements of T which are almost integral over R forms a ring which is called the ***complete integral closure*** of R in T. The reference to T is omitted when $T = T(R)$ and R is said to be ***completely integrally closed*** when it equals its own complete integral closure. Note that if we have a pair of rings $T_1 \subset T_2$ (both containing R) and an element $s \in T_1$, Gilmer's definition of almost integral allows for s to be almost integral over R as an element of T_2 but not as an element of T_1.

For an element $t \in T(R)$ it is not hard to show that t is almost integral over R if and only if there exists a regular element $r \in R$ such that $rt^n \in R$ for each $n > 0$.

EXAMPLE 10. *Let K be a field and let $D_3 = K[\{UY^n : n \geq 0\}, \{VY^n : n \geq 0\}]$. Let $\mathcal{P}$ be the set of prime ideals of D_3 which do not contain both U and V and form the ring $R_3 = D_3 + B$. As in the first two examples, $S = K$ so that $T(R_3) = R_3$. Hence R_3 is trivially completely integrally closed. However, $R_3[X]$ is not completely integrally closed.*

Just like the ring R_1 in Example 1, we can represent the element y as an element of $T(R_3[X])$. Since no prime ideal in $\mathcal{P}$ contains (U, V), $uX + v$ is a regular element of $T(R_3[X])$. Hence $y = (uyX + vy)/(uX + v) \in T(R_3[X])$. Since $UY^n, VY^n \in D_3$ for each $n > 0$, $y^n(uX + v) \in R_3[X]$ for each $n > 0$. Thus y is almost integral over $R_3[X]$. Moreover, multiplication by y defines an R_3-module endomorphism on the semi-regular ideal J generated by the set $\{uy^n, vy^n : n \geq 0\}$. Note that even though y is an element of $Q_0(R_3)$ it is not almost integral over R_3 as an element of $Q_0(R_3)$. However $R_3[y] = Q_0(R_3)$ is the complete integral closure of R_3 in $T(R_3[X])$.

Recall from above that $\mathcal{F}$ is the set of all semi-regular ideals of the ring R. Just like $\bigcup\{(I : I) : I$ is a nonzero ideal of $D\}$ is the complete integral closure of the integral domain D, when R is reduced $R^\sharp = \varinjlim\{Hom_R(I, I) : I \in \mathcal{P}\}$ is the complete integral closure of R in $T(R[X])$. Moreover, the complete integral closure of $R[X]$ is $R^\sharp[X]$. Whence $R[X]$ is completely integrally closed if and only if (R is reduced and) R is completely integrally closed in $T(R[X])$. For a proof of these remarks see [20].

For an integral domain D, it is possible for D to be completely integrally closed domain while some localization D_S is not. The ring of entire functions is one such example (see [8, Exercises 16 and 21, pages 147 and 148]). Example 2.4 of [20] shows that it is possible for $R[X]$ to be completely integrally closed while $T(R)[X]$ is not.

3 PRÜFER RINGS

In the late 1950's, P. Samuel gave a number of different ways to extend the notion of a valuation domain to arbitrary pairs of rings $R \subset T$ (see [24]). One definition is in terms of a "valuation" map. Specifically a ***valuation*** on T is a function ϕ from T onto a totally ordered abelian group G (plus an element ∞) such that for all $r, s \in T$

(*i*) $$\phi(rs) = \phi(r)\phi(s),$$

(*ii*) $$\phi(r + s) \geq min\{\phi(r), \phi(s)\},$$

(*iii*) $$\phi(1) = 0 \text{ and } \phi(0) = \infty.$$

The ring R is said to be a ***valuation ring*** of T if there is a valuation ϕ on T such that $R = \{r \in T : \phi(r) \geq 0\}$. An alternate definition is in terms of "valuation pairs". Specifically, a pair (R,P) is a ***valuation pair*** of T if P is a (prime) ideal of R and for each $t \in T \backslash R$ there is an element $r \in P$ such that $rt \in R \backslash P$. In the late 1960's, M. Manis introduced valuation pairs and proved the following result [22].

THEOREM 11. *Let $R \subset T$ be a pair of rings and let P be a prime ideal of R. Then the following are equivalent.*

1. *(R, P) is a valuation pair of T.*
2. *There is a valuation ϕ on T such that $R = \{r \in T : \phi(r) \geq 0\}$.*
3. *The only ring between R and T with an ideal which contracts to P (in R) is R itself.*

Following the standard practice, when $T = T(R)$ the reference to T is dropped. When $T = Q_0(R)$, we refer to R as a Q_0- valuation ring and the pair (R, P) as a Q_0-valuation pair.

In [6], Davis proved that for an integral domain D, D is a Prüfer domain if and only if every overring of D is integrally closed. Thus one way to extend the notion of Prüfer domains to arbitrary pairs of rings $R \subset T$, is to say that R is a *T-Prüfer ring* if every ring between R and T is integrally closed in T. For $T = Q_0(R)$ this is saying that every *Q_0-overring* is integrally closed in $Q_0(R)$. When $T = T(R)$ this is equivalent to every finitely generated regular ideal of R being invertible. As with valuation rings, we call R a Prüfer ring or Q_0-Prüfer ring when T is either $T(R)$ or $Q_0(R)$. It is not the case that every valuation ring is a Prüfer ring (see for example [12] or [13, Example 7, page 182]). However it is true that Prüfer rings are in a way "locally" valuation rings.

Let P be a prime ideal of R. The *large quotient ring* of R with respect to P is the ring $R_{[P]} = \{t \in T(R) : rt \in R$ for some $r \in R\backslash P\}$ and the *Q_0-quotient ring* of R with respect to P is the ring $R_{\{P\}} = \{t \in Q_0(R) : rt \in Q_0(R)$ for some $r \in R\backslash P\}$. To extend P to these rings of "quotients" we use $[P] = \{t \in T(R) : rt \in P$ for some $r \in R\backslash P\}$ and $\{P\} = \{t \in Q_0(R) : rt \in P$ for some $r \in R\backslash P\}$. Both of these extensions contract to P in R and each is a prime ideal. A proof of the following can be found in both [16] and [11].

THEOREM 12. *The following are equivalent for the ring R.*

(1) *R is a Prüfer ring.*
(2) *$(R_{[M]}, [M])$ is a valuation pair for each maximal ideal M of R.*
(3) *Every finitely generated regular ideal of R is invertible.*

For $f \in Q_0(R)$, we let $dom_R f = \{r \in R : fr \in R\}$ and $ran_R f = f dom_R f$.

THEOREM 13. *The following are equivalent for the ring R.*

(1) *R is a Q_0-Prüfer ring.*
(2) *$(R_{\{P\}}, \{P\})$ is a Q_0-valuation pair for each maximal ideal P of R.*
(3) *For each $f \in Q_0(R)$, $dom_R f + ran_R f = R$.*
(4) *R is integrally closed in $Q_0(R)$ and each Q_0-overring of R is an intersection of Q_0-quotient rings of R.*

PROOF: The proof that (1) implies (2) is essentially the same as that given for Theorem 13, (5) implies (1) of [11].

For a proof of (2) implies (3) see the proof of Theorem 2, (ii) implies (iii) of [7].

The proof of (4) implies (1) follows from the following lemma.

LEMMA 14.

(A) *Let P be a prime ideal of R. If R is integrally closed in $Q_0(R)$, then $R_{\{P\}}$ is also integally closed in $Q_0(R)$.*
(B) *$R = \bigcap\{R_{\{M\}} : M \in MaxR\}$.*

To see that (3) implies (2), first note that for each $f \in Q_0(R)$ no prime ideal of R can contain both $dom_R f$ and $ran_R f$. Hence if P is a prime ideal of R and $f \in Q_0(R)\backslash R_{\{P\}}$, then there is an element $r \in P \subset \{P\}$ such that $rf \in R\backslash P \subset R_{\{P\}}\backslash\{P\}$. Whence for each maximal ideal (in fact each prime ideal) M of R, $(R_{\{M\}}, \{M\})$ is a Q_0-valuation pair.

To complete the proof we show that (2) implies (4). That R is integrally closed in $Q_0(R)$ follows from Lemma 14 and the fact that every Q_0-valuation ring is integrally closed in $Q_0(R)$. As (2) and (3) are equivalent, the proof of (3) implies (2) allows us to assert that for each prime ideal P of R, $(R_{\{P\}}, \{P\})$ is a Q_0-valuation pair.

Let V be a Q_0-overring of R and let M be a maximal ideal of V and let $P = M \cap R$. Moreover if $t \in \{M\} \cap R_{\{P\}}$, then there are elements $s \in V \backslash M$ and $r \in R \backslash P$ such that $st \in M$ and $tr \in R$. Thus $s(tr) \in M$ so that $tr \in M \cap R = P$. Whence $\{M\} \cap R_{\{P\}} = \{P\}$. By the above and Theorem 11 we have $V_{\{M\}} = R_{\{P\}}$. Therefore by Lemma 14, $V = \bigcap \{V_{\{M\}} : M \in MaxV\} = \bigcap \{R_{\{P\}} : P = M \cap R$ for some $M \in MaxV\}$.

A ring R is said to be an *I-ring* if every ring between R and $Q(R)$ is integrally closed in $Q(R)$. I-rings were named and first investigated by N. Eggert in [7]. Among other things, Eggert proved that if R is an I-ring, then for every $f \in Q(R)$ the set $\{r \in R : fr \in R\}$ is finitely generated [7, Theorem 6]. In other words, if R is an I-ring, then $Q_0(R) = Q(R)$. Hence R is an I-ring if and only if R is a Q_0-Prüfer ring and $Q_0(R) = Q(R)$. Results in [23] imply that if $R \subset T$ is a T-Prüfer ring, then in fact T must be a subring of $Q_0(R)$ (in the sense that for each element of $t \in T$, multiplication by t can be thought of as an R- module homomorphism on some semi-regular ideal of R).

Other variations on Prüfer domains include arithmetical rings (which won't be considered here) and strong Prüfer rings (which will). Obviously since we have an example of a total quotient ring R_1 for which the corresponding polynomial ring $R_1[X]$ is not integrally closed, $R[X]$ need not be Prüfer when R is. In fact Gilmer and T. Parker proved that $R[X]$ is a Prüfer ring if and only if R is von Neumann regular [10]. However if R is a Prüfer ring with property A, then $R(X) = R[X]_U$ is a Prüfer ring where U is the set of unit content polynomials of R [4]. A ring R is said to be a ***strong Prüfer ring*** if every finitely generated faithful ideal is locally principal—or equivalently, if $R(X)$ is Prüfer [4].

An ideal I is said to be Q_0-*invertible* if there is an R-module J of $Q_0(R)$ such that $IJ = R$. Essentially the same proof that shows an invertible ideal is finitely generated and locally principal shows that a Q_0-invertible ideal is also finitely generated and locally principal. However, while an invertible ideal must be regular, a Q_0-invertible ideal need only be faithful (see for example [8, Exercise 10, page 456]). Since $IHom_R(I, R) \subset R$ for any semiregular ideal I, if I is Q_0-invertible, then $IHom_R(I, R) = R$. It follows that a finitely generated faithful ideal I is Q_0-invertible if and only if I is locally principal since for each maximal ideal M, $Hom_{R_M}(I_M, R_M)$ is naturally isomorphic to $(Hom_R(I, R))_M$ (see also [14, Theorem 3.4.1]). Hence Q_0-invertible ideals are the same as finitely generated cancellation ideals (see [8, Exercises 5, 6 and 7, pages 66 and 67] and [3, Theorem 1]).

LEMMA 15. *Let I be a finitely generated faithful ideal of R. Then the following are equivalent.*

(1) *I is Q_0-invertible.*
(2) *I is locally principal.*
(3) *I is locally invertible.*
(4) *$IR(X)$ is invertible.*
(5) *$IR(X)$ is locally principal.*
(6) *$IR(X)$ is principal.*

PROOF: A proof of the equivalence of (1) and (2) appears above. For the others see either [2] or [13, Theorems 15.1 and 15.2].

THEOREM 16. *The following are equivalent for a ring R.*

(1) *R is a strong Prüfer ring.*
(2) *Every finitely generated faithful ideal of R is Q_0- invertible.*
(3) *$R(X)$ is a Prüfer ring.*
(4) *R is Q_0-Prüfer and $Q_0(R)$ has property A.*
(5) *R is integrally closed in $Q_0(R)$ and $T(R[X]) = Q_0(R)[X]_U$.*

PROOF: The equivalence of (1) and (2) follows from Lemma 15.

For the proof of (1)⇔(3) see [4, Theorem 3.2].

((1)⇒(4)) Let I be a finitely generated faithful ideal of R. Then I is locally principal and so by Lemma 15, $IHom_R(I, R) = R$.

If J is finitely generated faithful ideal of $Q_0(R)$, then there is a finitely generated faithful ideal I of R such that $IJ \subset R$. Thus $IJHom_R(IJ, R) = R$ and so $J = Q_0(R)$. Hence $Q_0(R)$ has property A.

Let $R \subset S \subset Q_0(R)$. Then $R(X) \subset S[X]_U \subset Q_0(R)[X]_U \subset T(R[X])$. Since $R(X)$ is Prüfer, $S[X]_U$ is integrally closed in $Q_0(R)[X]_U$. Thus $S = S[X]_U \cap Q_0(R)$ is integrally closed in $Q_0(R)$ and R is a Q_0-Prüfer ring.

((4)⇒(2)) Let I be finitely generated faithful ideal of R. Since $Q_0(R)$ has property A, $IQ_0(R) = Q_0(R)$. Let $I = (a_1, a_2, \ldots, a_n)$ and let $q_1, q_2, \ldots, q_n \in Q_0(R)$ be such that $a_1q_1 + a_2q_2 + \cdots + a_nq_n = 1$. Let M be a maximal ideal of R and let ϕ be the valuation corresponding to the Q_0-valuation pair $(R_{\{M\}}, \{M\})$. Since the set of products $\{q_ka_m\}$ is finite, the set $\{\phi(q_ka_m)\}$ has a minimum value say $-b \leq 0$. Let $r \in \{M\}$ be such that $\phi(r) = b$. Then $\phi(rq_ka_m) \geq 0$ for each k and m and is equal to 0 for some k and m. Again from finiteness, there is a element $t \in R\backslash M$ such that $trq_ka_m \in R$ for each k and m and $trq_ka_m \in R\backslash M$ for some k and m. Thus $trq_k \in Hom_R(I, R)$ for each k and $IHom_R(I, R) \not\subset M$. Therefore no maximal ideal contains $IHom_R(I, R)$ and I is Q_0-invertible.

((4)⇒(5)) If R is a Q_0-Prüfer ring, then R is integrally closed in $Q_0(R)$.

Let $a(X) = a_nX^n + \cdots + a_1X + a_0$ be a regular element of $R[X]$. It suffices to show that $1/a(X) \in Q_0(R)[X]_U$. Since $a(X)$ is regular, $c(a)$ is a finitely generated faithful ideal of R and therefore $c(a)Q_0(R) = Q_0(R)$ since $Q_0(R)$ has property A. Whence, as above, $c(a)$ is Q_0-invertible. Pick $b_0, b_1, \ldots, b_n \in Hom_R(c(a), R)$ such that $a_nb_0 + \ldots a_0b_n = 1$ and set $b(X) = b_nX^n + \cdots + b_0$. Then the coefficient on X^n in the expansion of $a(X)b(X)$ is 1. Consequently, $1/a(X) = b(X)/a(X)b(X) \in Q_0(R)[X]_U$.

((5)⇒(4)) Let $I = (a_0, a_1, \ldots, a_n)$ be a faithful ideal of $Q_0(R)$ and let $a(X) = a_nX^n + \cdots + a_0$. Since $a(X)$ is a regular element of $Q_0(R)[X]$, there are polynomials $d(X) \in Q_0(R)[X]$ and $u(X) \in U$ such that $a(X)d(X) = u(X)$. As $c(u) = R$, $IQ_0(R) = Q_0(R)$. Whence $Q_0(R)$ has property A.

Let M be a maximal ideal of R. By Theorem 13 it suffices to show $(R_{\{M\}}, \{M\})$ is a Q_0-valuation pair. Let $t \in Q_0(R)\backslash R_{\{M\}}$. The polynomial $X - t$ is a regular element of $Q_0(R)[X]$. Hence $1/(X - t) \in T(R[X]) = Q_0(R)[X]_U$. Thus there are polynomials $d(X) \in Q_0(R)[X]$ and $u(X) \in R[X]\backslash M[X]$ with the degree of $u(X)$ minimal such that $(X - t)d(X) = u(X)$. Write $u(X) = u_nX^n + \cdots + u_0$. Since $u(t) = 0$ and $R_{\{M\}}$ is integrally closed in $Q_0(R)$, $(u_n)^{n-1}u(t) = (u_nt)^n \cdots + u_0(u_n)^{n-1} = 0$ implies

$u_n t \in R$. As $t \notin R_{\{M\}}$, we must have $u_n \in M$. Whence, by the minimality of $u(X)$, the equation $(u_n t + u_{n-1})t^{n-1} + \cdots + u_0 = 0$ implies $u_n t \in R \backslash M$ (and $n = 1$). Thus $(R_{\{M\}}, \{M\})$ is a Q_0-valuation pair.

The result above shows that every strong Prüfer ring is a Q_0-Prüfer ring. Hence: Prüfer + property A $\Rightarrow$ strongly Prüfer $\Rightarrow$ Q_0-Prüfer $\Rightarrow$ Prüfer. Examples exist to show that each converse is false (see [13, § 27] and [21]).

A result of Gilmer and J. Hoffman [9, Theorem 3] shows that the saturation of U in $R^b[X]$ (where R^b is the integral closure of R is $Q_0(R)$) is the set, U', of unit content polynomials over R^b.

COROLLARY 17. *The integral closure of R in $Q_0(R)$ is a strong Prüfer ring if and only if $Q_0(R)[X]_U = T(R[X])$.*

PROOF: Since U' is the saturation of U in $R^b[X]$, $Q_0(R)[X]_U = Q_0(R)[X]_{U'}$. The result now follows from the equivalence of (1) and (5) in Theorem 16.

COROLLARY 18. *For an integral domain D with quotient field K, the integral closure of D is a Prüfer domain if and only $K(X) = K[X]_U$.*

ACKNOWLEDGEMENTS

Unreferenced results above are from the following sources. Examples 1 and 2 first appeared in [17]. Lemmas 4 and 5 and Corollaries 8 and 9 are from [18]. Theorem 7 combines results of [18] and [20]. Lemma 6 and Example 10 are in [20]. Theorems 13 and 16 and Corollaries 17 and 18 are from [21].

REFERENCES

1. T. Akiba, *Integrally-closedness of polynomial rings*, Jap. J. Math. **6** (1980), 67–75.
2. D. D. Anderson, *Some remarks on the ring $R(X)$*, Comment. Math. Univ. St. Pauli **26** (1977), 137–140.
3. D. D. Anderson and D. F. Anderson, *Some remarks on cancellation ideals*, Math. Japonica **29** (1984), 879–886.
4. D. D. Anderson, D. F. Anderson and R. Markanda, *The rings $R(X)$ and $R\langle X\rangle$*, J. Algebra **95** (1985), 96–115.
5. J. Brewer, D. L. Costa and K. McCrimmon, *Seminormality and root closure in polynomial rings and algebraic curves*, J. Algebra **58** (1979), 217–226.
6. E. Davis, *Overrings of commutative rings, II*, Trans. Amer. Math. Soc. **110** (1964), 196–222.
7. N. Eggert, *Rings whose overrings are integrally closed in their complete quotient rings*, J. reine angew. Math. **282** (1976), 88–95.
8. R. Gilmer, Multiplicative ideal theory, Marcel Dekker, New York, (1972).
9. R. Gilmer and J. Hoffman, *A characterization of Prüfer domains in terms of polynomial rings*, Pac. J. Math. **60** (1975), 81–85.
10. R. Gilmer and T. Parker, *Semigroup rings as Prüfer rings*, Duke Math. J. **41** (1974), 219–230.
11. M. Griffin, *Prüfer rings with zero divisors*, J. reine angew. Math. **239** (1969), 55–67.
12. M. Griffin, *Valuations and Prüfer rings*, Can. J. Math. **26** (1974), 412–429.

13. J. Huckaba, Commutative rings with zero divisors, Marcel Dekker, New York, (1988).
14. J. T. Knight, Commutative algebra, London Mathematical Society Lecture Note Series, 5, Cambridge University Press, London, (1971).
15. J. Lambek, Lectures on rings and modules, Chelsea, New York, (1986).
16. M. Larsen and P. McCarthy, Multiplicative theory of ideals, Academic Press, New York and London, (1971).
17. T. G. Lucas, *Characterizing when* $R[X]$ *is integrally closed*, Proc. Amer. Math. Soc. **105** (1989), 861–867.
18. T. G. Lucas, *Characterizing when* $R[X]$ *is integrally closed, II*, J. Pure Appl. Algabra **61** (1989), 49–52.
19. T. G. Lucas, *Root closure and* $R[X]$, Comm. Algebra **17** (1989), 2393–2414.
20. T. G. Lucas, *The complete integral closure of* $R[X]$, Trans. Amer. Math. Soc. **330** (1992), 757–768.
21. T. G. Lucas,*Strong Prüfer rings and the ring of finite fractions*, J. Pure Appl. Algebra **84** (1993), 59–71.
22. M. Manis, *Valuations on a commutative ring*, Proc. Amer. Math. Soc. **20** (1969), 193–198.
23. C. P. L. Rhodes, *Relatively Prüfer pairs of commutative rings*, Comm. Algebra **19** (1991), 3423–3445.
24. P. Samuel, *La notion de place dans un anneau*, Bull. Soc. Math. France **85** (1957), 123–133.

19
Graded Rings and Modules

Hideyuki Matsumura Nagoya University, Nagoya, Japan

In this article all rings are assumed to be commutative with a unit element. In section 1 we shall consider graded rings of type G, where G is an abelian group, and discuss some properties which behave well for all G and also some others which depend on G. section 2 will be a brief summary of some of the main results of Goto-Watanabe [1978a]. In section 3 we shall consider the following topics: projective embeddings, graded modules defined by regular differential forms, the canonical ring of a curve.

1 G-graded rings

Let G be an arbitrary abelian group. A graded ring of type G (G-graded ring for short) is a ring R with a direct decomposition (as $\mathbb{Z}$-module)

$$R = \bigoplus_{\gamma \in G} R_\gamma \quad \text{such that} \quad R_\gamma R_\delta \subset R_{\gamma+\delta}.$$

(Since we are dealing with commutative rings only, the group G is necessarily abelian. The book Năstănescu-Oystaeyen [1982] considers non-commutative graded rings with non-commutative group G also.)

A graded R-module is an R-module M with a direct decomposition

$$M = \bigoplus_{\gamma \in G} M_\gamma \quad \text{such that} \quad R_\gamma M_\delta \subset M_{\gamma+\delta}.$$

An element of R_γ (or M_γ) is called a homogeneous element of degree γ. The set of homogeneous elements of R and M will be denoted by $h(R)$ and $h(M)$, respectively. Thus, $h(R) = \bigcup_\gamma R_\gamma$.

For a graded module M and for $\gamma \in G$, define the **shifted module** $M(\gamma)$ as the same R-module with grading $M(\gamma)_\delta = M_{\gamma+\delta}$.

A morphism of graded R-modules $f : M \to N$ is an R-linear map which preserves grade, i.e., such that $f(M_\gamma) \subset N_\gamma$. Graded R-modules and their morphisms form an abelian category which we denote by $\underline{\mathfrak{M}}(R)$, while the category of usual (ungraded) R-modules is denoted by $\mathfrak{M}(R)$. We shall put $*$ in front of technical terms concerning graded objects. For instance, a $*$ideal of R is a graded (or homogeneous) ideal, and a $*$injective module is a graded module which is an injective object in $\underline{\mathfrak{M}}(R)$.

An R-linear map $f : M \to N$ such that $f(M_\delta) \subset N_{\delta+\gamma}$ $(\forall \delta \in G)$ is said to be of degree γ, and the set of such maps is denoted by $\underline{\mathrm{Hom}}_R(M, N)_\gamma$. Their direct sum

$$\underline{\mathrm{Hom}}_R(M, N) := \bigoplus_{\gamma \in G} \underline{\mathrm{Hom}}_R(M, N)_\gamma$$

is a graded R-module, and is a submodule of the usual R-module $\mathrm{Hom}_R(M, N)$ of all R-linear maps from (ungraded) M to N. It is easy to see that if M is finitely generated or if G is finite, then we have

$$\underline{\mathrm{Hom}}_R(M, N) = \mathrm{Hom}_R(M, N),$$

but in general these two modules are different. It is also clear that, for any graded modules M, N and for any $\gamma \in G$, we have

$$\underline{\mathrm{Hom}}_R(M(-\gamma), N) = \underline{\mathrm{Hom}}_R(M, N(\gamma)) = \underline{\mathrm{Hom}}_R(M, N)(\gamma).$$

A graded module N is $*$injective, by definition, if the functor $M \mapsto \underline{\mathrm{Hom}}_R(M, N)_0$ from $\underline{\mathfrak{M}}(R)$ to the category of abelian groups is exact. But if $0 \to L \to M$ is exact in $\underline{\mathfrak{M}}(R)$, then $0 \to L(-\gamma) \to M(-\gamma)$ is also exact and so $\underline{\mathrm{Hom}}_R(-, N)$ is also an exact functor. Thus, N is $*$injective iff $\underline{\mathrm{Hom}}_R(-, N)$ is exact. Similarly for $*$projectivity.

Graded modules of the form $\bigoplus_i R(\gamma_i)$ are called graded free (or $*$free) modules. If M is a graded module generated by $\{\xi_i\}$, where ξ_i is a homogeneous element of degree γ_i, then there is a surjective morphism $\bigoplus R(-\gamma_i) \to M$ in $\underline{\mathfrak{M}}(R)$.

A. Where life is easy (results which are valid for all G)

(A1) *A graded module M is $*$projective iff it is projective.* Năstănescu-Oystaeyen [1982]

In fact, if M is $*$projective then it is a direct summand of a $*$free module F, and F is certainly free, hence M is projective in the usual sense. Conversely, suppose M is projective, and let $f : M \to N$ and $g : L \to N$ be morphisms in $\underline{\mathfrak{M}}(R)$ with g surjective. Then there is an R-linear map $\varphi : M \to L$ such that $g\varphi = f$. Let φ_0 be the "homogeneous component of φ of degree 0", i.e., for $x \in h(M)$ define $\varphi_0(x)$ to be the homogeneous component of $\varphi(x)$ of the same degree as x. Then it is easy to check that φ_0 is a morphism in $\underline{\mathfrak{M}}(R)$ and $g\varphi_0 = f$. Therefore M is $*$projective.

Remark. Similarly, a graded module which is injective is $*$injective. But a $*$injective module is not necessarily injective. Example: Let $R = k[x]$, a polynomial ring over a field k in one variable x with the usual $\mathbb{Z}$-grading. Then $k[x, x^{-1}]$ is its $*$injective hull in $\underline{\mathfrak{M}}(R)$, whereas its injective hull is the quotient field $k(x)$. We shall denote the $*$injective hull of a graded module M by $\underline{E}_R(M)$ to distinguish it from the ordinary injective hull $E_R(M)$ (in $\mathfrak{M}(R)$).

(A2) *The abelian category $\underline{\mathfrak{M}}(R)$ has enough projectives and enough injectives.*

Every object of $\underline{\mathfrak{M}}(R)$ is a homomorphic image of a $*$free module, and $*$free modules are certainly $*$projective. The existence of enough injectives in $\underline{\mathfrak{M}}(R)$ is a special case of a general existence theorem of the case of Grothendieck [1957] Th. 1.10.1) to the effect that an abelian category which has a generator (in the present case $\bigoplus_{\gamma \in G} R(\gamma)$

will do) and satisfies the axiom (AB5) has enough injectives. One can also prove the existence of $\underline{E}_R(M)$ directly by constructing it in $E_R(M)$, *cf.* [Bruns-Herzog].

(A3) *Let M be a graded module and N be a graded submodule of M. Then M is a *essential extension of N (i.e., any non-zero graded submodule of M has a non-zero intersection with N) iff it is an essential extension of N.* (Năstănescu-Oystaeyen [1982] p.9)

proof is easy.

(A4) Let M and N be graded modules. Then $M \otimes_R N$ has a natural structure of graded R-module such that if $x \in M_\gamma$ and $y \in N_\delta$, then $x \otimes y \in (M \otimes N)_{\gamma+\delta}$. The Tor modules $\mathbf{Tor}_i^R(M, N)$ can be calculated by using *free resolution of M (or of N), hence they are also naturally graded (and are the left-derived functors of $\otimes$ in $\mathfrak{M}(R)$). *A graded module M is flat iff it is *flat, i.e., iff the functor $N \mapsto M \otimes N$ is exact in $\mathfrak{M}(R)$.* This can be proved either by using (the graded version of) D. Lazard's theorem that a flat module is a direct limit of free modules (Fossum-Foxby [1974], *cf.* Lazard [1969]) or by the criterion of flatness in terms of linear relations (Herrmann-Orbanz [1982] or Matsumura [1982], *cf.* Bourbaki [1961a] p.43, Cor.1). From this follows the local criterion of flatness: If I is an ideal (not necessarily graded) and M be a graded R-module such that

(1) M/IM is flat over R/I,

(2) $\mathrm{Tor}_1^R(M, R/I) = 0$,

(3) for every graded ideal H of R, the R-module $H \otimes_R M$ is I-adically separated.

Then M is flat over R. (Herrman-Orbanz [1982] or Matsumura [1982]).

(A5) Suppose that our graded ring R is noetherian and M is a graded finite R-module. Then M has a resolution

$$\to F_i \to F_{i-1} \to \cdots \to F_0 \to M \to 0$$

by finite *free modules, hence

$$\underline{\mathbf{Ext}}_R^i(M, N) := H^i(\underline{\mathrm{Hom}}(F_\bullet, N)) = H^i(\mathrm{Hom}(F_\bullet, N)) = \mathrm{Ext}_R^i(M, N)$$

for any graded module N. Note that the graded Ext modules $\mathrm{Ext}_R^i(M, N)$ can be also calculated by means of *injective resolution of N.

Let I be a graded ideal of R and M be a graded module (not necessarily finite). Let $\Gamma_I(M)$ denote, as usual, the set of those elements of M which are killed by some power of I. This is a graded submodule of M, and the right derived functors of the left-exact functor $M \mapsto \Gamma_I(M)$ are called the graded local cohomology $\underline{\boldsymbol{H}}_I^i(\boldsymbol{M})$. Since these are calculated by means of *injective resolutions of M, a priori they are not equal to the ungraded local cohomology $H_I^i(M)$. But one can easily see that $H_I^i(M) \cong \varinjlim_{n\to\infty} \underline{\mathrm{Ext}}_R^i(R/I^n, M)$, hence by what we have seen about Ext we have $\underline{\boldsymbol{H}}_I^i(\boldsymbol{M}) \cong \boldsymbol{H}_I^i(\boldsymbol{M})$ **for all i and M.**

B. Where one must be careful (results depend on G)

(B1) Prime avoidance.

Consider the following statement:
⟨Let $I, P_1, \cdots, P_r$ be graded ideals. Suppose that I is not contained in any P_i, and that all P_i are prime. Then there exists $a \in h(I)$ which does not lie in any P_i.⟩

This is true if R is $\mathbb{N}$-graded (i.e., $G = \mathbb{Z}$ and $R_i = 0$ for $i < 0$) and $I \subset R_+ = \bigoplus_{i>0} R_i$, or more generally if R is $\mathbb{Z}$-graded and $P_i \not\supset R_+$ for all i.

Counterexamples.

(1) $R = k[X,Y]$ with $k =$ a field $\subset R_0$, $G = \mathbb{Z}$, $\deg X = 0$, $\deg Y = 1$. Let $I = (X^2, Y)$, $P_1 = (Y)$, $P_2 = (X)$. Then R is $\mathbb{N}$-graded but $h(I) \subset P_1 \cup P_2$.

(2) $R = k[YX, Y^{-1}] \subset k[X, Y, Y^{-1}]$ with $G = \mathbb{Z}$, $\deg X = 0$, $\deg Y = 1$. Thus R is the Rees algebra of the ideal XA of the ring $A = k[X]$. Then $R_i = AX^iY^i$ $(i > 0)$, $R_i = AY^i$ $(i \leq 0)$. Let $I = (Y^{-1}, YX)$, $P_1 = (XY)$, $P_2 = (Y^{-1})$. We have $R/P_1 \cong k[Y_{-1}], R/P_2 \cong k[YX]$, so that the P_i are primes. But $h(I) \subset P_1 \cup P_2$.

(3) $R = k[X,Y]$ with $G = \mathbb{Z}^2$, $\deg(X^\alpha Y^\beta) = (\alpha, \beta)$. The homogeneous elements are $cX^\alpha Y^\beta (c \in k)$, and the only graded prime ideals are $(X,Y), (X), (Y), (0)$. If we put $I = (X,Y)$, $P_1 = (X)$, $P_2 = (Y)$, then we have a counterexample to the statement.

(B2) Associated primes

Suppose the graded ring R is noetherian, and consider the following statement:
⟨If M is agraded R-module,the prime ideals associated to M are graded and are of the form $\operatorname{ann}(\xi)$ with $\xi \in h(M)$.⟩
This is true if G is torsion-free, but false for $G = \mathbb{Z}_r = \mathbb{Z}/r\mathbb{Z}$.
Proof. When G is torsion-free one can make it an ordered group, i.e., G has a total order such that $\alpha < \beta$ implies $\alpha + \gamma < \beta + \gamma$. (This is well known and can be seen by imbedding G into a vector space over $\mathbb{Q}$.) Let $P \in \operatorname{Ass}(M)$. Then $P = \operatorname{ann}(\xi)$ for some ξ. Let $\xi = \xi_1 + \cdots + \xi_n$, $\xi_i \in M_{\gamma_i}$, $\gamma_1 < \gamma_2 < \cdots < \gamma_n$. If $n = 1$ then P is clearly graded and we are done. We use induction on n. Let $a = a_1 + \cdots + a_r \in P$, $\deg(a_i) = \delta_i, \delta_1 < \cdots < \delta_r$. We claim that $a_i \in P$ for all i. From $a\xi = 0$ we have $a_1\xi_1 = 0$. If $a_1\xi \neq 0$ then $\operatorname{ann}(\xi) = \operatorname{ann}(a_1\xi)$ because P is a prime,and $a_1\xi$ has at most $n-1$ homogeneous components, hence by induction on n we are done.

A counterexample when $G = \mathbb{Z}_2$. Let $R = k[X]/(1 - X^2) = k + k\epsilon$ with $\deg \epsilon = 1, \epsilon^2 = 1$. Then $(1-\epsilon)$ and $(1+\epsilon)$ are the associated prime ideals of R and are non-graded.

(B3) *A graded ideal P is a prime iff $ab \in P$ implies $a \in P$ or $b \in P$ for homogeneous elements a, b.*

This is again true if G is torsion-free but false if G is of torsion. Proof is similar to the above.

Added in Feb. 1993: Kamoi [1992] showed that, by calling a graded ideal I a G-prime when it satisfies $ab \in I \Rightarrow a \in I$ or $b \in I$ for homogeneous elements and using G-prime ideals instead of (or besides) usual prime ideals, one can remedy some of the pathologies such as discussed above.

2 Noetherian $\mathbb{Z}$-graded (or $\mathbb{N}$-graded) rings

Theorem (2.1). (P. Samuel) *A $\mathbb{Z}$-graded ring $R = \bigoplus_{i \in \mathbb{Z}} R_i$ is noetherian iff the following conditions are satisfied:*

(1) *R_0 is a noetherian ring,*

(2) *R is finitely generated as R_0-algebra.*

Therefore a $\mathbb{Z}$-graded noetherian ring is of the form

$$R = R_0[X_1, \cdots, X_r, Y_1, \cdots, Y_s],\ \deg(X_i) > 0,\ \deg(Y_j) < 0,$$

where R_0 is noetherian, and consequently each R_i is afinite R_0-module. The theorem is not difficult to prove(cf. e.g. Bourbaki, Algebre Commutative, Chap.3, or D. Rees, Lectures on the Asymptotic Theory of Ideals.)

Let R be as above and P be a prime ideal. Let P^* be the ideal generated by the homogeneous elements of P. Then P^* is prime by (B3) of §2.

(2.2) If P is not graded then $\operatorname{ht} P^* = \operatorname{ht} P - 1$. To prove this, first we prove $\operatorname{ht}(P/P^*) = 1$ by observing that a $\mathbb{Z}$-graded ring in which every non-zero homogeneous element is invertible is either a field or of the form $K[t, t^{-1}]$ with K a field, and then take a prime ideal $Q \subset P$ such that $\operatorname{ht} Q = \operatorname{ht} P - 1$ and use induction on $\operatorname{ht} P$. Cf. Matijevic-Roberts [1974].
(Remark. In the case of $\mathbb{Z}^n$-graded rings we have $\operatorname{ht} P^* \geq \operatorname{ht} P - n$, cf. Goto-Watanabe [1978b].)

(2.3) If P is a graded prime and $\operatorname{ht} P = h$, then there is a chain of graded prime ideals $P = P_0 \supset P_1 \supset \cdots \supset P_h$ of length h. In other words, $*\operatorname{ht} P = \operatorname{ht} P$. (Therefore we do not need the notatin $*$ht.)

(2.4) $*\dim R \geq \dim R - 1$. (The left hand side is the length of the longest chain of graded primes in R.) When R is $\mathbb{N}$-graded we have $*\dim R = \dim R$. This is because every $*$ maximal ideal is maximal in this case.

(2.5) R is CM (=Cohen-Macaulay) iff for every $*$maximal ideal $\mathfrak{m}$, $R_\mathfrak{m}$ is CM. Similarly for regular, Gorenstein etc. (For a proof, cf. Matijevic [1975]).

* *

An $\mathbb{N}$-graded ring $R = \bigoplus_{n \geq 0} R_n$ with $R_0 = k$, where k is a field, is said to be $\mathbb{N}$-graded over k. We denote its unique $*$maximal ideal $\bigoplus_{n>0} R_n$ by $\mathfrak{m}$. It is well known that such a pair $(R, \mathfrak{m})$ behaves like a complete local ring. In particular, Matlis duality and local duality take the following simple forms (Goto-Watanabe [1978a]).

Let R be an $\mathbb{N}$-graded ring over k of Krull dimension d. For each graded R-module M, set $k = R/\mathfrak{m}$ (= the field k as a graded R-module concentrated at degree zero), and define the k-dual $M^\vee$ of M by $M^\vee = \underline{\operatorname{Hom}}_k(M, k)$. Therefore $(M^\vee)_n = \operatorname{Hom}_k(M_{-n}, k)$, the dual space of M_{-n}.

Theorem (2.6). *$R^\vee$ is the injective envelope of $k = R/\mathfrak{m}$:*

$$R^\vee = \underline{E}_R(\underline{k}) = E_R(\underline{k}).$$

Corollary. $M^\vee = \underline{\mathrm{Hom}}_k(M, \underline{k}) = \underline{\mathrm{Hom}}_R(M, R^\vee)$ *is the Matlis dual of* M.

Definition. We set $K_R := H^d_{\mathfrak{m}}(R)^\vee$, where $d = \dim R$ and call it the *canonical module* of R. Since $H^d_{\mathfrak{m}}(R)$ is graded and Artinian, by Matlis theory K_R is finitely generated and graded.

Theorem (2.7).

(a) *If* $M \in \mathfrak{M}(R)$ *is finitely generated, then*

$$H^d_{\mathfrak{m}}(R)^\vee \cong \underline{\mathrm{Hom}}_R(M, K_R).$$

(b) *If* R *is Cohen-Macaulay, then we have, for all* i,

$$H^{d-i}_{\mathfrak{m}}(R)^\vee \cong \underline{\mathrm{Ext}}^i_R(M, K_R)$$

Remarks.

(2.8) Let R be an $\mathbb{N}$-graded ring over k of Krull dimension d, and $\mathfrak{m}$ be its unique *maximal ideal. then

$$R \text{ is CM} \iff H^i_{\mathfrak{m}}(R) = 0 \text{ for } 0 \le i < d.$$

This follows from (2.5) and from the easy observation that $H^i_{\mathfrak{m}}(R) = H^i_{\mathfrak{m}}(R_{\mathfrak{m}})$ because $\mathfrak{m}$ is a maximal ideal.

(2.9) Let R be an arbitrary (not necessarily noetherian) $\mathbb{Z}$-graded ring, and t be an indeterminate over R. Then there is an injective homomorphism $\psi : R \to R[t]$ such that $\psi(a) = at^n$ for $a \in R_n$. This is very useful. For instance, it can be used to prove the following theorem (Bourbaki [1961b], Chap. 5, §1, no. 8):

Suppose that R is an integral domain with quotient field K, and let R' denote the integral closure of R in K. Let S be the set of non-zero homogenous elements of R. Then R_S is a $\mathbb{Z}$-graded ring in the obvious way, and R' is a graded subring of R_S. If R is $\mathbb{N}$-graded, then so is R'.

If R is noetherian and $R_0 = k =$ a field, then R is finitely generated over k by (2.1) and it is well known that R' is a finitely generated R-module for such a ring R. Therefore R' is again noetherian.

3 Connections with Algebraic Geometry

A. Projective embeddings

From the modern viewpoint, an algebraic variety V (over an algebraically closed field k) is a ringed space, i.e., a topological space with a sheaf of rings $\mathcal{O}_V$, satisfying certain conditions. In particular, the stalks of $\mathcal{O}_V$ are local rings, which are called the local rings of V. If all the local rings of V are regular (resp. normal, CM, Buchsbaum) then V is said to be non-singular or smooth (resp. normal, CM, Buchsbaum). The cohomology groups $H^q(V, \mathcal{O}_V)$ are important invariants of V.

If V is isomorphic (as ringed space) to a closed subvariety V' of a projective space $\mathbb{P}^r$, then V' is called a projective embedding of V, and V is called a projective variety.

Let $I(V')$ be the ideal of $S = k[T_0, \ldots, T_r]$ defining V', and let $R(V') = S/I(V')$. Then $R(V')$ is called the homogeneous coordinate ring of V'. For simplicity we shall consider here irreducible (and reduced) varieties only. Then $I(V')$ is a graded prime ideal of S, and $R(V')$ is a noetherian domain $\mathbb{N}$-graded over k. The homogeneous coordinate ring $R(V')$ depends on the projective embeddings. If it is normal (resp. CM, etc.) then we say that V' is arithmetically normal (resp. arithmetically CM, etc.). Therefore we can ask the following question:

Question (3.1). *When V has a good property (e.g. non-singular), does there exist a good (e.g. arithmetically CM) projective embedding?*

The following theorem is classical.

Theorem (3.2). *If V is normal, then it has an arithmetically normal projective embedding.*

To prove this, take an arbitrary projective embedding V'. Then the derived normal ring (i.e., the integral closure in the quotient field) of $R(V')$, say S, is again a noetherian $\mathbb{N}$-graded ring over k by (2.9), but it is not necessarily generated by elements of degree 1, and then we cannot write it $R(V'')$ for some $V'' \subset \mathbf{P}^s$. But if V is normal then the conductor of $R(V')$ is an irrelevant ideal (i.e., contains some power of the $*$maximal ideal). This means that $R(V')_n = S_n$ for n large.

Let $R = R(V') = \bigoplus_{n \geq 0} R_n$ and let d be a positive integer. Set $R^{(d)} = \bigoplus_{n \geq 0} R_{nd}$ and view it a $\mathbb{Z}$-graded ring by defining $R^{(d)}_n = R_{nd}$. Then $R^{(d)}$ is of the form $R(W)$ for some projective embedding $W \subset \mathbf{P}^q$ of V, which is called the Veronese transform of V' of degree d.

Since we have $R^{(d)} = S^{(d)}$ for d sufficiently large, W is an arithmetically normal embedding of V.

(3.3) Next we assume that V is CM. Does it have an arithmetically CM embedding? The answer is no, even if we assume that V is non-singular. To see that we need some preparation from cohomology theory.

A projective space $\mathbf{P}^r$ has an invertible sheaf denoted by $\mathcal{O}_{\mathbf{P}}(1)$, which corresponds to the (linear equivalence class of) hyperplanes. For $n \in \mathbb{Z}$ we set $\mathcal{O}_{\mathbf{P}}(n) = \mathcal{O}_{\mathbf{P}}(1)^{\otimes n}$, and for a quasi-coherent sheaf $\mathcal{F}$ on $\mathbf{P}^r$ we define $\mathcal{F}(n) = \mathcal{F} \otimes \mathcal{O}(n)$, $\Gamma_*(\mathcal{F}) = \bigoplus_{n \in \mathbb{Z}} \Gamma(\mathbf{P}^r, \mathcal{F}(n))$. The homogeneous coordinate ring of $\mathbf{P}^r$ is $S = k[T_0, \ldots, T_r]$, and $\Gamma_*(\mathcal{F})$ is a graded S-module. When V is a closed subvariety of $\mathbf{P}^r$ and $\mathcal{F}$ is a quasi-coherent sheaf on V, we can view $\mathcal{F}$ as a sheaf on $\mathbf{P}^r$ and form $\Gamma_*(\mathcal{F})$. Then $\Gamma_*(\mathcal{F})$ is a graded $R(V)$-module and also a $\Gamma_*(\mathcal{O}_V)$-module, and one can show that $\Gamma_*(\mathcal{O}_V)$ coincides (at least when V is normal) with the derived normal ring of $R(V)$, see Hartshorne [1977] p.126. Conversely, if M is a graded $R(V)$-module, then one can construct a quasi-coherent sheaf $\tilde{M}$ on V such that $\Gamma(U_i, \tilde{M}) = M_{(t_i)}$, where t_i is the image of T_i in $R(V)$ and U_i is the open set of V defined by $t_i \neq 0$, and $M_{(t_i)}$ is the degree zero part of the localization of M with respect to the multiplicative set $\{t_i^\nu \mid \nu \geq 0\}$. One can show that $\Gamma_*(\mathcal{F})\tilde{} \cong \mathcal{F}$. (For all these, see Hartshorne [1977] Chap.II §5.) Moreover, the local cohomology modules $H^i_{\mathfrak{m}}(M)$ and the sheaf cohomology $H^i(V, \tilde{M})$, where M is a finitely generated graded $R(V)$-modules and $\mathfrak{m}$ is the $*$-maximal ideal of $R(V)$, are related as follows:

$$0 \to H^0_{\mathfrak{m}}(M) \to M \to \Gamma_*(\tilde{M}) \to H^1_{\mathfrak{m}}(M) \to 0 \text{ (exact)}$$

$$H^i_{\mathfrak{m}}(M)_n \cong H^{i-1}(V, \tilde{M}(n)) \quad (i \geq 2).$$

(For these, see e.g. Stückrad-Vogel [1986] pp. 36-38, or Goto-Watanabe [1978a] (5.1.6)).

Serre [1955] proved that if $\mathcal{F}$ is a coherent sheaf on $V \subset \mathbf{P}^r$, then

i) $\Gamma(V, \mathcal{F}(n))$ generates $\mathcal{F}$ and $H^i(V, \mathcal{F}(n)) = 0$ $(i > 0)$ for $n \gg 0$,

and

ii) if V is a CM variety of dimension d, and $\mathcal{F}$ is locally free, then

$$H^i(V, \mathcal{F}(n)) = 0 \ (i < d) \text{ for } n \ll 0.$$

Therefore, if V is a CM and normal projective variety of dimension d in $\mathbf{P}^r$ and R is its homogeneous coordinate ring, then we have

$$H^0_{\mathfrak{m}}(R) = 0 \text{ (since } R \text{ is a domain)}, \ H^1_{\mathfrak{m}}(R) \cong R'/R,$$

where R' is the derived normal ring of R, and

$$H^i_{\mathfrak{m}}(R)_n = 0 \text{ for } 0 < i \leq d \text{ and } |n| > n_0 \text{ for some integer } n_0.$$

(Note that $(R'/R)_n = 0$ for $n < 0$ or $n \gg 0$, and that $\dim R = d + 1$.) It follows that, if we take Veronese transform of sufficiently high degree, then we get a projective embedding such that

$$(*) \qquad H^i_{\mathfrak{m}}(R)_n = 0 \text{ for } 0 \leq i \leq d \text{ and } n \neq 0, \text{ and } R = R'.$$

But $H^i_{\mathfrak{m}}(R)_0 = H^{i-1}(V, \mathcal{O}_V)$ $(i \geq 2)$ are invariants of V and do not depend on the projective embedding. On the other hand, there is a theorem (independently proved by Goto and Stückrad-Vogel) which says that a noetherian graded module of dimension $d+1$ over a polynomial ring $S = k[T_0, \ldots, T_r]$ such that $H^i_{\mathfrak{m}}(M)_n = 0$ for $n \neq 0$ and for all $i \leq d$ is a Buchsbaum module. (Cf. Stückrad-Vogel [1986] p.98). This means that the coordinate ring R satisfying $(*)$ is a Buchsbaum ring. Therefore we get the following conclusion:

Theorem (3.4). (i) *Let V be a CM normal projective variety. Then it has an arithmetically normal and arithmetically Buchsbaum embedding.*
(ii) *There are non-singular projective varieties of dimension $d > 1$ with $H^i(V, \mathcal{O}) \neq 0$ for some $0 < i < d$ (e.g. abelian varieties of dimension > 1), and such a variety has no arithmetically CM embeddings.*

B. Graded modules associated with sheaves of differential forms

Let $V \subset \mathbf{P}^r$ be a non-singular projective variety of dimension $d > 1$, and let R be its homogeneous coordinate ring. For simplicity we assume that V is arithmetically normal, i.e., R is normal. Let Ω^p be the sheaf of regular differential p-forms on V, $0 \leq p \leq d$. Then Ω^p is a locally free sheaf of rank $\binom{d}{p}$. Set $W^p := \Gamma_*(\Omega^p)$. Then $W^0 = R$ since V is arithmetically normal, and the W^p are graded R-modules.

Recall that Serre duality states:

(1) $H^d(V, \Omega^d) \cong k$,

(2) If $\mathcal{F}$ is a locally free sheaf and $\mathcal{F}^* = \mathcal{H}om(\mathcal{F}, \mathcal{O})$ is the dual sheaf, then the cup-product pairings

$$H^i(V, \mathcal{F}) \times H^{d-i}(V, \mathcal{F}^* \otimes \Omega^d) \to H^d(V, \Omega^d) \cong k$$

are non-degenerate. (See Hartshorne [1977] Chap. III Cor. 7.7).

For $\mathcal{F} = \Omega^p$ we have $\mathcal{F}^* \otimes \Omega^d \cong \Omega^{d-p}$, hence we can identify the k-dual space of

$$\left[H^{d+1}_{\mathfrak{m}}(R)\right]_{-\nu} = H^d(V, \mathcal{O}(-\nu))$$

with $H^0(V, \Omega^d(\nu)) = [W^d]_\nu$. This means that $W^d = [H^{d+1}_{\mathfrak{m}}(R)]^* \cong K_R$. It is easy to check that the isomorphism $W^d \cong K_R$ is an isomorphism of graded R-modules. Similarly, $[H^{q+1}_{\mathfrak{m}}(W^p)]^* \cong H^{d-q+1}_{\mathfrak{m}}(W^{d-p})$ $(q > 1)$.

In the case $V = \mathbf{P}^r$, the graded modules W^p $(p > 0)$ appear as syzygies in the minimal free resolution of $k = S/\mathfrak{m}$. In fact, we have $S = k[T_0, \ldots, T_r]$, $\mathfrak{m} = (T_0, \ldots, T_r)$, and the minimal free resolution of k is nothing but the Koszul complex with respect to $T_0, \ldots, T_r$. Therefore it has the following form:

$$0 \to S(-r-1) \to \cdots \to S(-p)^{a_p} \xrightarrow{d_p} \cdots \xrightarrow{d_1} S \to k \to 0,$$

where $a_p = \binom{r+1}{p}$, and by an easy calculation one can show that W^p is isomorphic to $\mathrm{Ker}(d_p)$. It follows that $\mathrm{pd}\, W^p = r - p$, $\dim W^p = r + 1$, $\mathrm{depth}\, W^p = (r+1) - (r-p) = p + 1$. Moreover, it is easy to see that the W^p are Buchsbaum S-modules. I don't know whether this last statement is true for any non-singular variety V. The example of an abelian variety (where all Ω^p are free and so all W^p are free modules) show that the equality $\mathrm{depth}\, W^p = p + 1$ is not true in general.

C. The canonical ring of a curve

Let V be a non-singular projective variety of dimension d, and $K = K_V$ be a canonical divisor of V. Thus the sheaf $\mathcal{O}(K)$ is isomorphic to Ω^d and is an invertible sheaf. Set

$$R(V, K) = \bigoplus_{n \geq 0} \Gamma(V, \mathcal{O}(nK)).$$

This is an $\mathbf{N}$-graded ring over k, and is called the canonical ring of V. This is known to be a birational invariant of V. Whether it is finitely generated over k is a delicate question for $d > 1$, and has been answered affirmatively when $d = 2$ (Zariski and Mumford) and when $d = 3$ and $\mathrm{char}(k) = 0$ (by joint effort of Benveniste, Fujita, Kawamata, Mori and Shokurov. Cf. Kawamata [1990].) Here we consider the case $d = 1$ only. Therefore V is a non-singular curve. Let g be its genus. When $g = 0$ the canonical ring reduces to k, and when $g = 1$ we may take $K = 0$, so that the canonical ring is isomorphic to the polynomial ring in one variable: $R(V, K) \cong k[T]$. When $g > 1$, V is hyperelliptic (i.e., if there is a finite morphism $V \to \mathbf{P}^1$ of degree 2) if and only if $|K|$ is not very ample (see Hartshorne [1977] Chap. IV Prop. 5.2). Since $\deg K = 2g - 2 > 0$, $|nK|$ is very ample for large n. But in the hyperelliptic case the image of the morphism defined by $|K|$ is a rational curve (Hartshorne [1977] Chap IV Prop. 5.3), hence $R(V, K)$ is not generated by elements of degree one. When V is non-hyperelliptic (and so necessarily g is ≥ 3), the following is known:

Theorem (3.5). *Suppose V is a non-singular curve of genus $g \geq 2$ which is not hyperelliptic. Then $R = R(V, K)$ is generated by R_1 over $R_0 = k$, and $V = \mathrm{Proj}(R)$.*

Proof is not so easy and we refer the reader to Saint-Donat [1973]. Also cf. Hartshorne [1977] Chap. IV. This theorem means that R coincides with the homogeneous coordinate ring of the image of the projective embedding (called the canonical embedding) $V \to \mathbf{P}^{g-1}$. Therefore the canonical embedding is arithmetically normal, and consequently arithmetically CM since $\dim R = 2$. In this embedding, $\mathcal{O}(1) = \mathcal{O}(K) = \Omega^1$ and so $\mathcal{O}(n+1) = \Omega^1(n)$, hence $W^1 = K_R = R(1)$ where $R = R(V, K)$. This means R is Gorenstein with $a(R) = 1$ (cf. Goto-Watanabe [1978a]). According to Eisenbud [1990], the canonical rings of curves are characterized as S/I, where $S = k[X_0, \ldots, X_{g-1}]$ and I is a graded prime ideal, such that a) S/I is normal, b) $\dim S/I = 2$, c) S/I is Gorenstein, d) degree $S/I = 2g - 2$.

On the other hand, one can show that a hyperelliptic curve cannot be a complete intersection in any projective embedding (Hartshorne [1977] Chap. IV Ex. 5.1).

References

Bourbaki, N. (1961a). *Algèbre Commutative, Chap. 1,2,* Hermann, Paris.

— (1961b). *Algèbre Commutative, Chap. 5,6,* Hermann, Paris.

Bruns, W. and Herzog, J. *Cohen-Macaulay Rings and Modules,* Cambridge Univ.Press, forthcoming.

Demazure, M. (1988). *Anneaux gradués normaux,* Travaux en cours, **37** Introduction à la théorie des singlarités II (Lê Dũng Tráng ed.).

Eisenbud, D. (1990). *Green's Conjecture: An Orientation for Algebraists,* preprint.

Eisenbud, D. and Goto, S. (1984). *Linear free resolution and minimal multiplicity,* J. Alg. **88**, 89-133.

Fossum, R. and Foxby, H.-B. (1974). *The Category of graded modules,* Math. Scand. **35**, 288-300.

Goto, S. and Watanabe, K. (1978a). *On graded rings, I,* J. Math.Soc.Japan, **30**, 179-213.

— (1978b). *On graded rings, II ($\mathbb{Z}$-graded rings),* Tokyo J. Math. **1**, 237-261.

Grothendieck, R. (1957). *Sur quelques points d'algèbre homologique,* Tôhoku Math. J. **9**, 119-221.

Hartshorne, R. (1977). *Algebraic Geometry,* Springer.

Herrmann, M., Ikeda, S. and Orbanz, U. (1988). *Equimultiplicity and Blowing-up,* Springer.

Herrmann, M. and Orbanz, U. (1982). *Two Notes on flatness,* Manuscripta math. **40**, 104-133.

Kamoi, Y. (1992). *On graded rings with grading by an abelian group,* preprint.

Kawamata, Y. (1990), *Canonical and Minimal modules of Algebraic Varieties,* Proc. International Congress of Mathematicians 1990, Vol. I, Springer.

Lazard, D. (1969). *Autour de la platitude,* Bull. Soc. Math. France, **97**, 81-128.

Matsumura, H. (1982). *A remark on flatness over a graded ring,* The 4th symposium on Commutative Algebra in Japan, Karuizawa, 93-95.

— (1991). *On graded modules associated with sheaves of differential forms,* to appear on Ann. Univ. Bucharest.

Matijevic, J. (1975). *Three local conditions on a graded ring,* Trans. Amer. Math. Soc. **205**, 275-284.

Matijevic, J. and Roberts, P. (1974). *A Conjecture of Nagata on graded Cohen-Macaulay rings,* J. Math. Kyoto Univ. **14**, 125-128.

Năstănescu, C. and Oystaeyen, F. van (1982). *"Graded Ring Theory,"* North-Holland Publ. Co., Amsterdam.

Saint-Donat, B. (1973). *On Petri's analysis of the linear system of quadrics through a canonical curve,* Math. Ann. **206**, 157-175.

Serre, J.-P. (1955). *Faisceaux Algébriques Cohérents,* Ann. of Math. **61**, 197-278.

Stückrad, J. and Vogel, W. (1986). *Buchsbaum Rings and Applications,* VEB Deutcher Verlag der Wissenschaften.

Tomari, M. and Watanabe, K. *Normal cyclic covers, I (Normal $\mathbb{Z}_r$-graded rings),* to appear.

20
Symbolic Powers, Rees Algebras and Applications

Jun-ichi NISHIMURA
Department of Mathematics
Faculty of Science, Kyoto University
Kyoto, 606, JAPAN

This is a survey of one of the examples constructed in my paper entitled "A few examples of local rings, I". So, we give here the results without proof. Throughout this paper, all rings are commutative with identity. We fully use the notation and terminology of EGA [EGA], Matsumura [Mat 80] and Nagata [Nag 75]. The set of natural numbers and that of non-negative integers are denoted respectively by $\mathbf{N}$ and $\mathbf{N}_0$.

We say that a local ring A is unmixed, or quasi-unmixed if $\dim \hat{A}/\hat{\mathfrak{p}} = \dim A$, respectively, for any $\hat{\mathfrak{p}} \in \mathrm{Ass}(\hat{A})$, or for any $\hat{\mathfrak{p}} \in \mathrm{Min}(\hat{A})$. A famous theorem of Nagata–Ratliff tells us that a local domain A is universally catenary if and only if A is quasi-unmixed. Hence, if A is (quasi-)unmixed, A/P is quasi-unmixed for any $P \in \mathrm{Spec}(A)$. So, the question is: If furthermore A is unmixed, then is A/P also unmixed for any $P \in \mathrm{Spec}(A)$?

Indeed, this has been a longstanding unsolved problem. Then, Brodmann–Rotthaus [BR 83] and Ogoma [Ogo 82] independently and almost at the same time found counter-examples, which we shall reconstruct below.

Now we summarize the content of this note. In section 1, we first recall a theorem of Hochster which gives criteria for the equality of ordinal and symbolic powers of primes. Then we revisit one of his examples of prime ideals whose ordinal and symbolic powers are equal, which plays an important role in the last section.

Next, we review the fundamental construction of local domains with *reasonably* odd completion, following Rotthaus [Rot 79], Ogoma [Ogo 80], Heitmann [Hei 82], and Brodmann–Rotthaus [BR 82].

Further, we sketch a sort of generalization of the construction above, which produces local domains with peculiar non-zero *prime* elements. It might be worth noting here that, to get such examples, the consideration of two (extended) Rees Algebras is crucial. For the details of these two sections, we refer the reader to [Nis 92].

In the final section, as applications of the observations above, we give a simplified version of Brodmann–Rotthaus construction of an example of an unmixed local domain A with a prime ideal P such that A/P is not unmixed. We end this article by remarking that Ogoma's example can be obtained in the same manner.

1 SYMBOLIC POWERS, HOCHSTER'S THEOREM AND HIS EXAMPLE

Throughout this section, R denotes a Noetherian domain, $P = (p_1, \ldots, p_k)$ a prime ideal of R and $\boldsymbol{p}$ the k-tuple $p_1, \ldots, p_k$. As mentioned in Introduction, we first recall Hochster's criteria for the equality of ordinal and symbolic powers of a prime ideal P of a Noetherian domain R, namely, criteria for the equality $P^n = P^{(n)} = P^n R_P \cap R$ for any n. Then, for the later use, we review Hochster's third example of a height 2 prime ideal P of a polynomial ring R in 4 variables over a field K such that R/P is not Cohen–Macaulay but $P^n = P^{(n)}$ for any n.

To state Hochster's criteria, we fix notation. If $P = (p_1, \ldots, p_k)$ is a prime ideal of a Noetherian domain R, taking $k+1$ algebraically independent indeterminates $t_1, \ldots, t_k, q$ over R, we set $S = R[t_1, \ldots, t_k]$. We define an increasing sequence of ideals of S recursively as follows:

$$J_0(\boldsymbol{p}) = (0),\ J_{n+1}(\boldsymbol{p}) = \{\textstyle\sum_{i=1}^k s_i t_i \mid s_i \in S,\ \sum_{i=1}^k s_i p_i \in J_n(\boldsymbol{p})\} + J_n(\boldsymbol{p}),$$
$$J(\boldsymbol{p}) = \textstyle\bigcup_n J_n(\boldsymbol{p}).$$

THEOREM 1.1 (Hochster [Hoc 73]) *The following conditions on a prime ideal $P = (p_1, \ldots, p_k)$ of a Noetherian domain R are equivalent:*

(1.1.1) *$P^n = P^{(n)}$ for every positive integer n, and the associated graded ring of R_P is a domain.*

(1.1.2) *$PS + J(\boldsymbol{p})$ is prime.*

(1.1.3) *For some integer $n > 0$, $PS + J_n(\boldsymbol{p})$ is a prime of height k. In this case, $PS + J_n(\boldsymbol{p}) = PS + J(\boldsymbol{p})$.*

(1.1.4) *There is a height k prime Q of S such that $Q \subset PS + J(\boldsymbol{p})$. In this case, $Q = PS + J(\boldsymbol{p})$.*

(1.1.5) *q is a prime element in the subring $R[q, p_1/q, \ldots, p_k/q]$ of $R[q, 1/q]$.*

As applications of the theorem above, Hochster observed three examples of prime ideals whose ordinal and symbolic powers are equal. The first example is a prime ideal generated by an R-sequence, the second one is the prime ideal generated by the k by k minors of a k by $k+1$ matrix of indeterminates over a field. Though they are interesting as well, we shall only look close at the third one for our later purpose.

Let X, Y be indeterminates over a field K. Set $A = K[X, XY, Y^2, Y^3]$, which is not Cohen–Macaulay. Let x, z_1, z_2, z_3 be indeterminates over K and set $R = K[x, z_1, z_2, z_3]$. Let $\phi: R \to A$ be the K-homomorphism which maps x, z_1, z_2, z_3 to X, XY, Y^2, Y^3, respectively. Let $P = \operatorname{Ker}\phi$. Then

EXAMPLE 1.2 (cf.[Hoc 73, p.61]) $P = (z_2^3 - z_3^2, z_2 x^2 - z_1^2, z_2 z_1 - x z_3, z_2^2 x - z_3 z_1) = (p_1, p_2, p_3, p_4)$ is a height 2 prime ideal of R where R/P is not Cohen–Macaulay but $P^n = P^{(n)}$ for every positive integer n.

Indeed, $J_1(\boldsymbol{p})$ contains $a = xt_1 - z_3t_3 - z_2t_4, b = z_1t_1 - z_2^2t_3 - z_3t_4, c = z_2t_2 + z_1t_3 - xt_4$ and $d = z_3t_2 + z_2xt_3 - z_1t_4$. Then, $e = t_1t_2 + z_2t_3^2 - t_4^2 \in J_2(\boldsymbol{p})$. Hence, $Q = (p_1, p_2, p_3, p_4, a, b, c, d, e)S \subset PS + J_2(\boldsymbol{p}) \subset PS + J(\boldsymbol{p})$. Hochster shows that Q is a height 4 prime ideal of $S = R[t_1, t_2, t_3, t_4]$.

We end the first section by notifying that the five relations a, b, c, d, e above will appear again in a very crucial step of our reconstruction of Brodmann–Rotthaus peculiar unmixed local domain given below.

2 CONSTRUCTION OF PECULIAR LOCAL DOMAINS

2.0 Notation

Let K_0 be a countable field, for example, $\mathbf{Q}$ the field of rational numbers, $\mathbf{F}_q$ the finite field with q elements, or $\overline{\mathbf{F}}_p$ the algebraic closure of the prime field of characteristic $p > 0$, etc..., and let K be a purely transcendental extension field of countable degree over K_0, that is, $K = K_0(a_{ik})$ with transcendental basis $\{a_{ik} \mid i = 1, \ldots, n;\ k = 1, 2, \ldots\}$. Further, for any $k \in \mathbf{N}$, we write:

$$K_k = K_{k-1}(a_{1k}, \ldots, a_{nk}) = K_{k-1}(a_{ik}) \text{ and } K = \bigcup_k K_k \tag{2.0.1}$$

Taking $n+1$ indeterminates u, $z_1, \ldots, z_n$ over K, we set:

$$S_0 = K_0[u, z_1, \ldots, z_n] \text{ and } \mathfrak{N}_0 = (u, z_1, \ldots, z_n)S_0,$$
$$S_k = S_{k-1}[a_{1k}, \ldots, a_{nk}] \text{ and } \mathfrak{N}_k = (u, z_1, \ldots, z_n)S_k.$$

Then, $S_k = K_0[a_{ih}, u, z_1, \ldots, z_n]$ with $i = 1, \ldots, n$ and $1 \le h \le k$. We shall write:

$$S = \bigcup_k S_k = K_0[a_{ik}, u, z_1, \ldots, z_n] \text{ and } \mathfrak{N} = (u, z_1, \ldots, z_n)S \tag{2.0.2}$$

where $i = 1, \ldots, n;\ k = 1, 2, \ldots$. Moreover, let

$$R_0 = (S_0)_{\mathfrak{N}_0} = K_0[u, z_1, \ldots, z_n]_{(u, z_1, \ldots, z_n)} \text{ with } \mathfrak{n}_0 = (u, z)R_0,$$

$$R = S_{\mathfrak{N}} = K[u, z_1, \ldots, z_n]_{(u, z_1, \ldots, z_n)} \text{ with } \mathfrak{n} = (u, z_1, \ldots, z_n)R.$$

Then R is a countable regular local ring.

Notation being as above, we take a subset of S with the following property:

$$\mathcal{P} = \left\{ p \in S \;\middle|\; \begin{array}{l} \text{for each height one } \mathfrak{p} \in \operatorname{Spec}(R), \text{ there is} \\ \text{at least one element } p \text{ such that } p \in \mathfrak{p} \end{array} \right\} \tag{2.0.3}$$

Then, $\mathcal{P}$ is a *countable* set and we may assume that $u \in \mathcal{P}$ and that $\mathcal{P}$ contains an infinite number of elements of S_0.

2.1 Numbering

Now we come to an important lemma due to [Hei 82], which guarantees a *good* enumeration on $\mathcal{P}$.

With notation and assumptions above, we fix a surjective mapping $\rho: \mathbf{N} \to \mathcal{P}$, called a *numbering* on $\mathcal{P}$, which, if we set $\rho(i) = p_i$, satisfies the following:

$$p_1 = u, \text{ and } p_i \in S_{i-2} \text{ for any } i \geq 2 \tag{2.1.1}$$

We define:

$$z_{i0} = z_i \text{ and } z_{ik} = z_i + a_{i1}q_1 + \cdots + a_{ik}q_k^k \text{ with } q_k = p_1 \cdots p_k \tag{2.1.2}$$

Then $P_k = (z_{1k}, \ldots, z_{nk})R$ is a prime ideal of height n for any $k \geq 0$.

LEMMA 2.2 (Heitmann's Lemma) *With notation as above, let ρ be a numbering on $\mathcal{P}$ which satisfies* (2.1.1). *Then*

(2.2.1) $p_h \notin P_k$ *whenever* $k \geq h - 1$, *and*

(2.2.2) $(z_{1k}, \ldots, z_{mk})S_k$ *is a prime ideal, generated by an S_k-regular sequence* $z_{1k}, \ldots, z_{mk}$ *for any* m $(1 \leq m \leq n)$.

2.3 Relations

Taking polynomials in n variables over K_0 without constant term: $F_1(Z), \ldots, F_r(Z) \in K_0[Z_1, \ldots, Z_n]$, we set

$$\alpha_{jk} = \frac{1}{q_k^k} F_j(z_{1k}, \ldots, z_{nk}) \in L = Q(R),\ j = 1, \ldots, r \tag{2.3.1}$$

Then, by definition

$$\begin{aligned} \alpha_{j(k+1)} &= \frac{1}{q_{k+1}^{k+1}} F_j(z_{1(k+1)}, \ldots, z_{n(k+1)}) \\ &= \frac{1}{q_{k+1}^{k+1}} F_j(z_{1k} + a_{1(k+1)}q_{k+1}^{k+1}, \ldots, z_{nk} + a_{n(k+1)}q_{k+1}^{k+1}). \end{aligned}$$

Thus

$$\alpha_{jk} = \frac{q_{k+1}^{k+1}}{q_k^k}\alpha_{j(k+1)} + \frac{q_{k+1}^{k+1}}{q_k^k} r_{jk} \text{ with } r_{jk} \in R \tag{2.3.2}$$

Let $B = \bigcup_k R[\alpha_{jk}] \subset L$ with $j = 1, \ldots, r$ and $k = 1, 2, \ldots$. Then

LEMMA 2.4 *With notation as above, let* $M = (u, z_1, \ldots, z_n)B$. *Then* M *is a maximal ideal of* B.

Notation being as above, for $i = 1, \ldots, n$ and $j = 1, \ldots, r$, we set:

$$\zeta_i = z_i + a_{i1}q_1 + \ldots + a_{ik}q_k^k + \ldots = z_i + \textstyle\sum_{k=1}^{\infty} a_{ik}q_k^k,$$
$$f_j = F_j(\zeta_1, \ldots, \zeta_n) \in \hat{R} = K[[u, z_1, \ldots, z_n]] = K[[u, \zeta_1, \ldots, \zeta_n]].$$

Let $A = B_M \subset L$ be a quasi-local domain with maximal ideal $\mathfrak{m} = MA$. Then

THEOREM 2.5 $(A, \mathfrak{m})$ *is a* Noetherian *local domain with the following conditions:*

(2.5.1) $\tilde{\imath}\colon K[[u, \zeta_1, \ldots, \zeta_n]]/(F_1(\zeta), \ldots, F_r(\zeta)) \cong \hat{R}/(f_1, \ldots, f_r) \tilde{\to} \hat{A}$,

(2.5.2) $\hat{\mathfrak{p}} = (\tilde{\imath}(\zeta_1), \ldots, \tilde{\imath}(\zeta_n))\hat{A}$ *is a prime ideal of* $\hat{A}$ *and* $\hat{\mathfrak{p}} \cap A = (0)$,

(2.5.3) $A/\mathfrak{p}$ *is essentially of finite type over* K *for any non-zero prime* $\mathfrak{p} \in \mathrm{Spec}(A)$.

3 REES ALGEBRAS AND LOCAL DOMAINS WITH ODD NON-ZERO PRIMES

To begin with, we make a minor change of the previous notation.

3.0 Notation

Let K_0, K and K_k be as in (2.0.1). Taking $n+2$ indeterminates u, x, $z_1, \ldots, z_n$ over K, we set

$$S_0 = K_0[u, x, z_1, \ldots, z_n] \text{ and } \mathfrak{N}_0 = (u, x, z_1, \ldots, z_n)S_0,$$
$$S_k = S_{k-1}[a_{1k}, \ldots, a_{nk}] \text{ and } \mathfrak{N}_k = (u, x, z_1, \ldots, z_n)S_k.$$

Then, $S_k = K_0[a_{ih}, u, x, z_1, \ldots, z_n]$ with $i = 1, \ldots, n$ and $1 \le h \le k$. Further, we set

$$S = \bigcup_k S_k = K_0[a_{ik}, u, x, z_1, \ldots, z_n] \text{ and } \mathfrak{N} = (u, x, z_1, \ldots, z_n)S \qquad (3.0.1)$$

where $i = 1, \ldots, n$; $k = 1, 2, \ldots$. Moreover, let

$$R_0 = (S_0)_{\mathfrak{N}_0} = K_0[u, x, z_1, \ldots, z_n]_{(u,x,z)} \text{ with } \mathfrak{n}_0 = (u, x, z)R_0,$$

$$R = S_{\mathfrak{N}} = K[u, x, z_1, \ldots, z_n]_{(u,x,z)} \text{ with } \mathfrak{n} = (u, x, z_1, \ldots, z_n)R.$$

Then R is a countable regular local ring.

3.1 Numbering

Putting $\mathrm{Spec}(R)^* = \mathrm{Spec}(R) \setminus \{xR\}$, we take a subset of S with the following property:

$$\mathcal{P}^* = \left\{ p \in S \;\middle|\; \begin{array}{l} \text{for each height one } \mathfrak{p} \in \mathrm{Spec}(R)^*, \text{ there is} \\ \text{at least one element } p \text{ such that } p \in \mathfrak{p} \text{ and} \\ p \notin xR \end{array} \right\} \tag{3.1.1}$$

Then, $\mathcal{P}^*$ is a *countable* set and we may assume that $u \in \mathcal{P}^*$.

Denoting by $\overline{s}$ the image of $s \in S$ in $\overline{S} = S/xS$ (or in $\overline{R} = R/xR$), we may further assume that $\overline{\mathcal{P}} = \{\overline{p} \in \overline{S} \mid p \in \mathcal{P}^*\}$ satisfies the same condition as in (2.0.3), namely,

$$\overline{\mathcal{P}} = \left\{ \overline{p} \in \overline{S} \;\middle|\; \begin{array}{l} \text{for each height one } \overline{\mathfrak{p}} \in \mathrm{Spec}(\overline{R}), \text{ there is} \\ \text{at least one element } \overline{p} \text{ such that } \overline{p} \in \overline{\mathfrak{p}} \end{array} \right\} \tag{3.1.2}$$

Next, we fix a surjective mapping $\rho^*: \mathbf{N} \to \mathcal{P}^*$ with $\rho^*(i) = p_i$, a *numbering* on $\mathcal{P}^*$, which satisfies the following:

$$p_1 = u, \text{ and } p_i \in S_{i-2} \text{ for any } i \geq 2 \tag{3.1.3}$$

Moreover, we remark that, if ρ^* is the numbering above, then the induced mapping $\overline{\rho}: \mathbf{N} \to \overline{\mathcal{P}}$, which maps i to $\overline{p}_i$, is also an enumeration on $\overline{\mathcal{P}}$ satisfying:

$$\overline{p}_1 = \overline{u}, \text{ and } \overline{p}_i \in \overline{S}_{i-2} = S_{i-2}/xS_{i-2} \text{ for any } i \geq 2 \tag{3.1.4}$$

Hence, $\overline{\rho}$ becomes a numbering on $\overline{\mathcal{P}}$. As in Section 2, we define:

$$z_{i0} = z_i \text{ and } z_{ik} = z_i + a_{i1}q_1 + \cdots + a_{ik}q_k^k \text{ with } q_k = p_1 \cdots p_k \tag{3.1.5}$$

Then $Q_k = (x, z_{1k}, \ldots, z_{nk})R$ is a prime ideal of height $n+1$ for any $k \geq 0$. And the same proof as in Lemma 2.2 shows:

LEMMA 3.2 *With notation as above, if ρ^* is a numbering on $\mathcal{P}^*$ which satisfies* (3.1.3), *then*

(3.2.1) $p_h \notin Q_k$ *whenever* $k \geq h - 1$,

(3.2.2) $(x, z_{1k}, \ldots, z_{mk})S_k$ *is a prime ideal, generated by an S_k-regular sequence x, z_{1k}, $\ldots$, z_{mk} for any m $(1 \leq m \leq n)$.*

3.3 Relations

Taking polynomials in $n+1$ variables over K_0 without constant term: $G_1(X, Z)$, $\ldots$, $G_r(X, Z) \in K_0[X, Z_1, \ldots, Z_n]$, we set $F_j(Z) = G_j(0, Z) \in K_0[Z_1, \ldots, Z_n]$. Let

$$\beta_{jk} = \frac{1}{q_k^k} G_j(x, z_{1k}, \ldots, z_{nk}) \in L = Q(R), \; j = 1, \ldots, r \tag{3.3.1}$$

Then, by definition

$$\begin{aligned}\beta_{j(k+1)} &= \frac{1}{q_{k+1}^{k+1}} G_j(x, z_{1(k+1)}, \ldots, z_{n(k+1)}) \\ &= \frac{1}{q_{k+1}^{k+1}} G_j(x, z_{1k} + a_{1(k+1)} q_{k+1}^{k+1}, \ldots, z_{nk} + a_{n(k+1)} q_{k+1}^{k+1}).\end{aligned}$$

Thus

$$\beta_{jk} = \frac{q_{k+1}^{k+1}}{q_k^k} \beta_{j(k+1)} + \frac{q_{k+1}^{k+1}}{q_k^k} s_{jk} \text{ with } s_{jk} \in R \tag{3.3.2}$$

Let $B = \bigcup_k R[\beta_{jk}] \subset L$ with $j = 1, \ldots, r$ and $k = 1, 2, \ldots$. Then, the same reasoning as in Lemma 2.4 gives:

LEMMA 3.4 *Notation being as above, let $M = (u, x, z_1, \ldots, z_n)B$. Then M is a maximal ideal of B and $B/M \cong R/\mathfrak{n} \cong K$.*

With notation as above, for $i = 1, \ldots, n$ and $j = 1, \ldots, r$, we set:

$$\begin{aligned}\zeta_i &= z_i + a_{i1} q_1 + \ldots + a_{ik} q_k^k + \ldots = z_i + \textstyle\sum_{k=1}^{\infty} a_{ik} q_k^k, \\ g_j &= G_j(x, \zeta_1, \ldots, \zeta_n) \in \hat{R} = K[[u, x, z_1, \ldots, z_n]] = K[[u, x, \zeta_1, \ldots, \zeta_n]], \\ f_j &= F_j(\zeta_1, \ldots, \zeta_n) \in \hat{R}/x\hat{R} = K[[u, z_1, \ldots, z_n]] = K[[u, \zeta_1, \ldots, \zeta_n]].\end{aligned}$$

Let $A = B_M \subset L$ be a quasi-local domain with its maximal ideal $\mathfrak{m} = MA$. On the other hand, let $\tilde{\phi}$, ϕ be ring homomorphisms:

$$\begin{aligned}&\tilde{\phi}\colon K_0[X, Z, Q][T_1, \ldots, T_r] \to K_0[X, Z, Q][\frac{G_1}{Q}, \ldots, \frac{G_r}{Q}] \text{ with } T_j \mapsto G_j/Q, \\ &\phi\colon K_0[Z, Q][T_1, \ldots, T_r] \to K_0[Z, Q][\frac{F_1}{Q}, \ldots, \frac{F_r}{Q}] \text{ with } T_j \mapsto F_j/Q.\end{aligned}$$

Under the circumstances, we get:

THEOREM 3.5 *Regarding $K_0[Z, Q]$ as $K_0[X, Z, Q]/XK_0[X, Z, Q]$, suppose that*

(3.5.0) $\operatorname{Ker}\phi = K_0[Z, Q] \otimes_{K_0[X,Z,Q]} \operatorname{Ker}\tilde{\phi}$.

Then, $(A, \mathfrak{m})$ is a Noetherian local domain with prime *element x which satisfies the following conditions:*

(3.5.1) $\imath\colon K[[u, x, \zeta]]/(G_1(x, \zeta), \ldots, G_r(x, \zeta)) \cong \hat{R}/(g_1, \ldots, g_r) \xrightarrow{\sim} \hat{A}$,

(3.5.2) $\tilde{\imath}\colon K[[u, \zeta]]/(F_1(\zeta), \ldots, F_r(\zeta)) \cong (R/xR)^\wedge/(f_1, \ldots, f_r) \xrightarrow{\sim} \hat{A}/x\hat{A}$,

(3.5.3) $\hat{\mathfrak{q}} = (\tilde{\imath}(x), \tilde{\imath}(\zeta_1), \ldots, \tilde{\imath}(\zeta_n))\hat{A}$ *is a prime ideal of $\hat{A}$ and $\hat{\mathfrak{q}} \cap A = xA$,*

(3.5.4) *$A/\mathfrak{p}$ is essentially of finite type over K for any non-zero prime $\mathfrak{p} \in$* $\operatorname{Spec}(A) \setminus \{xA\}$.

4 BRODMANN–ROTTHAUS EXAMPLE AND OGOMA'S EXAMPLE

In this section, we start with showing that Brodmann–Rotthaus example (Example 4.1) can be gained as a joint application of Theorem 3.5 and Example 1.2. Next, we shall remark that Ogoma's example (Example 4.2) can be reproduced in the same manner. The crucial point in our reconstruction of these examples is to check the condition (3.5.0). Of course, even though they mentioned implicitly, the same is one of the essential and hard parts of their original work. Namely, Brodmann–Rotthaus use Hochster's relations a, b, c, d, e in Example 1.2 to get (3.5.0). As well, Ogoma wisely calculates the *Kernels* by hand. Nevertheless, it should be interesting that MACAULAY [Mac 89] gives us the same result automatically.

EXAMPLE 4.1 ([BR 83]) Three dimensional analytically irreducible local domain A, hence unmixed, but has $\mathfrak{p} \in \mathrm{Spec}(A)$ such that $A/\mathfrak{p}$ is *not* unmixed.

Construction. With notation as in Theorem 3.5, take

$$G_1(X,Z) = Z_2^3 - Z_3^2, \qquad G_2(X,Z) = Z_2X^2 - Z_1^2,$$
$$G_3(X,Z) = Z_2Z_1 - XZ_3, \quad G_4(X,Z) = Z_2^2X - Z_3Z_1.$$

Here, MACAULAY gives us (cf. Example 1.2):

$$\begin{aligned}\mathrm{Ker}\,\tilde{\phi} = (&QT_1 - G_1,\, QT_2 - G_2,\, QT_3 - G_3,\, QT_4 - G_4,\\ &XT_1 - Z_3T_3 - Z_2T_4,\, Z_1T_1 - Z_2^2T_3 - Z_3T_4,\, Z_2T_2 + Z_1T_3 - XT_4,\\ &Z_3T_2 + Z_2XT_3 - Z_1T_4,\, T_1T_2 + Z_2T_3^2 - T_4^2),\\ \mathrm{Ker}\,\phi = (&QT_1 - F_1,\, QT_2 - F_2,\, QT_3 - F_3,\, QT_4 - F_4,\\ &- Z_3T_3 - Z_2T_4,\, Z_1T_1 - Z_2^2T_3 - Z_3T_4,\, Z_2T_2 + Z_1T_3,\\ &Z_3T_2 - Z_1T_4,\, T_1T_2 + Z_2T_3^2 - T_4^2).\end{aligned}$$

Consequently, $\mathrm{Ker}\,\phi = K_0[Z,Q] \otimes_{K_0[X,Z,Q]} \mathrm{Ker}\,\tilde{\phi}$. Therefore, by Theorem 3.5, we get a local domain $(A, \mathfrak{m})$ with *prime* element x which satisfies the following conditions:

(4.1.1) $\quad \hat{A} \xrightarrow{\sim} K[[u,x,\zeta]]/(G_1(x,\zeta),\ldots,G_r(x,\zeta)) \cong K[[u,x,xy,y^2,y^3]],$

(4.1.2) $\quad \hat{A}/x\hat{A} \cong K[[u,x,xy,y^2,y^3]]/xK[[u,x,xy,y^2,y^3]].$

EXAMPLE 4.2 ([Ogo 82]) Three dimensional analytically unramified unmixed local domain A, which has $\mathfrak{p} \in \mathrm{Spec}(A)$ such that $A/\mathfrak{p}$ is *not* unmixed.

Construction. Notation being as in Theorem 3.5, take

$$G_1(X,Z) = Z_1Z_3, \quad G_2(X,Z) = Z_1(X + Z_2),$$
$$G_3(X,Z) = Z_2Z_3, \quad G_4(X,Z) = Z_2(X + Z_2).$$

Then, by MACAULAY, we get:

$$\begin{aligned}\operatorname{Ker}\tilde{\phi} = (&QT_1 - G_1,\ QT_2 - G_2,\ QT_3 - G_3,\ QT_4 - G_4,\\ &(X+Z_2)T_1 - Z_3T_2,\ (X+Z_2)T_1 - Z_1T_3,\\ &(X+Z_2)T_2 - Z_1T_4,\ (X+Z_2)T_3 - Z_3T_4,\ T_1T_4 - T_2T_3),\\ \operatorname{Ker}\phi = (&QT_1 - F_1,\ QT_2 - F_2,\ QT_3 - F_3,\ QT_4 - F_4,\\ &Z_2T_1 - Z_3T_2,\ Z_2T_1 - Z_1T_3,\ Z_2T_2 - Z_1T_4,\\ &Z_2T_3 - Z_3T_4,\ T_1T_4 - T_2T_3).\end{aligned}$$

Consequently, $\operatorname{Ker}\phi = K_0[Z,Q]\otimes_{K_0[X,Z,Q]}\operatorname{Ker}\tilde{\phi}$. Therefore, by Theorem 3.5, we get a local domain $(A,\mathfrak{m})$ with *prime* element x which enjoys the following:

(4.2.1) $\hat{A} \cong K[[u,x,\zeta_1,\zeta_2,\zeta_3]]/(\zeta_1,\zeta_2)\cap(\zeta_3, x+\zeta_2)$,

(4.2.2) $\hat{A}/x\hat{A} \cong K[[u,\zeta_1,\zeta_2,\zeta_3]]/(\zeta_1\zeta_3,\zeta_1\zeta_2,\zeta_2\zeta_3,\zeta_2^2)$.

References

[BR 82] Brodmann, M. and Rotthaus, C. (1982). Local domains with bad sets of formal prime divisors, J. Algebra **75**, 386–394.

[BR 83] Brodmann, M. and Rotthaus, C. (1983). A peculiar unmixed domain, Proc. Amer. Math. Soc. **87**, 596–600.

[EGA] EGA chapitre IV, IHES Publ. Math. **20** (1964), **24** (1965).

[Hei 82] Heitmann, R. C. (1982). A non-catenary, normal, local domain, Rocky Mountain J. Math. **12**, 145–148.

[Hoc 73] Hochster, M. (1973). Criteria for equality of ordinary and symbolic powers of primes, Math. Z. **133**, 53–65.

[Mac 89] MACAULAY, Macaulay User Manual, May 25, 1989.

[Mat 80] Matsumura, H. (1980). Commutative Algebra, Benjamin (second ed.).

[Nag 75] Nagata, M. (1975). Local Rings, John Wiley (reprt. ed. Krieger).

[Nis 92] Nishimura, J. (1992) A few examples of local rings, I, preprint.

[Ogo 80] Ogoma, T. (1980). Non-catenary pseudo-geometric normal rings, Japan. J. Math. **6**, 147–163.

[Ogo 82] Ogoma, T. (1982). Construction of an unmixed domain A with $A/\mathfrak{p}$ mixed for some prime ideal $\mathfrak{p}$ of A ; making use of Poincaré Series, preprint.

[Rot 79] Rotthaus, C. (1979). Universell Japanische Ringe mit nicht offenem regulärem Ort, Nagoya Math. J. **74**, 123–135.

21
Soundable Subsets of a Spectrum and Depth

Gabriel PICAVET

Département de Mathématiques
Université BLAISE PASCAL (Clermont II)
U.F.R. AUBIERE
63177 AUBIERE -CEDEX FRANCE

This note represents a summary of the author's talk given at the Fes algebra meeting. The results contained in this paper will be published in a forthcoming paper together with their proofs and various other developments.

0 - PRELIMINARIES ON REGULAR SEQUENCES AND FLAT TOPOLOGY

1 - The following result is standard :

PROPOSITION 1 : Let A be a Noetherian ring, M a finitely generated A-module and I an ideal of A. Then I contains a M-regular sequence of length $n+1$ if and only if $Ext^i_A(A/I, M) = 0$, for all $i \leq n$.

The proof can be done by induction on n. When $n = 0$, the inductive hypothesis is neither but the following condition :

$$(0) \qquad \begin{cases} \textit{for all finitely generated ideals } I \textit{ of } A,\ Hom_A(A/I, M) = 0 \\ \textit{implies that } I \textit{ contains a } M-\textit{regular element} \end{cases}$$

After a careful scrutiny of the proof, one can observe that proposition 1 is still valid, under the hypothesis : A is any ring , I is a finitely generated ideal of A and M is an A-module of which every quotient module satisfies condition (0).

This condition can be translated into a property of the (weak) assassin, as we shall see later. We then say that the assassin is a soundable subset of the spectrum of A. Our main purpose is the study of such soundable subsets, as well as closely related ideas.

2 - The flat topology on $Spec(A)$ has been studied in [3] by us : an open base is given by the sets $V(I)$, where I is a finitely generated ideal of A. The closure of a subset X of $Spec(A)$ is $X^p = \cap[D(I); I \in I_f(A) \text{ and } X \subset D(I)]$, where $I_f(A)$ is the set of finitely generated ideals of A. We recall that the content of a polynomial $f(t) = a_o + \ldots + a_n t^n \in A[t]$ is the ideal $C(f(t)) = (a_0, \ldots, a_n)$ of A. We have shown in [3] the following results :

$\Sigma_X = \{f(t) \in A[t] \quad / \quad X \subset D(C(f(t)))\}$ is a multiplicative subset of $A[t]$; denote by $X(A)$ the ring $A[t]_{\Sigma_X}$, then the canonical morphism $A \to X(A)$ is flat and its spectral image is X^p. We consider now in A a multiplicative subset $S_X = \{a \in A \quad / \quad X \subset D(a)\}$. Then we obtain a factorization $A \to A_X \to X(A)$.

DEFINITION 2 : Let A be a ring and $X \subset Spec(A)$, we define :

a) the size of X as $U(X) = \cup[P; P \in X] = A - S_X$

b) a Moore's closure u on $Spec(A)$ by $P \in X^u$ if $P \subset U(X)$.

If X^g is the generization of X, one has $X^g \subset X^p \subset X^u$:

observe that $X^u = \cap[D(a); a \in A \text{ and } X \subset D(a)]$, then the last inclusion follows easily.

DEFINITION 3 : Let A be a ring and $X \subset Spec(A)$, we say that :

1) X is *soundable* if $X^p = X^u$

2) X is *expanded* if $X^g = X^u$

3) X is *primal* (resp. *of finite size*) if $U(X)$ belongs to $Spec(A)$ (resp. $U(X)$ is equal to a finite union of prime ideals of A).

Moreover, X is said to verify absolutely one of the above properties if all the subsets of X verify this property.

It is clear that expanded implies soundable.

Let I be an ideal of A, one can observe that $I \subset U(X)$ if and only if I does not meet S_X ; hence I lifts in a proper ideal of A_X. Therefore $I \subset U(X)$ if and only if there exists a prime P such that $I \subset P \subset U(X)$. Using this remark, the proof of the following proposition is straightforward.

PROPOSITION 4 : Let A be a ring and $X \subset Spec(A)$, then

1) X is soundable if and only if every finitely generated ideal I contained in $U(X)$ is contained in some prime ideal P of X.

2) X is expanded if and only if every ideal I (resp. every prime ideal I) contained in $U(X)$ is contained in some prime ideal P of X.

1 - SOUNDABLE AND EXPANDED MODULES

If M is an A-module, the (weak) assassin of N. Bourbaki $Ass(M)$ is well known. The set $Att(M)$ is perhaps not so well known : a prime ideal P belongs to $Att(M)$ if for all $I \in I_f(A)$ such that $I \subset P$, there exists $x \in M$, $x \neq 0$, such that $I \subset 0 : x \subset P$. One says that P is attached to M. This notion comes from papers of D. G. Northcott and P. Dutton, see [1].

We denote by $Z(M)$ the set of zero divisors on the A-module M. Then, we have :

PROPOSITION 5 : Let M be an A-module , then $Z(M) = U(Ass(M)) = U(Att(M))$ and $Ass(M)^p = Att(M)^p$. Moreover, if $I \in I_f(A)$, then $Hom_A(A/I, M) = 0$ is equivalent to $Ass(M) \cap V(I) = \emptyset$ (resp. $Att(M) \cap V(I) = \emptyset$).

Thus , condition (0) is valid for M if and only if $Ass(M)$ (resp. $Att(M)$) is a soundable subset of $Spec(A)$.

Consequently, we are led to give the following definition :

DEFINITION 6 : An A-module M is said soundable (resp. expanded, primal) if $Ass(M)$ is soundable (resp. expanded , primal).

PROPOSITION 7 : Let M be an A-module, the following statements are equivalent :

1) The A-module M is soundable

2) For all $I \in I_f(A)$, the inclusion $I \subset Z(M)$ implies that there exists an element $x \neq 0$ in M, such that $Ix = 0$.

We observe that an A-module M is soundable and primal if and only if for all finite subsets $\{a_1 \ldots, a_n\}$ of $Z(M)$, there exists an element $x \neq 0$ in M, such that $a_1x = \ldots = a_nx = 0$.

EXAMPLES :

A) Let M be an A-module, then the $A[t]$-module $M[t]$ is soundable.

B) A finite Goldie dimensional A-module is soundable.

This can be deduced from the examples below.

C) An A-module is soundable if and only if its injective envelope is soundable.

D) An A-module with an irreducible zero submodule is soundable and primal.

For example, consider an indecomposable injective module.

E) Let M be an A-module such that $Ass(M/N)$ has a finite size for all submodules $N \neq M$, then M is soundable.

Various properties of ascent and descent by ring morphisms can be proved. We don't go further here.

One recovers the Noetherian case :

F) Let M be an A-module , with Krull dimension, see [4], then all quotient modules of M have finite Goldie's dimension. Such a module is soundable.

2 - CHARACTERIZATIONS OF SOUNDABLE OR EXPANDED SUBSETS OF A SPECTRUM

We first observe :

PROPOSITION 8 : A subset X of $Spec(A)$ is expanded if and only if it is soundable and quasi-compact .

Another remark is :

PROPOSITION 9 : Let X be a subset of $Spec(A)$, the following statements on X are equivalent :

1) X is soundable and primal

2) X^p is a directed set

3) X is irreducible for the flat topology

4) X^p is soundable and primal.

The following proposition is immediate.

PROPOSITION 10 : Let A be a ring, I an ideal of A and S a multiplicative subset of A, then $V(I)$ and $Im(Spec(A_S))$ are expanded.

PROPOSITION 11 : A subset X of $Spec(A)$ is absolutely soundable , if X is absolutely of finite size.

DEFINITION 12 : Let A be a ring. We say that A is an absolutely soundable (resp. expanded , of finite size) ring if $Spec(A)$ is absolutely soundable (resp. expanded , of finite size).

Then, we have :

PROPOSITION 13 : Let A be a ring, then A is an absolutely soundable (resp. expanded, of finite size) ring if and only if every A module M is soundable (resp. expanded, of finite size).

We recover in two cases some well known rings :

PROPOSITION 14 : Let A be a ring , then :

1) A is absolutely of finite size if and only if A is an universally zero-divisor ring (for every A-module M and all submodules $N \neq M$, the size of $Ass(M/N)$ is finite , see [5]).

2) A is absolutely expanded if and only if A is a compactly packed ring. This last kind of rings is characterized as rings in which every semi-prime ideal is the radical of a principal ideal.

3) A is absolutely soundable if and only if the radical of every finitely generated ideal of A is the radical of a principal ideal.

Clearly, a Bézout domain is absolutely soundable. One can also show that a ring of continuous real functions is absolutely soundable.

The three classes of rings, considered in the definition 12, are different. To end, we add that these classes of rings are stable under quotient and localization.

REFERENCES

1. P. Dutton. Prime ideals attached to a module . *Quart. J. Math. Oxford* , (2) , 29 (1978), 403 - 413.
2. D.G. Northcott. Finite free resolutions . *Cambridge University Press*, 71 . Cambridge, London, New York, Melbourne (1976).
3. G. Picavet. Propriétés et applications de la notion de contenu . *Comm. Algebra*, 13, (1985), 2231 -2265.
4. R. Gordon and J.C. Robson. Krull dimension. *Memoirs of the American Mathematical Society*, n° 113, (1973).
5. S. Visweswaran. A note on universally zero-divisor rings. *Bull. Austral. Math. Soc.*, 43 , (1991) , 233 - 239.

22
t-Closed Rings

Martine PICAVET-L'HERMITTE

Département de Mathématiques
Université Blaise Pascal (Clermont II)
U.F.R. AUBIERE
63177 AUBIERE-CEDEX FRANCE

In a recent paper, see [4], we made a study of t-closed morphisms : the definition of this kind of morphisms is given by N. Onoda, T. Sugatani and K. Yoshida in [3], with restrictive hypothesis ; we have generalized in a suitable way their definition and thus obtained a theory paralleled with the one made by R. G. Swan, about seminormal morphisms, see [6].

Let us recall some definitions :

An injective rings morphism $A \to B$ is said to be t-closed in case an element b of B, satisfying both $b^2 - rb$ and $b^3 - rb^2 \in A$ for some $r \in A$, belongs to the ring A.

An injective rings morphism $A \to B$ is said seminormal if, whenever $b \in B$ satisfies $b^2, b^3 \in A$, then $b \in A$.

Now, R.G. Swan calls a ring seminormal if whenever $a, c \in A$ satisfy the relation $a^3 = c^2$, there exists $b \in A$ such that $a = b^2$ and $c = b^3$.

A swift thought convinces us that the relation $a^3 = c^2$ is nothing else but the annihilation condition for the resultant of the polynomials $X^2 - a$ and $X^3 - c$. We are thus led to mimic the definition of a seminormal ring, which give rise to t-closed rings.

1 - t-CLOSED RINGS

Let A be a ring and let a, r, c be elements of A such that there exists $b \in A$, with $b^2 - rb = a$ and $b^3 - rb^2 = c$. The last two relations are nothing but the condition : the polynomials $X^2 - rX - a$ and $X^3 - rX^2 - c$ of $A[X]$ have a common zero ; so their resultant $a^3 + arc - c^2$ is zero.

DEFINITION 1.1 : Let A be a ring. We say that A is a t-closed ring if for every element $(a, r, c) \in A^3$, such that $a^3 + arc - c^2 = 0$, there exists an element $b \in A$ such that $a = b^2 - rb$ and $c = b^3 - rb^2$.

REMARK : A t-closed ring is a seminormal one and thus is reduced, see [7].

Here are two stability properties of t-closed rings :

PROPOSITION 1.2 : Let $\{A_i\}_{i \in I}$ be a family of t-closed rings, then $A = \prod_{i \in I} A_i$ is a t-closed ring.

PROPOSITION 1.3 : Let $\{A_i\}_{i \in I}$ be a family of underrings of an integral domain B. If every A_i is a t-closed ring, so is $\bigcap_{i \in I} A_i$.

PROPOSITION 1.4 : Let A be a t-closed ring and S a multiplicative subset of A $(1 \in S, 0 \notin S)$, then A_S is a t-closed ring.

PROPOSITION 1.5 : Every absolutely flat ring is a t-closed ring and is thus seminormal.

PROPOSITION 1.6 : Let A be an integrally closed ring, with an absolutely flat total quotient ring. Then A is a t-closed ring and particularly a semi-normal one.

Absolute flatness for total quotient ring is reached for reduced rings, with compact minimal spectrum, satisfying the condition (A) : every faithful finitely generated ideal of the ring contains a free element.

This condition is always fulfilled by a polynomial ring.

REMARK : An integrally closed integral domain (for instance, a valuation ring) is a t-closed ring.

2 - LINK WITH t-CLOSED MORPHISMS

THEOREM 2.1 : Let $A \to B$ be an injective rings morphism.

1) if B is a t-closed ring and $A \to B$ a t-closed morphism, then A is a t-closed ring.

2) If B is an integral domain and if A is a t-closed ring, then $A \to B$ is a t-closed morphism.

REMARKS :

1) Let A be an integral t-closed domain and let $(a, r, c) \in A^3$ be such that we have $a^3 + arc - c^2 = 0$. When $(a, c) \neq (0, 0)$, there exists an unique element b in A, such that $a = b^2 - rb, c = b^3 - rb^2$. Now, if $(a, c) = (0, 0)$, these two relations are fulfilled by $b = 0$ and $b = r$; thus, there is no more uniqueness, contrary to what happens with seminormality.

2) Part 2) of the theorem above is no longer true when B is not an integral domain : let $A = \mathbb{Z}$ and $B = \mathbb{Z}^2$; these two rings are t-closed ; let $A \to B$ be the canonical morphism; the element $b = (1, 0)$ of B verifies $b^2 - b = 0$ and $b^3 - b^2 = 0$; but b does not belong to A, so that $A \to B$ is not t-closed.

These two remarks show that t-closed rings and seminormal rings have a different behaviour.

COROLLARY 2.2 : Let A be a ring, then :

1) If the total quotient ring $Tot(A)$ is absolutely flat and if $A \to Tot(A)$ is t-closed, then A is a t-closed ring.

2) If A is an integral domain, $A \to Tot(A)$ is a t-closed morphism if and only if A is a t-closed ring.

REMARK : Thus we recover the definition of a t-closed ring given in [2].

COROLLARY 2.3 : Let A be an integral domain, with quotient field K. Then A is t-closed if and only if the following condition is fulfilled :

an element b of K belongs to A if there exist $r \in A, m \in \mathbb{N}^*$ such that

$$b^{n+1} - rb^n \in A, \quad \text{for all} \quad n \geq m$$

COROLLARY 2.4 : Let A be a ring and let $\overline{A}$ be its integral closure in $Tot(A)$.

1) When $Tot(A)$ is absolutely flat and $A \to \overline{A}$ is a t-closed morphism, then A is a t-closed ring.

2) If A is an integral domain, then A is t-closed if and only if $A \to \overline{A}$ is a t-closed morphism.

COROLLARY 2.5 : Let A be a ring with an absolutely flat total quotient ring and let $\overline{A}$ be its integral closure. If the conductor of the morphism $A \to \overline{A}$ is a maximal ideal of $\overline{A}$, then A is a t-closed ring.

3 - THE CASE OF AN INTEGRAL DOMAIN

Theorem 2.1 gives a necessary and sufficient condition for a ring to be t-closed, in the integral case. We are going to infer further results from it.

PROPOSITION 3.1 : Let A be an integral domain. Then A is a t-closed ring if and only if A_M is a t-closed ring, for all maximal ideals M of A. The same is true when prime ideals are used.

PROPOSITION 3.2 : Let A be a ring. Then A is a t-closed ring provided that $A[X]$ is a t-closed ring. The converse is true if A is an integral domain.

When A is a one-dimensional domain, we obtain more acute results.

PROPOSITION 3.3 : Let A be a one-dimensional domain, then A is a t-closed ring if and only if A is a seminormal ring and the map $Spec(\overline{A}) \to Spec(A)$ is bijective.

COROLLARY 3.4 : Let A be a one-dimensional local domain such that $A \to \overline{A}$ is a finite morphism. Then A is a t-closed ring if and only if the maximal ideal of A is the only maximal ideal in $\overline{A}$.

Now, we introduce tools for constructing t-closed rings.

4 - THE t-CLOSURE OF AN INTEGRAL DOMAIN

Let $f : A \to B$ be an injective rings morphism and let $\overline{B}$ be the integral closure of A in B. If Q is a prime ideal of $\overline{B}$, over a prime ideal P of A, we denote by C_Q the image of the canonical morphism $A_P \to \overline{B}_Q$.

We defined in [4] the t-closure t_BA of A in B, in the following way : elements b of t_BA are elements $b \in \overline{B}$ such that, for all $Q \in Spec(\overline{B})$, the image of b in $\overline{B}_Q$ belongs to $C_Q + Q\overline{B}_Q$. Then we proved that t_BA is the smallest subring C of B such that $C \to B$ is t-closed. We emphasize that the results above are free of any Noetherian or integral hypothesis, given in the paper of T. Sugatani and K. Yoshida, cf. [5].

PROPOSITION 4.1 : Let $A \to B$ be an injective rings morphism such that B is a t-closed ring. Then the t-closure t_BA of A in B is a t-closed ring.

PROPOSITION 4.2 : Let A be an integral domain. Then $t_{\overline{A}}A$ is the smallest t-closed integral domain containing A.

We call it the t-closure of A and denote it by ${}^tA$.

COROLLARY 4.3 : Let A be an integral domain. Then A is a t-closed ring if and only if $A = {}^tA$.

COROLLARY 4.4 : Let A be an integral domain, then :

1) If S is a multiplicative subset of A, we have $({}^tA)_S = {}^t(A_S)$.

2) ${}^t(A[X]) = ({}^tA)[X]$.

5 - t-CLOSED RINGS AND QUASINORMAL RINGS

We showed in [4] a link between t-closed morphisms and quasinormal morphisms, for one-dimensional domains. We recall that an integral domain A is said to be quasinormal if the map $Pic(A) \to Pic(A[X, X^{-1}])$ is bijective. This last condition is equivalent to the following one : $A \to \overline{A}$ is a quasinormal morphism. Recall also that an injective morphism $A \to B$ is said to be quasinormal if the map $MPic(A) \to MPic(B)$ is injective, where $MPic(A)$ is the cokernel of the map $Pic(A) \to Pic(A[X, X^{-1}])$.

PROPOSITION 5.1 : Let A be a one-dimensional domain. Then, A is t-closed if and only if it is a locally quasinormal ring. This being so, A is a quasinormal ring.

COROLLARY 5.2 : A one-dimensional Noetherian domain is t-closed if and only if it is a quasinormal ring.

COROLLARY 5.3 : The coordinate ring A of an irreducible affine curve over an algebraically closed field is t-closed if and only if the curve is non-singular. In this case, A is integrally closed.

6 - t-CLOSED OVERRINGS AND SUBRINGS

D. Anderson, D. Dobbs and J. Huckaba gave in [1] conditions for subrings and overrings of an integral domain being all seminormal. We obtain similar results with t-closed rings.

PROPOSITION 6.1 : Let A be an integral domain. Every subring of A is a t-closed ring if and only if every subring of A is seminormal.

In particular, see [1], these conditions are equivalent to every subring of A is an Euclidian domain.

PROPOSITION 6.2 : Let A be a subring of a field L, with quotient field K. Every ring between A and L is t-closed if and only if one of the following conditions is fulfilled :

1) $A = K$ and the morphism $K \to L$ is an algebraic one.

2) $L = K$, every overring of A, integral over A is a t-closed ring and the integral closure $\overline{A}$ of A is a Prüfer ring.

PROPOSITION 6.3 : Let A be a one-dimensional t-closed domain ; every ring between A and $\overline{A}$ is a t-closed one .

PROPOSITION 6.4 : Let A be an integral domain with quotient field K, such that every overring B of A satisfies the following condition :

$$\text{(C)} \quad : \quad \forall b \in K, \text{ such that } \exists a, r \in B \text{ with } b^2 - rb = a, \text{ then } b \in B.$$

Then, every overring of A is a t-closed ring.

REMARK : A ring fulfilling the condition (C) is said to be quadratically integrally closed.

Proofs of the results above will be published in a forthcoming paper [8].

REFERENCES

1. D.F. Anderson, D. Dobbs and J.A. Huckaba, On seminormal overrings, *Comm. Algebra*, 10, (13), (1982), 1421-1448.

2. Y. Koyama, T. Sugatani and K. Yoshida, Some remarks on divisorial and seminormal overrings, *Comm. Algebra* 13, (4), (1985), 795-810.

3. N. Onoda, T. Sugatani and K. Yoshida, Local quasinormality and closedness type criteria, *Houston J. Math.*, 11, (1985), 247-256.

4. G. Picavet et M. Picavet-L'Hermitte, Morphismes t-clos, *Comm. Algebra*, 21, (1), (1993), 179-219.

5. T. Sugatani and K. Yoshida, On t-closure, *C.R. Math. Rep. Acad. Sci. Canada*, 6, (1984), 55-59.

6. R.G. Swan, On seminormality, *J. Algebra*, 67, (1980), 210-229.

7. D.L. Costa, Seminormality and projective modules, in Séminaire d'Algèbre Paul Dubreuil et Marie-Paule Malliavin, (Ed. M.P. Malliavin), *Lectures Notes in Mathematics*, Vol. 924, Springer-Verlag, Berlin, Heidelberg, New York, (1981), 400-412.

8. G. Picavet et M. Picavet-L'Hermitte, Anneaux t-clos, to appear.

23
Some Aspects of the Asymptotic Theory of Ideals. Generalization to Filtrations

DAOUDA SANGARE Département de Mathématiques, FAST, Université d'Abidjan, 22 B.P. 582, Abidjan, Côte d'Ivoire

INTRODUCTION

The asymptotic theory of ideals originated with the paper "Some asymptotic properties of powers of ideals "by Pierre SAMUEL [13] . This theory is based on the fact that in a nœtherian ring, large powers of an ideal are well behaved in some sense. It gives useful tools in commutative algebra and also in algebraic geometry. This theory has been developed at the beginning by D. Rees and M. Nagata and more recently by many other authors. It deals with topics like Samuel numbers of two ideals, homogeneous pseudo valuations, Rees valuation theorem, integral dependence over an ideal, reduction of ideals, Hilbert-Samuel polynomials, multiplicities of ideals, asymptotic prime divisors of an ideal. This paper is part of a current trend to extend to filtrations some classical results among the above topics (see for instance (D. Rees, [12]), (W. Bishop, [2]) , (L.J. Ratliff, Jr, [11])).

Indeed we intend to give a rapid introduction to the theory of multiplicity for filtrations following (W. Bishop, [2]) and to the notions of integral dependence over a filtration, reduction of filtrations and generalized Samuel numbers introduced by P. Ayegnon, H. Dichi, M. Soumaré and D. Sangaré from the University of Abidjan. So, this paper is rather expository in nature, the only new result here being Theorem 4.4. and its corollaries.

Throughout this paper, A will denote a commutative ring and I,J, arbitrary ideals of A. In [13], Samuel has shown that if A is a nœtherian ring, large powers of ideals of A are well behaved in some sense. One of his ideas was, in different language, to examine the numbers $\overline{v}_I(J)$ and $\overline{w}_I(J)$, depending on the two ideals I and J and defined as follows :

$$v_I(J) = \operatorname{Sup}\{r \in \mathbb{N} \ ; J \subseteq I^r\}$$

$$w_I(J) = \operatorname{Inf}\{r \in \mathbb{N} \ ; J \supseteq I^r\}$$

$$\overline{v}_I(J) = \lim_{n\to\infty} \frac{1}{n} v_I(J^n)$$

$$\overline{w}_I(J) = \lim_{n\to\infty} \frac{1}{n} w_I(J^n) .$$

The numbers $\overline{v}_I(J)$ and $\overline{w}_I(J)$ are called the Samuel numbers of I and J. Samuel proved that these numbers exist under the following conditions : A is nœtherian, I and J are non-nilpotent, $\sqrt{I} = \sqrt{J}$ and

$$\bigcap_n I^n = (o) = \bigcap_n J^n .$$

If x is an element of A, we put $v_I(x) = v_I(xA)$ and

$$\overline{v}_I(x) = \overline{v}_I(xA) = \lim_{n\to\infty} \frac{1}{n} v_I(x^n).$$

The mappings $x \mapsto v_I(x)$ and $x \mapsto \overline{v}_I(x)$ are pseudo valuations on A. Furthermore $\overline{v}_I$ is homogeneous i.e $\overline{v}_I(x^n) = n\,\overline{v}_I(x)$ for all $x \in A$ and for all integers $n \geqslant 1$.

The ideals I and J are termed <u>asymptotically equivalent</u> following Samuel if $\overline{v}_I = \overline{v}_J$.

An element x∈A is said to be <u>integral over the ideal I</u> if x satisfies an equation of the form :

$$x^m + a_1 x^{m-1} + \cdots + a_j x^{m-j} + \cdots + a_m = 0, \text{ where } a_j \in I^j \text{ for all } j.$$

The set I' of all elements which are integral over I, is an ideal of A called the <u>integral closure</u> of I. The ideal J is asymptotically equivalent to the ideal I if and only if J' = I'.

The concept of integral dependence over an ideal is closely related to that of reduction of an ideal, where <u>an ideal I is called a reduction of the ideal</u> J if $I \subseteq J$ and if there exists an integer $r \geqslant 0$ such that $J^{r+1} = I J^r$. It is easy to show that if A is nœtherian and if $I \subseteq J$, then I is a reduction of J if and only if $J \subseteq I'$.

Recently many authors have given in the literature an extension to filtrations of some classical results of the asymptotic theory of ideals (see for instance (D. Rees, [12]), (W. Bishop, [2]) , (L. J. Ratliff Jr., [11])). Here we intend to give a rapid introduction to some notions like integral dependence over a filtration, reduction of filtrations, generalized Samuel numbers of two filtrations which have been introduced by H. Dichi, M. Soumaré, P. Ayegnon and D. Sangaré ([4] , [6] and [1]).

In §1 we give a list of filtrations that are not I-adic and that have some of the nice properties of I-adic filtrations. In §2 we recall for a given finitely generated module M over a nœtherian ring A, the definition and some properties of the multiplicity of M with respect to a given filtration f on the ring A, that we need later. The definition was introduced by (Bishop, [2]) .

In § 3 we give a generalization to filtrations of the concept of integral dependence over an ideal and that of reduction of ideals. We will see that there are many possibilities for the extension of this last concept. In § 4 we extend to filtrations the notion of Samuel numbers of two ideals. Theorem 4.4 and his corollaries are generalizations of Theorem 3 of (Samuel, [13]) relating multiplicities, Samuel numbers and Krull dimension of the ring A.

1. FILTRATIONS, REES RINGS

1.1. By a filtration on A we mean any family $f = (I_n)_{n\in\mathbb{Z}}$ of ideals of A such that $I_0 = A$, $I_{n+1} \subseteq I_n$ and $I_n I_m \subseteq I_{n+m}$ for all m, $n\in\mathbb{Z}$. It follows that $I_n = A$ for all $n\leqslant o$. For any ideal I of A, the filtration $f_I = (I^n)_{n\in\mathbb{Z}}$, where $I^n = A$ for all $n\leqslant o$, is called I-adic.

For any real number $\lambda > o$, let $\{\lambda\}$ be the least integer greater than or equal to λ, and for any filtration $f = (I_n)_{n\in\mathbb{Z}}$, let $f^{(\lambda)} = (J_n)_{n\in\mathbb{Z}}$, where $J_n = I_{\{n\lambda\}}$ if $n\geqslant o$ and $J_n = A$ for all $n\leqslant o$.

It is easily shown that $f^{(\lambda)}$ is a filtration on A.

The set of all filtrations on A is ordered by

$$f = (I_n) \leqslant g = (J_n) \Longleftrightarrow I_n \subseteq J_n \text{ for all n.}$$

1.2. Let I be an ideal of A. The following two auxiliary graded rings associated to I and introduced by Rees have proved to be useful in many questions in the asymptotic theory of ideals :

$R(A, I) = \sum_{n\geq o} I^n X^n$ and $\mathcal{R}(A, I) = \sum_{n\in\mathbb{Z}} I^n X^n$, where $I^n = A$ for all $n\leq o$ and where X is an indeterminate. R(A, I) and $\mathcal{R}(A, I)$ are called the Rees rings of the ideal I.

R(A, I) is a graded subring of the polynomial ring A[X], while $\mathcal{R}(A, I)$ is a graded subring of $A[u, X]$, where we put $u = X^{-1}$.

The above construction can easily be extended to any filtration

$f = (I_n)$ on A to get the <u>Rees rings of f</u>

$$R(A, f) = \sum_{n \geq 0} I_n X^n$$

$$\text{and } \mathcal{R}(A, f) = \sum_{n \in \mathbb{Z}} I_n X^n .$$

In particular if $f = f_I$ is the I-adic filtration, we recover the above rings $R(A, I)$ and $\mathcal{R}(A, I)$.

Several properties of the Rees rings of an ideal can be extended to the Rees rings of a filtration. However it should be noted that if A is nœtherian, then $R(A, f)$ and $\mathcal{R}(A, f)$ need not be, whereas $R(A, I)$ and $\mathcal{R}(A, I)$ are.

It is useful to remark that $\mathcal{R}(A, f) = R(A, f)[u]$, where $u = X^{-1}$.

1.3 Here we give a list of filtrations that are not I-adic and that have some of the nice properties of I-adic filtrations. Most of them have been introduced by Bishop in different language. We end this section by an implication diagram between these filtrations.

Let $f = (I_n)$ be a filtration on A

a) f is said to be <u>I-good</u>, where I is an ideal of A if $I I_n \subseteq I_{n+1}$ for all n and if $I I_n = I_{n+1}$ $\forall$ n >> o, this last notation meaning "for all large n".

b) f is called <u>approximatable by powers</u> (A P for short) if there exists a sequence of positive integers (k_n) such that

$$I_{k_n m} \subseteq I_n^m \text{ for all } m, n \text{ and } \lim_{n \to \infty} \frac{k_n}{n} = 1 .$$

c) f is called strongly A P with rank k if there exists an integer $k \geqslant 1$ such that $f^{(k)}$ is the I_k-adic filtration i.e $I_{nk} = I_k^n$ for all n.

The following implications hold in any ring A.

f I-adic ⟹ f I-good ⟹ f strongly A P ⟹ f A P.

d) f is called nœtherian if the Rees ring $\mathcal{R}(A, f) = \sum_{n \in \mathbb{Z}} I_n X^n$ is nœtherian.

If A is nœtherian then

f I-good ⟹ f nœtherian ⟹ f strongly A P.

Consequently in a nœtherian ring, among the filtrations which behave like I-adic filtrations the most general ones are A P filtrations. This explains the role played by A P filtrations in the sequel.

2. MULTIPLICITY OF A FILTRATION IN A NŒTHERIAN RING

We recall the definition of the multiplicity of a finitely generated module M over a nœtherian ring A with respect to an ideal I, starting from the Hilbert-Samuel polynomial of M with respect to I. This multiplicity is a positive integer. It is generalized to an additive function on the category of finitely generated A-modules before being extended to filtrations in the sense of Bishop. As opposed to the I-adic case, any positive irrational number can be shown to be the multiplicity of a suitable filtration on a suitable ring.

2.1 Multiplicity of an ideal

A function $\varphi : \mathbb{Z} \longrightarrow \mathbb{Z}$ is called a polynomial function if there exists a polynomial $P(X) \in \mathbb{Q}[X]$ such that $\varphi(n) = P(n)$ for all large n. Such a polynomial is unique and is called the polynomial associated with φ.

For any A-module M, let $\ell(M)$ denote the length of M.

In the sequel of this section A is a nœtherian ring, I an ideal of A and M a finitely generated A-module. Suppose that $\ell(M/IM)$ is finite. Then $\ell(M/I^n M)$ is finite for all $n \geq 1$ and it is well known that, under this condition, the function $n \longrightarrow \ell(M/I^n M)$ is a polynomial function. The polynomial associated with this polynomial function is called the Hilbert-Samuel polynomial of M with respect to I. We denote by χ_I^M this polynomial. Then $\chi_I^M \in \mathbb{Q}[X]$ and

$$\chi_I^M (n) = \ell(M/I^n M) \text{ for all } n >> o.$$

Suppose in addition that $M \neq IM$. Then χ_I^M is not the null polynomial. Let $d = d_I(M)$ its degree. We may assume that

$$\chi_I^M = a_d X^d + a_{d-1} X^{d-1} + \cdots + a_o, \text{ where } a_j \in \mathbb{Q} \text{ for all } j \text{ and } a_d \neq o.$$

The multiplicity of M with respect to I is the number $e_I(M) = d!\, a_d$. It follows from the definition of χ_I^M that :

2.1.1. $$e_I(M) = \lim_{n\to\infty} \frac{d!}{n^d} \ell(M/I^n M).$$

This last formula is called the asymptotic formula of Samuel. It can be shown easily that $e_I(M)$ is a positive integer.

Unfortunately the multiplicity function $M \longmapsto e_I(M)$ is not an additive function on the category of finitely generated A-modules, i.e for any exact sequence $o \longrightarrow M' \longrightarrow M \longrightarrow M'' \longrightarrow o$ in this category, it is not necessarily true that $e_I(M) = e_I(M') + e_I(M'')$. But this multiplicity function can be extended to an additive multiplicity function as follows.

Suppose in addition that the coheight of the ideal I, i.e the Krull dimension of A/I, is null. Then $\ell(M/I M)$ is finite, hence the Hilbert-Samuel polynomial χ_I^M is well defined. Let s be the altitude of I, i.e the supremum of the heights of the minimal prime ideals over I. Following a result of Nagata ([9], Theorem 22.7),

$$d = d_I(M) = \text{alt}\left(\frac{I + \text{Ann } M}{\text{Ann } M}\right) \leqslant \text{alt } I = s.$$

Consequently we may write

$$\chi_I^M = a_s X^s + \cdots + a_d X^d + \cdots + a_0, \text{ where}$$

$$a_s = \begin{cases} 0 & \text{if } d_I(M) < s = \text{alt } I \\ a_d & \text{if } d_I(M) = s = \text{alt } I \end{cases}$$

The number $e_I^*(M) = s!\, a_s = \begin{cases} 0 & \text{if } d_I(M) < s = \text{alt } I \\ e_I(M) & \text{if } d_I(M) = s = \text{alt } I \end{cases}$

is still called the <u>multiplicity of M</u> with respect to I.
In particular if $M = A$, then $d_I(A) = \text{alt } I$, hence $e_I^*(A) = e_I(A)$. A second asymptotic formula of Samuel is still valid :

2.1.2. $$e_I^*(M) = \lim_{n \to \infty} \frac{s!}{n^s} \ell(M/I^n M),$$

where $s = \text{alt } I$.
It can be shown that the multiplicity function $M \longmapsto e_I^*(M)$ is additive on the category of finitely generated A-modules.

2.2 Multiplicity of a filtration

Starting from the second asymptotic formula (2.1.2) of Samuel, Bishop in [2], has introduced a concept of multiplicity for filtrations. Some preliminary definitions and remarks are necessary before defining this concept.

Let $f = (I_n)$ be a filtration on A. Then $I_1^n \subseteq I_n \subseteq I_1$ for all $n \geqslant 1$. Therefore $\sqrt{I_n} = \sqrt{I_1}$ for all $n \geqslant 1$. This constant ideal is called the <u>radical</u> of f and is denoted by $\sqrt{f}$. The <u>coheight</u> of f is by definition coht f = coht$\sqrt{f}$ = $\dim(\frac{A}{\sqrt{f}})$. Similarly the <u>altitude of f</u> is alt f = alt $\sqrt{f}$.

Let A be a nœtherian ring and let $f = (I_n)$ be a filtration on A such that coht f = o. Let s = alt f. Then for any finitely generated A-module M, $\ell(M/I_n M)$ is finite for all $n \geqslant 1$. Indeed $o = \dim \frac{A}{\sqrt{f}} = \dim \frac{A}{I_n}$. Therefore $V(I_n) \subseteq$ Max A, where $V(I_n)$ is the set of prime ideals which contain I_n and Max A the set of maximal ideals of A. Then by computing the support of the factor module M/I_nM we have

$$\operatorname{Supp}(M/I_nM) = V(I_n) \cap \operatorname{Supp}M \subseteq V(I_n) \subseteq \operatorname{Max}A.$$

Hence by a well known criterion, $\ell(M/I_n M)$ is finite. The <u>multiplicity of M with respect to f</u> is the number

$$e_f(M) = \lim_{n \to \infty} \frac{s\,!}{n^s}\, \ell(M/I_n M),$$

where s = alt f, whenever this limit exists.

2.3 Remarks

(i) Let I be an ideal of the nœtherian ring A such that coht I = o. Then for any finitely generated A-module M, the multiplicity $e_f(M)$ exists if $f = f_I$ is the I-adic filtration and $e_f(M) = e_I^*(M)$.

(ii) If f is not an I-adic filtration and if M is a finitely generated A-module, the multiplicity $e_f(M)$ need not exist and if it does, it need not be an integer nor a rational number as shown in the following example. Let A = $\mathbb{R}[X]$ the polynomial ring over the field $\mathbb{R}$ of real numbers and let λ be a positive real number. Consider the filtration $f = (X^{\{n\lambda\}}A)$. Then $e_f(A) = \lambda$ as easily shown.

(iii) However if coht f= o and if f is AP, then for any finitely generated module M over the nœtherian ring A, the multiplicity $e_f(M)$ exists and

$$e_f(M) = \lim_{n\to\infty} \frac{1}{n^s} e^*_{I_n}(M),$$

where s = alt f. For the proof of this statement, see [2], where it is also shown that for an AP filtration f such that coht f= o, the multiplicity function $M \longmapsto e_f(M)$ satisfies most of the properties of the I-adic case and that in particular it is an additive function on the category of finitely generated A-modules if A is nœtherian.

3. INTEGRAL DEPENDENCE OVER A FILTRATION, REDUCTION OF FILTRATIONS

It is well known that for ideals, the concepts of integral dependence and reduction are closely related when the ring A is nœtherian. These two concepts have recently given rise to various extensions to filtrations, e.g the concept of an f-good (resp. f-fine) filtration, of an integral (resp. a strongly integral) filtration over f and also of a reduction of f, for a given filtration f, as follows.

3.1 Let $f = (I_n)$ be a filtration on A. An element $x \in A$ is said to be <u>integral over f</u> if x satisfies an equation of the form :

$$x^m + a_1 x^{m-1} + \dots + a_j x^{m-j} + \dots + a_m = 0,$$

where m is an integer $\geqslant 1$ and $a_j \in I_j$ for all $j \geqslant 1$.

Let $y \in A$ and let $k \geq 1$ be an integer. Then the following statements are equivalent as easily shown :

(i) y is integral over $f^{(k)} = (I_{nk})_{n \in \mathbb{Z}}$

(ii) $y X^k$ is integral over the Rees ring $R(A,f)$.

(iii) $y X^k$ is integral over the Rees ring $\mathcal{R}(A,f)$.

An ideal J of A is said to be integral over f if each element of J is integral over f.

3.2. Let $f = (I_n)$ and $g = (J_n)$ be two filtrations on A.

a) g is said to be integral over f if J_n is integral over $f^{(n)}$ for all $n \geq 1$.

If $f \leq g$ then g is integral over f if and only if the ring $R(A,g)$ is integral over the ring $R(A,f)$ which is equivalent to saying that the ring $\mathcal{R}(A,g)$ is integral over the ring $\mathcal{R}(A,f)$. This follows from 3.1.

b) g is said to be strongly integral over f if $f \leq g$ and $R(A,g)$ is a finitely generated $R(A,f)$-module. It was shown in ([6], Corollary 2. 6) that this is equivalent to saying that $\mathcal{R}(A,g)$ is a finitely generated $\mathcal{R}(A,f)$-module.

3.3. We assume that $f \leq g$. Then g is said to be :

(i) f-fine if there exists an integer $N \geq 1$ such that

$$J_n = \sum_{p=1}^{N} I_p J_{n-p} \text{ for all } n > N.$$

(ii) Weakly f-good if there exists an integer $N \geq 1$ such that for all $n > N$,

$$J_n = \sum_{p=0}^{N} I_{n-p} J_p$$

(iii) f-good if there exists an integer $N \geqslant 1$ such that

$$J_n = \sum_{p=1}^{N} I_{n-p} J_p \text{ for all } n > N.$$

and f is called a reduction of g if there exists an integer $k \geqslant 1$ such that

$$J_{n+k} = J_k I_n \text{ for all } n \geqslant k.$$

3.4 Remarks

(i) Any filtration f is weakly f-good.

(ii) If a pair (f,g) of filtrations is such that g is f-good then necessarily g is g-good. Indeed let $N \geqslant 1$ such that

$$J_n = \sum_{p=1}^{N} I_{n-p} J_p \text{ for all } n > N.$$

Then

$$J_n \subseteq \sum_{p=1}^{N} J_{n-p} J_p \subseteq J_n ,$$

hence

$$J_n = \sum_{p=1}^{N} J_{n-p} J_p \text{ for all } n > N.$$

The above equality holds trivially for $o \leqslant n \leqslant N$. Therefore g is g-good.

(iii) If a pair (f,g) of filtrations is such that f is a reduction of g then necessarily g is f-good, hence g-good. In addition g is strongly AP and f is AP.

Indeed let $k \geqslant 1$ such that

$$J_{k+n} = J_k I_n \text{ for all } n \geqslant k.$$

Take N = 2k. If n>N then n-k>k and

$$J_n = J_k I_{n-k} \subseteq \sum_{p=1}^{2k} I_{n-p} J_p \subseteq J_n ,$$

hence

$$J_n = \sum_{p=1}^{2k} I_{n-p} J_p$$

and g is f-good. On the other hand, by hypothesis $J_{kn} = J_k^n$ for all n, therefore g is strongly AP.

As for f, write $n = q_n k + r_n$ with $0 \leqslant r_n < k$. Take $k_n = (q_{2k+n} + 1)k$.

Then

$$\lim_{n \to \infty} \frac{k_n}{n} = 1 ;$$

$$I_{k_n m} \subseteq J_{2k+n}^m = J_k^m I_{k+n}^m \subseteq I_n^m$$

for all n, m. Consequently f is AP.

(iv) g f-fine $\Rightarrow$ g f-good $\Rightarrow$ g weakly f-good.

(v) If g is strongly integral over f then g is integral over f and weakly f-good.

(vi) If A is nœtherian and if g is weakly f-good, then g is strongly integral over f.

3.5. PROPOSITION (Corollary 3.6 of [6]).

Let f,g be two filtrations on a nœtherian ring A. Assume that f$\leqslant$g and that f is nœtherian. Then the following statements are equivalent :

(i) g is f-fine

(ii) g is f-good

(iii) g is weakly f-good

(iv) g is strongly integral over f.

3.6 COROLLARY **Let f,g be two filtrations on A with $f \leqslant g$. Assume that the ring A is nœtherian.**

Then :

(i) If g is f-good, then f and g are nœtherian.

(ii) If f is a reduction of g then f and g are nœtherian and g is strongly integral over f.

Proof.

(i) g is g-good by 3.4, (ii). Since A is nœtherian, g is nœtherian (see [4]) On the other hand g is strongly integral over f by 3.4 (iv) and (vi). Then by Eakin's Theorem [7], f is nœtherian.

(ii) If f is a reduction of g, then g is f-good by 3.4, (iii) and following the first part, f and g are nœtherian. Therefore g is strongly integral over f by 3.5, (iv).

For further information, see [4] and [6] . The reader will find in ([6], Theorem 4.6) several criteria for a filtration f to be a reduction of a filtration g .

We will close this section by showing that strong integral dependence implies that multiplicities coincide whenever they exist.

3.7. PROPOSITION

Let A be a nœtherian ring and let $f = (I_n)$ and $g = (J_n)$ be two filtrations on A. Suppose that coht g = o and that g is strongly integral over f. Then for any finitely generated A-module M, if the multiplicity $e_g(M)$ exists then the multiplicity $e_f(M)$ exists and $e_f(M) = e_g(M)$.

Proof.

Since g is strongly integral over f , there exists an integer $N \geqslant 1$ such that

$$J_n = \sum_{p=0}^{N} J_{n-p} J_p.$$

In particular $J_{n+N} \subseteq I_n \subseteq J_n$ for all $n \geqslant o$. It follows that $\sqrt{g} = \sqrt{f}$, hence o = coht g = coht f and alt f = alt g = s. Therefore

$$\frac{s!}{n^s}\,\ell(M/J_n M) \leqslant \frac{s!}{n^s}\,\ell(M/I_n M) \leqslant \left(\frac{n+N}{n}\right)^s \frac{s!}{(n+N)^s}\,\ell(M/J_{n+N} M).$$

If $n\to\infty$ we obtain the above statements.

4. GENERALIZED SAMUEL NUMBERS

4.1 Let $f = (I_n)$ be a filtration on A. For any ideal J of A put

$$v_f(J) = \operatorname{Sup}\{r;\ J \subseteq I_r\},$$

$$w_f(J) = \operatorname{Inf}\{r;\ J \supseteq I_r\}$$

and

$$w_f(J) = \infty \text{ if the set } \{r;\ J \supseteq I_r\} \text{ is empty.}$$

Let $g=(J_n)$ be another filtration on A. By the generalized Samuel numbers of f and g, we mean the numbers

$$\overline{v}_f(g) = \lim_{n\to\infty} \frac{1}{n} v_f(J_n)$$

and

$$\overline{w}_f(g) = \lim_{n\to\infty} \frac{1}{n} w_f(J_n),$$

whenever they exist in $\overline{\mathbb{R}}_+ = \mathbb{R}_+ \cup (\infty)$.

In particular if I and J are ideals of A and if f_I (resp. f_J) is the I-adic (resp. J-adic) filtration on A, we have $v_{f_I}(J) = v_I(J)$; $w_{f_I}(J) = w_I(J)$, following the notations of the Introduction. We will write $\overline{v}_f(f_J) = \overline{v}_f(J)$; $\overline{w}_f(f_J) = \overline{w}_f(J)$; $\overline{v}_{f_I}(g) = \overline{v}_I(g)$; $\overline{w}_{f_I}(g) = \overline{w}_I(g)$. So, with the notations of the Introduction $\overline{v}_{f_I}(f_J) = \overline{v}_I(J)$; $\overline{w}_{f_I}(f_J) = \overline{w}_I(J)$.

4.2 Let $f = (I_n)$ and $g = (J_n)$ be two filtrations on A. Beside the case $f = f_I$ and $g = f_J$ studied by Samuel in [13], the sequence $\frac{1}{n} v_f(J_n)$ (resp. $\frac{1}{n} w_f(J_n)$) need not be convergent in $\overline{\mathbb{R}}_+$. However, it is convergent in $\overline{\mathbb{R}}_+$ if g (resp. f) is AP and in this case

$$\overline{v}_f(g) = \lim_{n\to\infty} \frac{1}{n} \overline{v}_f(J_n) \quad \text{([1], Proposition 2.6)}$$

$$\text{(resp. } \overline{w}_f(g)) = \lim_{n\to\infty} n\, \overline{w}_{I_n}(g), \quad \text{(H. Dichi, [5])}$$

If in addition f and g are A P , then

$$\overline{w}_f(g) = \lim_{n\to\infty} \frac{1}{n} \overline{w}_f(J_n) \quad \text{(H. Dichi, [5]).}$$

4.3 Several basic results concerned with the Samuel numbers of two ideals can be extended to generalized Samuel numbers of two filtrations. As a consequence, many properties of homogeneous I-adic pseudo valuations can be extended to homogeneous pseudo valuations associated with AP filtrations (See [5] and [2] for more information). We will end this paper by Theorem 4.4 and its corollaries which are generalizations of Theorem 3 of Samuel in [13].

The filtration $f = (I_n)$ is said to be <u>separated</u> if $\bigcap_n I_n = (o)$ and <u>nonnilpotent</u> if $I_n \neq (o)$ for all $n \geqslant 1$.

4.4 THEOREM.

Let $f = (I_n)$ and $g = (J_n)$ be two separated nonnilpotent AP filtrations on the nœtherian ring A and let s= altf. If coht f= o and if $\sqrt{f} = \sqrt{g}$, then for any finitely generated A-module M, we have :

$$(\overline{v}_f(g))^s e_f(M) \leqslant e_g(M) \leqslant (\overline{w}_f(g))^s e_f(M).$$

Proof. Put $v_n = v_f(J_n)$; $w_n = w_f(J_n)$. Then for all n>>o, $v_n \geqslant 1$, $w_n \geqslant 1$, $I_{w_n} \subseteq J_n \subseteq I_{v_n}$. Since $\sqrt{f} = \sqrt{g}$, we have o = coht f = coht g and s = alt f = alt g. On the other hand

$$\left(\frac{1}{n} v_n\right)^s \frac{s!}{v_n^s} \ell(M/I_{v_n} M) \leqslant \frac{s!}{n^s} \ell(M/J_n M) \leqslant \left(\frac{1}{n} w_n\right)^s \frac{s!}{w_n^s} \ell(M/I_{w_n} M).$$

If $n \to \infty$ then $v_n \to \infty$, $w_n \to \infty$ and the conclusion follows.

4.5 COROLLARY.

Let I, J be nonnilpotent proper ideals of the nœtherian ring A such that $\cap I^n = (o) = \cap J^n$, $\sqrt{I} = \sqrt{J}$ and coht I = o. Let s = alt I. Then for any finitely generated A-module M, the following asymptotic inequalities hold :

$$(\overline{v}_I(J))^s e_I^*(M) \leqslant e_J^*(M) \leqslant (\overline{w}_I(J))^s e_I^*(M).$$

Proof. This follows from Theorem 4.4 by taking $f = f_I$ and $g = f_J$.

4.6 COROLLARY.

Let A be a nœtherian semi local ring of dimension d≥1, and let I, J be two ideals of definition of A. Then for any finitely

generated A-module M,

$$(\overline{v}_I(J))^d\, e_I^*(M) \leqslant e_J^*(M) \leqslant (\overline{w}_I(J))^d\, e_I^*(M).$$

Proof.

Since I, J are ideals of definition of A, $\sqrt{I} = \sqrt{J} = r(A)$ the Jacobson radical of A. Therefore coht I = o. By Krull's intersection Theorem, $\cap I^n = o = \cap J^n$. I is nonnilpotent since dim A$\geqslant$1. Finally it is clear that s=alt I = d. To achieve the proof it suffices to apply 4.5.

4.7 COROLLARY (Theorem 3 of Samuel [13]). Let (A,$\mathfrak{M}$) be a nœtherian local ring of dimension d$\geqslant$1, I, J two $\mathfrak{M}$-primary ideals of A. Then :

$$\overline{w}_I(J) \leqslant (e_J(A)/e_I(A))^{1/d} \leqslant \overline{w}_I(J)$$

Proof. This follows from 4.6 with M=A.

ACKNOWLEDGEMENT

I am grateful to the referee for his careful reading of the earlier version of this paper and for his judicious remarks.

REFERENCES

1. P. Ayegnon, D. Sangaré ; Generalized Samuel numbers and AP Filtrations, *J. Pure and Applied Algebra*, 65 : 1-13, (1990).

2. W. Bishop, A theory of multiplicity for multiplicative filtrations, *J. Reine Angew. Math.*, 277 :8-26, (1975).

3. W. Bishop, J. W. Petro, L.J. Ratliff Jr. and D. Rush, Note on nœtherian filtrations, *Comm. in Algebra*, 17(2) : 471-485, (1989).

4. H. Dichi, Integral dependence over a filtration, *J. Pure Appl. Algebra* 58 : 7-18, (1989).

5. H. Dichi, Notes sur les nombres de Samuel généralisés, (to appear)

6. H. Dichi, D. Sangaré, M. Soumaré, Filtrations, Integral dependence, reduction, f-good filtrations, *Comm. in Algebra,* 20(8) : 2393-2418 (1992).

7. P. Eakin, The converse of a well known theorem on nœtherian rings, *Math. Ann.*, 177 : 278-282, (1968).

8. M. Nagata, Note on a paper of Samuel concerning asymptotic properties of ideals, *Memoirs Coll. Sci. Univ. Kyoto, Series A* , Vol. xxx, Mathematics n°2, (1957).

9. M. Nagata, Local rings, *Interscience Publisher,* New York, (1962).

10. D.G. Northcott and D. Rees, Reduction of ideals in local rings, *Proc. Cambridge philos. Soc.* 50 : 145-158, (1954).

11. L.J. Ratliff, Jr., Notes on essentially powers filtrations, *Michigan Math. J.*, 26 : 313-324, (1979).

12. D. Rees, FRS, Lectures on the asymptotic theory of ideals, *London Math. Soc. Lecture Note Series*, 113, Cambridge Univ. Press, (1988).

13. P. Samuel, Some asymptotic properties of powers of ideals, *Ann. Math.* 56 : 11-21, (1952).

24
Chain Conditions Arising from the Study of Non Finitely Generated Modules Over Commutative Rings

S. ZARZUELA Departament d'Àlgebra i Geometria, Universitat de Barcelona, Gran Via 585, E-08007 Barcelona, Spain

1 INTRODUCTION

In 1974 M. Hochster [5] introduced the big Cohen–Macaulay modules conjecture, and showed how to obtain from its validity the corresponding one for most of the local homological conjectures. Let us recall its statement:

BIG COHEN–MACAULAY MODULES CONJECTURE. *Let* $(A, \mathfrak{m})$ *be a (Noetherian) local ring and* $a_1, \dots, a_d$ *be a system of parameters for* A. *Then there exists an* A*-module* M *such that* $a_1, \dots, a_d$ *is a regular sequence for* M, *that is,* $\mathfrak{m}M \neq M$ *and* $a_{i+1} \notin Z_A(M/(a_1, \dots, a_i)M)$ *for any* i.

Then M is said to be a *big Cohen–Macaulay* A*-module with respect to the system of parameters* $a_1, \dots, a_d$.

M. Hochster himself proved that a big Cohen–Macaulay A–module with respect to a given system of parameters always exists if A contains a field. By this way he could extend the majority of the results about the local homological conjectures that were known at that moment. See the excellent book of J. R. Strooker [15] for more information about this and related topics.

Big Cohen–Macaulay modules are in general non finitely generated (the existence of a finitely generated one is claimed by the maximal Cohen–Macaulay modules conjecture, only formulated for complete rings and with stronger consequences than the big Cohen–Macaulay modules conjecture). It then happens that many of the properties satisfied by finitely generated modules over a local ring fail for even such special non finitely generated modules. The following example was given by H.–B. Foxby (see [3]):

Let $A = K[[X, Y]]$, *where* K *is a field and* X, Y *are two independent variables. Let* $M = A \oplus E_A(A/(Y))$. *Then* X, Y *is a regular sequence for* M *while* Y, X *is not.*

This example shows that if M is a big Cohen–Macaulay module then not any permutation of a regular sequence for M is a regular sequence for M, as well as that not any system of parameters for A is a regular sequence for M.

On the other hand (and for different purposes I'm not going to detail here) it is useful to handle with big Cohen–Macaulay A–modules M such that any system of parameters for A is a regular sequence for M. Under the hypothesis that big Cohen–Macaulay modules exist, there are several methods to obtain these particular big Cohen–Macaulay modules. The most general one comes from the following result of J. Bartjin and J. R. Strooker [2]:

If M *is a big Cohen–Macaulay* A*–module, then the separated completion of* M *with respect to the* $\mathfrak{m}$*–adic topology,* $\hat{M}$, *is a big Cohen–Macaulay* A*–module with respect to any system of parameters for* A.

See also [6] for another situation were these special big Cohen–Macaulay modules arise.

Simultaneously R. Y. Sharp called *balanced* these particular big Cohen–Macaulay modules, and initiated a systematic study of their properties in [11]. There is some kind of general philosophy in this study, namely, that the notion of balanced big Cohen–Macaulay module is a good generalization of the notion of finitely generated Cohen–Macaulay module, and that they satisfy many of the properties that hold in the finitely generated case.

The problems we are going to consider in this talk happen in this frame. They are mainly related with the following questions:

(a) *When the property of being a balanced big Cohen–Macaulay module is preserved by localization*, and
(b) *when the regular sequences for a balanced big Cohen–Macaulay module can be permuted.*

More specifically, we are going to relate both problems to some special chain conditions in a particular subset of the support that appears in a quite natural way when one deals with non finitely generated modules.

2 LOCALIZATION OF BALANCED BIG COHEN–MACAULAY MODULES

Let $(A, \mathfrak{m})$ be a (Noetherian) local ring and M be an A–module. The *small support* of M (introduced by H.–B. Foxby in the more general study of complexes of modules) is defined as:

$$supp_A(M) = \{\, \mathfrak{p} \in Spec(A) \mid \exists\, i \geq 0 \text{ with } \mu_A^i(\mathfrak{p}, M) \neq 0 \,\} \;,$$

where $\mu_A^i(\mathfrak{p}, M)$ is the i-th Bass number of M with respect to $\mathfrak{p}$. The prime ideals in the small support of M are those for which a copy of the injective envelope $E_A(A/\mathfrak{p})$ appears in some step of a minimal injective resolution of M. If M is finitely generated it is then well known that the usual support of M : $Supp_A(M)$, and the small support of M coincide, but as we will later see this is no more true for non finitely generated modules, even if they are balanced big Cohen–Macaulay.

Now let M be a balanced big Cohen–Macaulay (*bbCM* for short) A-module, that is, a (non necessarily finitely generated) A-module M such that $\mathfrak{m}M \neq M$ and any system of parameters for A is a regular sequence for M. In [12] R. Y. Sharp proved that the prime ideals in the small support of M satisfy very nice properties, and characterized them in different ways. In particular he showed that following formula holds:

PROPOSITION 2.1 ([12]) *Let M be a balanced big Cohen–Macaulay A-module and $\mathfrak{p} \in supp_A(M)$. Then $ht(\mathfrak{p}) + dim(A/\mathfrak{p}) = dim(A)$.*

By using this property is then easy to find a *bbCM* module M for which the support and the small support are not the same. Consider any local non catenary domain A containing a field, for instance any of the first ones constructed by M. Nagata. Then there exists a *bbCM* A-module M, and by proposition 2.1 $supp_A(M) \neq Spec(A) = Supp_A(M)$ since A is a domain.

From another point of view the following characterization for the prime ideals in the small support of a *bbCM* A-module M is given in [18]:

PROPOSITION 2.2 ([18]) *Let M be a balanced big Cohen–Macaulay A-module and $\mathfrak{p} \in Spec(A)$. Then $\mathfrak{p} \in supp_A(M)$ if and only if $\mathfrak{p}M_\mathfrak{p} \neq M_\mathfrak{p}$.*

It is well known that if N is a finitely generated Cohen–Macaulay A-module and $\mathfrak{p} \in Supp_A(M)$, then $N_\mathfrak{p}$ is a finitely generated Cohen–Macaulay $A_\mathfrak{p}$-module. As we have just seen in proposition 2.2, a similar result for *bbCM* A-modules is only possible if the prime ideal $\mathfrak{p}$ belongs to the small support of M. This is the question that R. Y. Sharp possed in [11]. Namely:

QUESTION 2.3 *Let M be a balanced big Cohen–Macaulay A-module and let $\mathfrak{p} \in supp_A(M)$. When is $M_\mathfrak{p}$ a balanced big Cohen–Macaulay $A_\mathfrak{p}$-module?*

R. Y. Sharp himself gave an affirmative answer to this question if A is a catenary local domain [12], and H.–B. Foxby (private comunication) and Y. Takeuchi [16] generalized the above result to the case when A is any catenary local ring. On the other hand T. Ogoma has shown that the general answer is negative by giving a counterexample in [9].

The following necessary and sufficient condition is proved in [18]:

PROPOSITION 2.4 ([18]) *Let M be a balanced big Cohen–Macaulay A-module and $\mathfrak{p} \in supp_A(M)$. Then the following are equivalent:*

(i) $M_\mathfrak{p}$ is a balanced big Cohen–Macaulay $A_\mathfrak{p}$-module.

(ii) For any prime ideal $\mathfrak{q} \in supp_A(M)$ *such that* $\mathfrak{q} \subset \mathfrak{p}, ht(\mathfrak{p}) = ht(\mathfrak{q}) + ht(\mathfrak{p}/\mathfrak{q})$.

Observe that any catenary local ring trivially satisfies condition (ii) above.

Now consider the following subset of $Spec(A)$:

$$spec(A) := \{ \mathfrak{p} \in Spec(A) \mid ht(\mathfrak{p}) + dim(A/\mathfrak{p}) = dim(A) \} \quad .$$

As we have seen in proposition 2.1, $supp_A(M) \subset spec(A)$ for any $bbCM$ A–module M. Furthermore, under the assumption that $bbCM$ A–modules exist, it is possible to find one with the bigest possible small support:

PROPOSITION 2.5 *There exists a balanced big Cohen–Macaulay* A*–module* M *such that* $supp_A(M) = spec(A)$.

Proof: Let $M := \bigoplus_{\mathfrak{p} \in sepc(A)} M^{\mathfrak{p}}$, where $M^{\mathfrak{p}}$ is a $bbCM$ $A_{\mathfrak{p}}$–module. Then by [18], (1.4) M is a $bbCM$ A–module and by construction $supp_A(M) = spec(M)$.

Now we have:

PROPOSITION 2.6 *Let* A *be a local ring. Then the following are equivalent:*

(i) For any balanced big Cohen–Macaulay A*–module* M *and for any* $\mathfrak{p} \in supp_A(M)$, $M_{\mathfrak{p}}$ *is a balanced big Cohen–Macaulay* $A_{\mathfrak{p}}$*–module.*

(ii) For any couple of prime ideals $\mathfrak{q} \subset \mathfrak{p}$ *in* $spec(A)$, $ht(\mathfrak{p}) = ht(\mathfrak{q}) + ht(\mathfrak{p}/\mathfrak{q})$.

(iii) For any chain of prime ideals in $Spec(A)$: $\mathfrak{p}_0 \subset \ldots \subset \mathfrak{p}_n$, *with* $\mathfrak{p}_0$, $\mathfrak{p}_n \in spec(A)$, $n = ht(\mathfrak{p}_n/\mathfrak{p}_0)$ *if and only if the chain is saturated and* $\mathfrak{p}_i \in spec(A)$ *for any* i.

Proof: (i) $\Longleftrightarrow$ (ii) follows propositions 2.4 and 2.5. (ii) $\Longrightarrow$ (iii) It is clear that if $n = ht(\mathfrak{p}_n/\mathfrak{p}_0)$ the chain is saturated. Moreover, for any $i : dim(A) \geq ht(\mathfrak{p}_i) + dim(A/\mathfrak{p}_i) \geq ht(\mathfrak{p}_0) + i + dim(A/\mathfrak{p}_n) + n - i = ht(\mathfrak{p}_0) + dim(A/\mathfrak{p}_n) + n = ht(\mathfrak{p}_0) + ht(\mathfrak{p}_n/\mathfrak{p}_0) + dim(A/\mathfrak{p}_n) = ht(\mathfrak{p}_n) + dim(A/\mathfrak{p}_n) = dim(A)$. Thus for any i, $ht(\mathfrak{p}_i) + dim(A/\mathfrak{p}_i) = dim(A)$, that is, $\mathfrak{p}_i \in spec(A)$.

Conversely: If the chain is saturated and $\mathfrak{p}_i \in spec(A)$ for any i, then $ht(\mathfrak{p}_{i+1}/\mathfrak{p}_i) = 1$ and $ht(\mathfrak{p}_n) = 1 + ht(\mathfrak{p}_{n-1}) = \ldots = n + ht(\mathfrak{p}_0)$, hence $ht(\mathfrak{p}_n/\mathfrak{p}_0) = n$.

(iii) $\Longrightarrow$ (ii) Let $\mathfrak{q} \subset \mathfrak{p}$ be prime ideals in $spec(A)$ and consider a saturated chain of prime ideals: $\overline{\mathfrak{q}}_0 \subset \ldots \subset \overline{\mathfrak{q}}_r = \mathfrak{q} = \mathfrak{q}_0 \subset \ldots \subset \mathfrak{q}_n = \mathfrak{p} = \overline{\mathfrak{p}}_0 \subset \ldots \subset \overline{\mathfrak{p}}_s = \mathfrak{m}$, where $r = ht(\mathfrak{q})$, $n = ht(\mathfrak{p}/\mathfrak{q})$ and $s = dim(A/\mathfrak{p})$. Since $\mathfrak{q} \in spec(A)$, $\overline{\mathfrak{q}}_0$ is a minimal prime ideal such that $dim(A/\overline{\mathfrak{q}}_0) = dim(A)$, that is, $\overline{\mathfrak{q}}_0 \in spec(A)$. Hence for any i, $\overline{\mathfrak{q}}_i \in spec(A)$. On the other hand, since $\mathfrak{m}, \mathfrak{p} \in spec(A)$ and $s = ht(\mathfrak{m}/\mathfrak{p})$, $\overline{\mathfrak{p}}_j \in spec(A)$ for any j. And similarly $\mathfrak{q}_t \in spec(A)$ for any t. Thus we have a saturated chain of prime ideals in $spec(A)$ and this gives by (iii) that $r + n + s = ht(\mathfrak{m}/\overline{\mathfrak{q}}_0) = dim(A)$. In particular $ht(\mathfrak{q}) + ht(\mathfrak{p}/\mathfrak{q}) + dim(A/\mathfrak{p}) = ht(\mathfrak{p}) + dim(A/\mathfrak{p})$, and $ht(\mathfrak{p}) = ht(\mathfrak{q}) + dim(A/\mathfrak{p})$.

As we have already remarked after proposition 2.4, it is clear that any catenary local ring satisfies the conditions of proposition 2.6. Next we give another family of examples.

EXAMPLE 2.7 *Let* A *be a local domain such that for any prime ideal* $\mathfrak{p} \in spec(A) - \{\mathfrak{m}\}$, $A_{\mathfrak{p}}$ *is catenary. Then* A *satisfies the conditions of proposition 2.5.* This holds if in particular A is any of the Nagata's counterexamples to the chain problem, since by [10], (B. 2.4) for any prime ideal $\mathfrak{p}$ different from the maximal one in A, $A_{\mathfrak{p}}$ is a regular local domain. It holds too if A is the non catenary normal local domain constructed by Ogoma in [8]: By [4], lemma 5, $A_{\mathfrak{p}}$ is also a regular local domain for any non maximal prime ideal $\mathfrak{p}$ in $Spec(A)$.

On the other hand it is clear that the counterexample given by Ogoma in [9] for the problem of the localization of *bbCM* A–modules shows that not any local ring satisfies the conditions of proposition 2.5. But the fact that this counterexample is not a domain leads to the following question:

QUESTION 2.8 *Does any local domain satisfy the conditions of proposition 2.5 ?*

Observe that if $spec(A) = Spec(A)$, an affirmative answer to the above question is given by Rattiff's characterization of catenary local domains: If A is a local domain, then A is catenary if and only if $ht(\mathfrak{p}) + dim(A/\mathfrak{p}) = dim(A)$ for any prime ideal $\mathfrak{p}$ in $Spec(A)$.

3 PERMUTATION OF REGULAR SEQUENCES FOR BALANCED BIG COHEN-MACAULAY MODULES

Regardless of whether the general conditions of proposition 2.6 are fulfilled, a particular balanced big Cohen–Macaulay A–module with special properties may localize. For instance, this would be the case if *M is a balanced big Cohen–Macaulay A–module such that* $supp_A(M) = Supp_A(M)$. For then $supp_A(M) = V(ann_A(M)) := \{ \mathfrak{p} \in Spec(A)$ such that $ann_A(M) \subset \mathfrak{p} \}$ ([19], (3.5)). By [7], proposition 7, $supp_A(M)$ is then catenary, hence by proposition 2.6 M localizes.

A balanced big Cohen–Macaulay A–module such that $supp_A(M) = Supp_A(M)$ satisfies an additional property that we want to consider now. Namely, that *any permutation of a regular sequence for* M *is again a regular sequence for* M. This happens since by [19], (3.3) and (3.6), a family of elements in $A : a_1, \ldots, a_r$ is a regular sequence for M if and only if $a_1, \ldots, a_r$ is part of a system of parameters for $A/ann_A(M)$.

For instance, if M is a balanced big Cohen–Macaulay module such that $sup\{ i \mid \mu^i_A(\mathfrak{m}, M) \neq 0 \} < \infty$ then $supp_A(M) = Supp_A(M)$ ([17], (2.5)). Furthermore and contrary to the finitely generated case, $sup\{ i \mid \mu^i_A(\mathfrak{m}, M) \neq 0 \} < \infty$ does not imply $id_A(M) < \infty$. Here there is an example:

EXAMPLE 3.1 ([17]). Let $S = K[[X_1, \ldots, X_n]]$, with K any field and $n \geq 2$. Let $t = X_1^2$ and $A = S/(t)$. If $\mathfrak{p} = (X_1)/(t)$, then one can find an $A_{\mathfrak{p}}$–module N with $id_{A_{\mathfrak{p}}}(N) = \infty$ ($A_{\mathfrak{p}}$ is not regular). Then $M = A \oplus N$ is a balanced big Cohen–Macaulay A–module such that $sup\{ i \mid \mu^i_A(\mathfrak{m}, M) \neq 0 \} < \infty$ but $id_A(M) = \infty$.

Given M an A–module the fact that any permutation of a regular sequence for

M is also a regular sequence for M holds under different circumstances. Next we list two of them.

EXAMPLE 3.2 *If M is complete with respect to the $\mathfrak{m}$-adic topology ([20], (8.2)).*

EXAMPLE 3.3 *If for any ideal $I \subset A$, IM is closed in M with respect to the $\mathfrak{m}$-adic topology ([13], proposition 1).*

Condition 3.2 is part of a more general set of ideas concerning complete modules, in the sense that they should satisfy the same properties as finitely generated ones. This line of thinking was initiated by J. Bartjin and J. R. Strooker (see [1] and [2]), and has been deeply stated by A. M. Simon (see [13] and [14]).

Unfortunately it is possible to give examples of balanced big Cohen–Macaulay A-modules for which not any permutation of a regular sequence is again a regular sequence.

EXAMPLE 3.4 Ogoma's counterexample of a balanced big Cohen–Macaulay A-modules M with a prime ideal $\mathfrak{p} \in supp_A(M)$ such that $M_\mathfrak{p}$ is not a balanced big Cohen–Macaulay $A_\mathfrak{p}$-module (see [20], remark 6).

EXAMPLE 3.5 Let $A = K[[X, Y, Z, T]]/(X, Y) \cap (Z, T)$, $\mathfrak{p} = (X, Y)/(X, Y) \cap (Z, T)$ and $\mathfrak{q} = (Z, T)/(X, Y) \cap (Z, T)$. Then $C = A/\mathfrak{p} \oplus E_A(A/\mathfrak{q})$ is a balanced big Cohen–Macaulay A-module for which $\overline{X} + \overline{Z}, \overline{T}$ is a regular sequence but $\overline{T} \in Z_A(C)$ (see [14], (3.4) for the details).

Observe that in example 3.5 A satisfies the conditions of proposition 2.6 since A is catenary.

To distinguish these balanced big Cohen–Macaulay A-modules for which regular sequences permute, A. M. Simon has introduced the notion of superbalanced big Cohen–Macaulay module. *A big Cohen–Macaulay A-module M is said to be superbalanced if any permutation of a regular sequence for M is again a regular sequence for M.* Superbalanced big Cohen–Macaulay A-modules are balanced but as we have already seen the converse is not true. Any complete big Cohen–Macaulay A-module is superbalanced.

A. M. Simon herself has related the property of being superbalanced with a chain condition in the small support in the following way:

PROPOSITION 3.6 ([14], 4.2) *Let M be a balanced big Cohen–Macaulay A-module. Then the following are equivalent:*

(i) M is superbalanced.

(ii) For any prime ideal $\mathfrak{p} \in supp_A(M)$ and for any element $a \in \mathfrak{m} - \mathfrak{p}$, there exists a prime ideal $\mathfrak{q}$ in $supp_A(M)$ such that $ht(\mathfrak{q}) = ht(\mathfrak{p}) + 1$ and $a \in \mathfrak{q}$.

Condition (ii) above in particular means that for any prime ideal $\mathfrak{p} \in supp_A(M)$ it is possible to find a saturated chain of prime ideals in $supp_A(M)$:

$$\mathfrak{p} = \mathfrak{p}_0 \subset \ldots \subset \mathfrak{p}_i \subset \ldots \subset \mathfrak{p}_r = \mathfrak{m} \quad ,$$

such that $r = ht(\mathfrak{m}/\mathfrak{p})$ (see [14], 4.4).

Our final question arises from the fact noted in example 3.5 that Ogoma's counterexample for the localization of balanced big Cohen–Macaulay modules is not superbalanced.

QUESTION 3.7 *Does any superbalanced big Cohen–Macaulay A–module localize?*

REFERENCES

1. J. Bartjin, Flatnes, completion, regular sequences: un ménage à trois, Thesis, Utrecht (1985).
2. J. Bartjin, J. R. Strooker, Modifications monomiales, in *Séminaire d'Algèbre P. Dubreil et M. P. Malliavin*, Paris 1982, Lecture Notes in Math. 1029, Springer–Verlag, 1983, 192–217.
3. P. Griffith, A representation theorem for complete local rings, *J. Pure Appl. Algebra* 7 (1976), 305–315.
4. R. C. Heitman, A non–catenary, normal, local domain, *Rocky Mountain J. of Math.* 12 (1) (1982), 145–148.
5. M. Hochster, Topics on the homological theory of modules over commutative rings, *C.B.M.S. Regional Conference Series in Mathematics* 24, Amer. Math. Soc., Providence, Ri, (1974).
6. M. Hochster, C. Huneke, Infinite integral extensions and big Cohen–Macaulay algebras, *Annals of Math.* 135 (1992), 53–89.
7. S. Mcadam, L. J. Rattiff, Semi–local taut rings, *Indiana University Math. J.* 26 (1) (1977), 73–79.
8. T. Ogoma, Non–catenary pseudo–geometric normal rings, *Japan J. Math.* 6 (1) (1980), 147–163.
9. T. Ogoma, Fibre product of Noetherian rings and their applications, *Math. Proc. Camb. Phil. Soc.* 97 (1985), 231–241.
10. L. J. Rattiff, Chain conjectures in Ring theory, *Lecture Notes in Math.* 647, Springer–Verlarg, 1978.
11. R. Y. Sharp, Cohen–Macaulay properties for balanced big Cohen–Macaulay modules, *Math. Proc. Camb. Phil. Soc.* 90 (1981), 229–238.
12. R. Y. Sharp, A Cousin–complex characterization of balanced big Cohen–Macaulay modules, *Quart. J. Math. Oxford* (2) 33 (1982), 471–485.
13. A. M. Simon, Some homological properties of complete modules, *Math. Proc. Camb. Phil. Soc.* 108 (1990), 231–246.
14. A. M. Simon, Modules complets et sutes régulières, *Bull. Soc. Math. de Belgique* XLIII (1991), 141–150.
15. J. R. Strooker, Homological questions in local algebra, *London Math. Soc. Lecture Notes Ser.* 145, Cambridge Univ. Press, 1990.
16. Y. Takeuchi, On localizations of a balanced big Cohen–Macaulay module, *Kobe J. Math.* 1 (1984), 43–46.
17. S. Zarzuela, Resolución injectiva de un gran módulo de Cohen–Macaulay, *Actas X*

Jornadas Hispano–Lusas de matemáticas, Sección 1, Universidad de Murcia, 1985, 185–193.

18. S. Zarzuela, Balanced big Cohen–Macaulay modules and flat extensions of rings, *Math. Proc. Camb. Phil. Soc.* 102 (1987), 203–209.
19. S. Zarzuela, Systems of parameters for non–finitely generated modules and big Cohen–Macaulay modules, *Mathematika* 35 (1988), 207–215.
20. S. Zarzuela, A generalization of a theorem of Sharp on big Cohen–Macaulay modules, *Math. Proc. Camb. Phil. Soc.* 108 (1990), 193–195.